MINDING NATURE
The Philosophers of Ecology

DEMOCRACY AND ECOLOGY

A Guilford Series

Published in conjunction with
the Center for Political Ecology

James O'Connor
Series Editor

IS CAPITALISM SUSTAINABLE?
Political Economy and the Politics of Ecology
Martin O'Connor, *Editor*

GREEN PRODUCTION
Toward an Environmental Rationality
Enrique Leff

MINDING NATURE
The Philosophers of Ecology
David Macauley, *Editor*

Forthcoming

MARXISM AND ECOLOGY
Ted Benton, *Editor*

RADICAL ENVIRONMENTAL MOVEMENTS TODAY
Daniel Faber

POLITICAL ECONOMY OF NATURE
James O'Connor

MINDING NATURE
The Philosophers of Ecology

Edited by
David Macauley

THE GUILFORD PRESS
New York London

A Division of Guilford Publications, Inc.
72 Spring Street, New York, NY 10012

Printed in the United States of America

This book is printed on acid-free paper.

Last digit is print number: 9 8 7 6 5 4 3 2 1

Library of Congress Cataloging-in-Publication Data

Minding nature : the philosophers of ecology / edited by
David Macauley.
p. cm. — (Democracy and ecology)
Includes bibliographical references and index.
ISBN 1-57230-058-2 — ISBN 1-57230-059-0 (pbk.)
1. Green movement. I. Macauley, David. II. Series.
JA75.8.M25 1996
363.7′001—dc20 95-39862
CIP

Contributors

David Abram is the author of *The Spell of the Sensuous: Perception and Language in a More-Than-Human World* (1996). He currently lives in northern New Mexico, where he writes on ecological and philosophical issues.

Henry T. Blanke is Reference Librarian at Marymount Manhattan College and is the coeditor of the *Progressive Librarian*.

Frank Coleman works for the Environmental Protection Agency and resides in Alexandria, Virginia. He is the author of *Hobbes and America* (1977) and *Politics, Policy and the Constitution* (1982).

John Ely, author of *The Greens of West Germany: An Alternative Modernity*, has written extensively on green politics, social ecology, and problems of relating socialism with ecology. He works at the University of San Francisco and is presently developing a "left-Aristotelian" and green view of "libertarian confederalism."

Andrew Feenberg is Professor of Philosophy at San Diego State University and author of *Critical Theory of Technology* (1991), *Lukács, Marx and the Sources of Critical Theory* (1986), and *Alternative Modernity* (1995).

Yaakov Garb's work focuses on the social and cultural aspects of environmental issues in the United States and Israel. His essay in this volume was completed during a year in the School of Social Science at the Institute for Advanced Study in Princeton.

Ramachandra Guha is an independent writer based in Bangalore in south India. He previously taught at Yale University and was a Fellow at the Wissenschaftcolleg zu Berlin. His books include *The Unquiet Woods* (1989) and, with Madhav Gadgil, *This Fissured Land* (1992). He is now working on a history of the Indian environmental movement.

Andrew Light is a Research Fellow in environmental health and Adjunct Professor of Philosophy at the University of Alberta. He is the coeditor, with Eric Katz, of *Environmental Pragmatism* (1995), the coeditor of the annual journal *Philosophy and Geography*, and the author of several articles on political ecology.

David Macauley teaches philosophy and literature classes at St. John's University and Marymount Manhattan College. He has published articles on ecology, political theory, and philosophy. He is presently completing his Ph.D. in philosophy at SUNY-Stony Brook and is planning a collection of essays on walking.

Joan Roelofs is Professor of Political Economy at Keene State College in New Hampshire, and is a member of the Boston Editorial Collective of *Capitalism, Nature, Socialism*.

Alan Rudy recently earned a doctorate in sociology from the University of California, Santa Cruz. In his dissertation he investigated the political economic history, ecological contradictions, and labor struggles associated with irrigated agriculture in the Imperial Valley of southeastern California.

Lawrence Vogel is Associate Professor of Philosophy at Connecticut College. He previously taught at Vassar and Yale. He is the author of *The Fragile "We": Ethical Implications of Heidegger's Being and Time* (1994) and editor of a forthcoming volume of Hans Jonas's later essays, *Mortality and Morality: A Search for the Good After Auschwitz* (1996).

Joel Whitebook teaches philosophy at The New School for Social Research and is the author most recently of *Perversion and Utopia: A Study in Pyschoanalysis and Critical Theory* (1995).

Michael E. Zimmerman is Professor of Philosophy at Tulane University and author of *Contesting Earth's Future: Radical Ecology and Postmodernity* (1994), *Heidegger's Confrontation with Modernity* (1990), and *Eclipse of the Self: The Development of Heidegger's Concept of Authenticity* (1981).

Contents

Introduction

Greening Philosophy and Democratizing Ecology

David Macauley

Greek Ideas and Green Ideals

Western philosophy begins as a meditation on nature in an attempt to discern the order of things and to speculate on its meaning, direction, and purpose. The first Greek *physiologoi*, or natural philosophers, reflected not simply upon the human *psyche* but directed themselves foremost toward the yawning heavens, the turnings and reversals of fire, the rhythm and play of water, and the outcroppings of rock and earth—more broadly, the four elements. In their search for a hidden *arche*, an underlying *logos*, or a guiding *telos*, they looked as well to the growth, movement, and relations of plants and animals. Science at this point was virtually indistinguishable from philosophy. Though moving away from the speculative insights of the Presocratics, Aristotle and his student Theophrastus can be seen as early forerunners of ecological thought—the Greek fathers of animal and plant ecology, respectively. To a remarkable extent, they integrated scientific observation of nature with philosophic justification and explanation.[1]

It is only later in history that such "theories of nature" are given over almost completely to "theories of mind" and to narrowed notions of *nous*, denatured conceptions of reason, and radically subjective perspectives on cosmology, aesthetics, and ethics.[2] It is the generous sensitivity, broad inclusiveness, and ontologically egalitarian orientation of this early Greek thought, rather than its scientific accuracy or correctness, which remains important because of its embeddedness and groundedness in a realm beyond the solely human.[3] One way to "green" or "ecologize" philosophy and to deepen and "democratize" ecology, then, would be to explore the first bold footsteps of the

Greek philosophers, observing their transitions from the interpretive framework of an intelligible, rational, and beautiful *kosmos*[4] to their movement away from myths and stories about the nonhuman, gradual severance from the natural world, and, finally, attempts to transcend this world altogether. One could inquire how they understood change, the elements, form and matter, and being and causality in relation to the natural and social realms, situating their thought in the context of social and political changes such as deforestation or domestication of plants and animals. These changes have helped to alter ecological and philosophical notions and allowed the process of dominating and controlling nature to become more complete.

Another approach to theorizing the relations of philosophy to ecology and, by extension, of society to nature, is to begin with contemporary aporias, problems, and debates and to understand them in relation to qualitative changes in philosophy, culture, and political economy over time. In this regard, the present collection of essays—many of which appeared in the journal *Capitalism, Nature, Socialism*—attempts to draw out, clarify, critique, and extend the insights of some of the most important philosophers, writers, and thinkers on nature and polity in this century, as well as several who antedated and influenced major twentieth-century figures. Each of the subjects of these essays has contributed deeply and significantly to contemporary discussion and to the historical understanding of political, epistemological, or social issues related to nature. With two exceptions, Hobbes and Heidegger, they stimulated constructive dialogue within progressive, democratic, and leftist circles. In various ways, they have helped to challenge the notion that conservation is inherently politically conservative or that our *oikos* (home) must be rendered uniformly economic where ecology (i.e., the household of nature) is concerned. In so doing, they have enabled us to rethink the possibility of creating a more democratic *and* ecological society.

One guiding thread through the thinking of many (though not all) of the figures considered in these essays is a certain commitment to or relation with critical social theory. Such theory can be characterized very loosely as a critical perspective on technology, power, scientism, and instrumental reason along with an opposition to exploitative capitalist social relations and a willingness to engage and reenvision the social and political world by examining the relationship between nature and society. In this sense, many of the philosophers and thinkers in this collection have relevance for understanding or critiquing emerging schools of thought and social movements such as social ecology, ecological feminism, bioregionalism, socialist ecology, and, perhaps, critical postmodern ecology (even if these perspectives do not always directly rely on or acknowledge critical social theory).

Important issues and ideas are at stake, many of which are debated in the following essays. These include (1) the respective claims and merits of anthropocentric, biocentric, and ecocentric positions; (2) the possibilities or pitfalls of modernity and, to a lesser extent, postmodernity and premodernity; (3) the nature, scope, and types of rationality and their relation to technology, language, social change, and forms of embodied knowledges[5]; and (4) the conceptions or constructions of nature and what they tell us about our vision of society. To these we can add (5) the degree to which existing political traditions or tendencies (socialist, anarchist, feminist, and liberal) are compatible with or antagonistic toward a new ecological sensibility, and (6) the sources for analysis and location of a viable ecological and political tradition (e.g., North American, European, Far Eastern, or Native American). Of central concern as well are (7) the specific and primary diagnostic critiques of the roots, forms, and manifestations of ecological and social problems. These range in starting point from capital formation, overpopulation, environmental pollution, and autonomous technology to social domination and hierarchy, repressive and instrumental rationality, the Judaic and Christian religious traditions, and historical forgetting of Being. Finally, at issue are (8) the proposals toward solutions, whether individualistic or collectivist, conflictual or consensual, libertarian or authoritarian, interventionist or "letting be."

In discussions of ecological philosophy and the mastery of nature, a common point of departure and historical foil has been either the empiricism of Francis Bacon or the rationalism of René Descartes,[6] two Renaissance thinkers who developed radically new methods for guiding scientific research and who characterized the natural world in terms of its use-value for or distinctness from humans. Marx himself, in fact, remarked, "That Descartes, like Bacon, anticipated an alteration in the form of production, and the practical subjugation of Nature by Man, as a result of the altered methods of thought, is plain from his 'Discours de la Méthode.' "[7] For Bacon, to know nature means to disturb and annoy it (*natura vexata*) through artisanry and technology so that it will disclose its secrets.[8] With Descartes, the understanding of nature and realization of certainty are achieved first by separation from the natural world, then its precise measurement. In both instances knowledge is premised upon detachment and a thoroughgoing control over the environment and experimental conditions. Coupled with this control is a privileging of the "how" over the "why," which eventuates in a utilitarian criterion of truth, a tendency to transform science into technique, and a reduction of the "object" of knowledge to an instrumental relation or quantifiable value.[9]

From Realism to Utopia: Hobbes and Fourier

The present collection of essays commences alternatively with an assessment of Thomas Hobbes (1588–1679), who was influenced by the Baconian equation of knowledge and power and likewise espoused a new epistemology, but who also built a political philosophy on his methodology and assumptions. That philosophy continues to exert a haunting philosophical and social influence. In contrast to Aristotle, whose model science is biology, and Marx, who proceeds by turning to history, Hobbes relies upon a mechanical physics that explains motion without recourse to the formal and final causes of Aristotelian science. And in contrast to Aristotle's conception of the political community as a natural association or Marx's depiction of it as an historical form, Hobbes conceives of community (which collapses into a commonwealth and all-powerful sovereign) as an assemblage of atomistic, self-interested individuals who band together out of fear of death (the *summum malum*) rather than desire for public freedom.[10]

Thus, for Hobbes the end of politics is simply self-preservation and security against a state of "slaughter, solitude, and the want of all things" rather than the good life of the community (Aristotle) or the attainment of species-being (Marx). Knowledge does not serve to attain *arete* (virtue or excellence) as in Aristotle or to overcome alienation as in Marx, but rather to construct a "commonwealth" or *Civitas* (artificial man) built upon a mechanistic psychology and physics. The contrast is vivid between Hobbes's artificial and mechanical political order, based upon *techne*, and established through a social contract, and the much earlier but still influential view of Aristotle that the *polis* (city or civic community), hence politics, is natural, in part because it is an extension of earlier organic forms. The significance of these comparisons becomes clear when one considers the new role of nature and its relation to society, a subject explored by Frank Coleman in the first essay.

Hobbes throws into perspective a great divide separating the modern or postmodern age from antiquity, where nature was endowed with and permeated by mind or intelligence and where the nature-society relation was organic. Current characterizations of and proposed solutions to ecological and social problems are often cloaked in Hobbesian language. If due to the gravity of environmental problems, the choice consists in near-absolute authority or ecological annihilation—in the words of William Ophuls, "Leviathan or oblivion"—this amounts to both a choice for Hobbes and a Hobson's choice (i.e., no choice at all). It leads Ophuls, for example, to hark back as well to Plato in proposing a class of "ecological guardians" to govern a steady-state society because "the only solution is a sufficient measure of coercion."[11]

Hobbesian assumptions also guide the thought of theorist William Catton and ecologist Garret Hardin, where they culminate similarly in analysis marked by scientific reductionism, neo-Malthusianism, and a lack of social critique.[12] In influential essays like "The Tragedy of the Commons" and "Living on a Lifeboat," Hardin grounds his egocentric ethic on the presupposition that humans are competitive by nature, that the commons functions like a marketplace, and that capitalism, in turn, is a completely natural form of association and exchange. It is no surprise, then, when he argues for the necessity of coercion and rule by a "strong and farsighted sovereign" (à la Hobbes), triage, and refusing aid to countries with incorrigible population problems.[13]

Put differently, some Cassandras of ecological doom may be offering us a false dilemma which conceals a very conservative or reactionary political agenda, notably some right-wing "ecological" parties in Europe and certain writers and political currents in the United States advocating, for example, the lifeboat ethic and survivalist, authoritarian, even ecofascistic solutions. It is vital to realize, then, that an image of nature offers us as well an image of society. In Lukács's terms, "nature is a social category," that is, "socially conditioned,"[14] or as Adorno writes, "In every perception of nature there is actually present the whole of society."[15] It makes great political and philosophical sense to examine representations of Earth, the environment, nature, and the wild for they may tell us as much as and often more about ourselves than about the nonhuman world around us. Historically significant characterizations can be found, for example, in Tennyson's image of nature as "red in tooth and claw"; Spinoza's conjunction of *Deus* with *Natura* and his identification of both with substance; the tendency in the West to gender matter, the Earth, and nature as feminine; Galileo's "mathematization of nature"; Emerson's transcendental spiritualization of nature; Nietzsche's injunction to "dehumanize nature" and "naturalize humanity"; and Rousseau's valorization of presocial "natural man." This kind of critical reading also throws light on issues related to teleology and evolution, cyclicality and progress, artifice and technique, and mechanism, organicism, and informationalism.

With Hobbes, the question of religion also arises, as Frank Coleman argues in exploring themes related to the "prosthetic God" and the location of the British philosopher at a busy historical intersection of liberalism, capitalism, and modern science. In Coleman's view, Hobbes derives his conception of "man" as artificer (*Homo faber*) from the Bible, where the infamous Leviathan is introduced in the Book of Job; consequently, he views nature as an artifact—either human or divine, but always and only an artifact. Thus, there are few to no limits placed on our capacities to transgress, destroy, or remake the natural world. The role of Judaism and Christianity in the ecological crisis has been the subject of some debate since Lynn White's 1967 essay on the subject,[16] and Coleman's discussion is in part an extension and deepening of that debate.

Bearing upon it, too, is the emergence from the desert landscape of a disembodied, omnipotent, and unknowable God, whose absolute transcendence and lack of concrete, phenomenal form can appear as empty and benighted as a lonely night beneath a starless Saharan sky. Among others, Paul Shepard, George Steiner, and David Miller have explored the relationship between the symbol of the desert and the idea of monotheism, and, correspondingly, of monotheism to society. Their work suggests that monotheism tends to promote a sense of abstract disconnectedness; a penchant toward authoritarianism, asceticism, and masculinism; and a failure to appreciate fully the particularities of landscape and place.[17] They surmise that the heightened experience of social and metaphysical alienation can result in periodic attacks upon those believed to have originated monotheism.[18] In contrast, like thinking itself, ecology is arguably a kind of polytheism (and pluralism); in the words of James Hillman, "both show interpenetrating and interlimiting patterns of autochthonous powers, each power a qualitative splendor, a presence that is both a unique example and a universal genus at once."[19]

Charles Fourier's (1772–1837) utopianism throws Hobbes's harsh realism into relief. Nature, which in Hobbes was relegated to a realm of chaos, disorder, and fear, reemerges as a sphere of potential freedom, pleasure, and spontaneity informing society. Further, human nature is no longer a dark constraint upon community. Unlike Hobbes's reductionism and mechanism, Fourier's psychology is complex, differentiated, pluralistic, and ecological in the sense of its emphasis on wholeness.[20] Because of Fourier's ecological vision of both nature and society, Murray Bookchin has written that "Fourier is in many ways the earliest social ecologist to surface in radical thought."[21]

One might question the apparent paradox of the *place* of utopian thought in political theory and, especially, in ecological philosophy. In other words, what has placeless and often groundless "idealism" or chiliasm to do with an ecological philosophy that is presumably inhabited, historical, and emplaced? It is helpful to recall that Thomas More, who coined the term *utopia* in 1516 for his novel of the same name, distinguished between *eu-topia* as "good place" and *outopia* as "no place." In the former sense, a utopia can serve as a kind of critical ethical and political guide toward the good and the good life rather than as a futuristic blueprint that all too frequently reinscribes or extends the realities of the present. The best utopian thought follows strong insights and tendencies to their logical conclusions; provides imaginative possibilities for association, communication, and organization; and is implicitly critical of existing conditions. In this sense, it is real insofar as it is a *realizable ideal*.

Following Lewis Mumford, one can distinguish between utopias of escape and utopias of reconstruction, where the latter (such as Fourier's) aim at new values, creative relationships, and liberatory institutions by blending idealism with practicality. Mumford's characterization of utopian thought illuminates as well the implicitly ecological dimensions to such thought. "Utopian thinking, as I came to regard it, then, was the opposite of one-sidedness, partisanship, partiality, provinciality, specialism. He [and she] who practiced the utopian method must view life synoptically and see it as an interrelated whole: not as a random mixture, but as an organic and increasingly organizable union of parts, whose balance it was important to maintain—as in any living organism."[22]

Many "utopias," of course, have been authoritarian models of rational planning gone awry, including Bacon's New Atlantis, Saint-Simon's industrial technocracy, B. F. Skinner's Walden Two, and Plato's Republic. Their stress on centralized, hierarchical, or absolute authority is in this regard unecological and antidemocratic in the extreme. Indeed, the twentieth century is witness to the dangers of utopian thought turned dystopic, where monsters have been born and bred from nightmarish dreams of reason, community, and harmony. This phenomenon has contributed (understandably) to its decline, as empiricism has supplanted speculative inquiry and Realpolitik has circumscribed imagination.

However, in the writings of William Morris, Ebenezer Howard, Paul Goodman, Gustav Landauer, Peter Kropotkin, Ernst Bloch, Herbert Marcuse, Theodore Roszak, and Murray Bookchin, utopian thinking (broadly defined) has also offered more fertile ecological and democratic visions. As Bookchin has observed insightfully:

> What marked the great utopians was not their lack of realism but their sensuousness, their passion for the concrete, their adoration of desire and pleasure. Their utopias were often exemplars of a qualitative "social science" written in seductive prose, a new kind of socialism that defied abstract intellectual conventions with their pedantry and icy practicality. Perhaps even more importantly, they defied the image that human beings were, in the last analysis, machines. . . . Their message of fecundity and reproduction thus rescued the image of humanity as an embodiment of the organic that had its place in the richly tinted world of nature, not in the workshop and the factory.[23]

These perspectives have fostered a kind of ecocommunalism predicated on such principles as mutual aid, simplicity, direct action, usufruct, respect for nature, decentralization, and appropriate technology. To this degree, they belong to an emerging *ecotopian* tradition which has found direct expression in cooperatives, collectives, and communes throughout the United States and elsewhere.[24]

As part of this tradition, Fourier offers both utopian ideas and practical proposals.[25] He envisions a world based on the *phalanstère*, a balanced community where the passions are allowed full play and human inclinations are set free. In Joan Roelofs's view, Fourier prefigures by more than a century the present red-green discussion in his critique of the nuclear family as perpetuating excess consumption, his early feminism, stress on sexual freedom, reorientation of work, and sensitivity to community, collective living, and the natural world.

The tension between "idealism" and "realism," between utopian sympathy and anti-utopian suspicion, is admittedly perennial. It has surfaced within the ecology movement itself, for example, between the more principled *Fundis* and the more pragmatic *Realos* in Germany and between radical ecologists and more mainstream environmentalists in the United States. Hence the value of critiques and visions like those of Fourier, which strive to overcome these oppositions and creatively to interlace play with work, freedom with necessity, and eventually society with nature.

The Ecology of Phenomenology: Heidegger and Merleau-Ponty

For critical and political ecology, the transition from Fourier to the Frankfurt School and phenomenology in the twentieth century entails a long, complex journey through Hegel, Marx, and Kropotkin, thinkers who are not represented in the present collection but who are integral to understanding it. Each has had a great influence on contemporary discussion.

Hegel (with Fichte and Schelling) restored organicism and teleology to nature, affirming with Aristotle (and contra Kant and Plato) that the "things themselves" can be known and that they possess *realitas* (though not *veritas*) and *nisus*, a striving to become definite. However, for Hegel, nature is developing into mind and in this sense is not complete. It is a phase in a process, the immediate, self-dirempted reflection of the Idea.[26]

Marx, who inverted and transformed Hegel so that nature or matter grounds consciousness, has been more directly adopted, adapted, and attacked by various schools of ecological thought. Support has been found in his work for positions ranging from a prefigurement of the science of ecology, an implicitly ecological critique of capitalism, and the inspiration for ecological socialism, on one end of the spectrum, to justification for the mastery of nature and an extension of productionism, technocratic thought, and Promethean ideals on the other end.[27]

Finally, Peter Kropotkin challenged the emphasis in evolutionary theory on notions of struggle and competition, calling attention to mutualistic, cooperative, and spontaneous aspects of animal and human societies. His works, such as *Mutual Aid* and *Fields, Factories and Workshops*, have influenced the development of social ecology, utopian criticism and eco-anarchism in the writings of Mumford, Martin Buber, and Bookchin, among others.[28]

Here, we turn to examine the work of Martin Heidegger (1889–1976) and Maurice Merleau-Ponty (1908–1961), two of this century's most profound philosophers and important phenomenologists. In the first instance, Michael Zimmerman reflects not only on Heidegger's contribution to ecological philosophy and deep ecology in particular, but also on his own sustained and influential efforts to introduce Heidegger to ecologically minded philosophers and activists. While still finding much of value in Heidegger's vast opus for ecological theory, Zimmerman warns of dangers and dead ends in his thought which are related to his antihumanism, resurrection of the "question of being," peculiar understanding of *physis* (nature), and political naiveté.

Heidegger was once rather awkwardly called the "metaphysician of ecologism" by George Steiner, and there is some truth to this claim. As Zimmerman has made clear, he seems to offer a kind of environmental ethos (as opposed to ethics). He speaks of "saving" the earth (meaning to free it into its own essence), stresses the primacy of dwelling, criticizes modern technology and science (as stripping nature of its significance), and presents humans as shepherds or stewards of the earth.

Nevertheless, Heidegger's thought is not without its share of problems. First, a residual or latent unexamined anthropocentrism *and* antihumanism are often present in his work. Second, *Dasein*, his unique term for existence, is not understood strictly as human although, in fact, it comes to mean no one and nothing other than the human. Third, there is a glaring lack of the *mit-Sein* (being-with) and intersubjectivity in his writing. Finally, he denies animals the capacity for language and establishes a sharp earth/world opposition which is usually described as a competitive battle or striving rather than as cooperative or symbiotic.

Heidegger grappled with the loss, forgetting, and recession of ground as *Boden* (soil), *Heimat* (home), and *Ort* (place), on the one hand, and, on the other, with the problem of grounding as in philosophical *Grund* (ground), *Gründung* (foundation), *Ur-grund* (primal ground), or *Ab-grund* (abyss). In both *An Introduction to Metaphysics* and his most important work, *Being and Time*, he is concerned with the most basic or fundamental question (*Grundfrage*), which deals with the nature and ground of Being. He repeatedly seeks to uncover and recover that which lies in depth or at the source and, in attempting to bringing this darkness to light, feels compelled to unearth and pile before us a mountain of words sharing the same *root*, in particular *Grund*. For example, "This question with its Why does not move on one level or surface only but penetrates the underlying realms to their ultimate reaches, to the limits; it is opposed to all surfaces and all shallowness and strives for the depth; as the widest question it is also of all deep questions the deepest."[29]

Heidegger's appeal to a *radical* solution (i.e., one that goes to the roots) is, of course, not immune to problems, in particular the charge of nostalgia and a privileging or mythologizing of origins, apparent in both the ontological and existential dimensions of *Being and Time*. He repeatedly instructs us to return, to go back, and to go home again in order to uncover the primordial (*ursprünglich*) "meaning of Being" which has been concealed by the tradition he hopes to dismantle and destroy. However, in counseling a going under (to roots) as a going back (return) in this way, he implies that some unmediated, pure, and authentic experiences and meanings in fact exist or at one time existed, a position which has been put into question effectively by Jacques Derrida, Theodor Adorno, and Jürgen Habermas.[30]

To his credit, Heidegger seeks to avoid *Ursprungphilosophie* in his later thought, and it is here perhaps that an ecophilosophical connection can be made more fully. His notion of *Geviert*, as One with fourfold polyvalence (earth, sky, divinities, and mortals), avoids grounding, explanation, and causality in any traditional sense. In other words, the world is not based on a ground external to it, and no single "element" of the four can provide such a foundation. The concealing-revealing *in* the world is a groundless ground since it is *of* this world.[31]

Merleau-Ponty, on the other hand, developed phenomenology in a different direction. He routed it through and connected it with the lived body, which Descartes had distinguished sharply from the mind (denigrating it in the process) and which Heidegger had avoided or at best marginalized.

Merleau-Ponty's starting point is that we find ourselves in the world and that we are embodied. Moreover, the self is not necessarily defined by the boundaries of the skin but is more regional, spread over a field (what Heidegger had called a "field of care") extending as far as one's concern. Unlike Descartes, who withdraws *from* the world in an attempt to find certainty, phenomenology begins again *in* the world, and consciousness is always *of* the world, not of an isolated mind. The shift is also to the meaning of the world; among other things, the result is a breakdown of the subject/object distinction. In the hands of Husserl, Heidegger, and Merleau-Ponty, phenomenology seeks to bring the things of the world to light and to allow them to be understood directly through an avoidance of presuppositions, a return to a world prior to knowledge yet basic to it, and a rigorous, methodological description. David Abram's essay looks at issues related to language, perception, and embodiment in Merleau-Ponty's late work, finding an unfinished gesture toward the natural world and ecology on which he speculates and creatively expands.

In the body of his work running from *Phenomenology of Perception* (1944) to *The Visible and Invisible* (1964) and passing through a deep encounter with Marxism, Merleau-Ponty focused upon, elucidated, and developed a wealth of ideas and concepts that lend themselves to an ecological sensibility, even if they were not so intended. Among them is *phenomenal field*, a term which deconstructs an opposition between earth and world and perhaps also between nature and culture. It is described as preobjective and like an atmosphere or a *Gestalt*, bearing resemblances to the quasi field theory of ecological science.[32] Similarly, *landscape* as "intentional tissue" and the "setting of our life," which serves as a middle term between earth and world, comes to replace phenomenal field, while his use of *inhabitation* implies a place-based understanding. Merleau-Ponty also values *ambiguity* (in contrast to Cartesian certainty), *experience* (the world as lived through, not simply what one thinks), and *expression* (an alternative to representational thought). Finally, he speculates about what he terms *transcendental geology*, a suggestive attempt perhaps to overcome the metaphysics of grounding.

With Merleau-Ponty, too, the Presocratic notion of the elements with their ecological implications makes a cryptic return, once again in the form of surviving "fragments" of writing, and, in particular, the working notes to *The Visible and the Invisible*. One of the central notions in his later work is that of *la chair* (the flesh), a general name for Being and the bond between body and world. According to Merleau-Ponty, "The flesh is not matter, is not mind, is not substance. To designate it, we should need the old term 'element' in the sense it was used to speak of water, air, earth, and fire."[33] Later he writes: "What we are calling flesh, this intensely worked over mass, has no name in any philosophy. As the formative medium of the object and subject, it is not the atom of being, the hard in itself. . . . We must think of it, as we said, as an element, as the concrete emblem of a general manner of being."[34]

Merleau-Ponty's presentation of flesh in elemental terms is understandable, for he seeks a medium between the body and the world and the visible and the invisible, hoping to express "our living bond with nature," as he puts it in his later lectures at the Collège de France. Connected closely with Merleau-Ponty's idea of the flesh are his interpretation of the earth (especially via Husserl's later writings) and his lectures on the "The Concept of Nature." For example, he writes of the "forgetfulness of the earth" just as Heidegger had explored the forgetting of being (*Seinsvergessenheit*), and claims that the earth serves as the "ground [*sol*] of experience" and "the root of our spatiality, our common homeland."[35]

Neo-Aristotelians and Nature: Arendt, Bloch, and Jonas

In my essay, I use Hannah Arendt (1906–1975) as a starting point to examine the phenomenon of earth alienation, meaning our attempts to transcend or escape our given home and what Arendt calls the "human condition," a term that stands in contrast to an essentializing notion of human nature. Philosophers have arguably been tempted to flee this-world ever since early thinkers like Anaximenes (who described air as originative and divine matter) contemplated the heavens and later ones like Kant placed stress on the *transcendental* dimensions of *universal* knowledge. With the advent of space technology and the subsequent planetary orbits by Soviet cosmonauts Yuri Gagarin and Gherman Titov, we have succeeded literally in leaving the earth. Such speculations, strivings, and successes have transformed us in the process (as have our flights *into* the self, the subatomic, and the psychological), altering our conceptions of nature and the cosmos and our image of humanity. In this sense we are, in the words of Werner Heisenberg, like the "astrophysicists [who] . . . must reckon with . . . the possibility that their outer world is only our inner world turned inside out."[36] Or, as Arendt puts it in interpreting a passage of Plato's *Theaetetus:* "As the rainbow connecting the sky with the earth brings its message to men, so thinking or philosophy, responding in wonder to the daughter of the Wonderer, connects the earth with the sky."[37]

Representing a minority tradition in philosophy, Arendt like Nietzsche—despite the latter's status as a poet and philosopher of air, height, verticality, levity, lightness, and the sublime—enjoins us to "remain faithful to the earth."[38]

In this regard, ecological philosophy counsels an attentiveness to lived place (rather than abstract space), body and matter (not only spirit and mind), changing appearances and phenomena (not simply static essences and noumena), and immanence (along with possible transcendence). It seeks to guide the flight of thought and the thought of flight gently back *down to earth* without cordoning it there unconditionally.

Arendt's concern with home and homelessness, earth and world alienation, and (in *The Life of the Mind*) appearance, surface, semblance, and the (no)where of thought retains an implicitly ecological dimension. In this final book, she moves to consider the work of Merleau-Ponty and biologist Adolph Portmann in order to evince the close relation between eye and environment and animality and appearance, returning to the approaches and insights of phenomenological investigation and *Zu den Sachen selbst* (to the things themselves). Her thought, however, also carries with it ideas and insights for a viable ecological politics since it is concerned with communitarian aims, participatory democracy, direct political action, meaningful work, unnecessary consumption and production, and a critique of modern science and technology.

Ernst Bloch (1885–1977), like Arendt and Merleau-Ponty, is not a name immediately familiar to many ecological theorists and activists. However, John Ely (following suggestions from Bookchin) makes a strong case for his relevance by exploring Bloch's unusual adherence to both Aristotelianism and Marxism as well as his commitments to the communitarian, natural law, and critical utopian traditions. In his interpretation of Feuerbach's eleventh thesis, Bloch speaks of the "resurrection of nature," where the goal is not a blind or regressive identification of humanity with nature but rather the recovery of an objective intentionality in the natural world. Nature is on the horizon, at the frontier, as a not-yet (*noch nicht*), implying the presence of a teleology (dependent ultimately on an ontology), but one more open and flexible than, for instance, in Hegel. "Nature," he writes in *The Principle of Hope*, "is not something that can be consigned to the past. Rather it is the construction-site that has not yet been cleared."[39] Like Arendt and Bookchin, Bloch offers the possibilities of incorporating neo- or left-Aristotelian and civic republican insights into ecological and political thought, expanding and integrating a sense of social place with given ecological niches, limiting and reversing the baleful effects of industrial capitalism and unchecked consumerism, and providing, as Ely argues, not only an embodied democracy but an embedded ethics.

In stark contrast to Bloch's principle of hope and its utopian promise, Hans Jonas (1903–1993) advocates a sober imperative of responsibility, an ethic of restraint, and a heuristic of fear.[40] Like Arendt and Bloch, Jonas is a neo-Aristotelian but, unlike Bloch and like Arendt, he is skeptical of the Marxist tradition and more elusive in his political leanings. Jonas also develops a philosophy of nature which seeks to bridge the perceived chasm between "is" and "ought," "objective" and "subjective," and to bring nature into the realm of human responsibility through an ontology and ethics based "in an objective assignment by the nature of things."[41] Along with other thinkers in the present collection of essays, Jonas sees that "the raping of nature and the civilizing of man go hand in hand" and that "natural science may not tell the whole story about Nature."[42]

Lawrence Vogel explores Jonas's response to Heidegger and the abyss of nihilism (which Nietzsche had diagnosed much earlier as the rejection and annihilation of the earth). In Vogel's view, Jonas attempts to overcome nihilism and to ground an ecological ethic which bypasses the horns of anthropocentrism and biocentrism. This project proceeds through three successive stages: (1) beginning with an existential interpretation of biology and a defense of purposiveness in nature, (2) passing through a metaphysical anchoring of ethical obligation, and (3) culminating in a theology of divine creation that Vogel finds consistent with Jonas's earlier views, but ultimately an unnecessary undertaking.

In addition to his search for a viable ecological ethic, Jonas bequeathed us a systematic philosophical analysis of modern technology, as did Mumford, Heidegger, and Marcuse in different ways. In looking at genetic engineering, cloning, medical support systems, and cybernetics, he voiced with urgency the Pandoran dangers that press threateningly and without public discussion upon the environment, our image of humanity, species survival, and the dignity of the person.[43] Toward historical clarification, he distinguished between classical *technique*—essentially premodern, static, possessable, and existing within well-defined and balanced limits—and modern *technology*, which is dynamic, process-oriented, self-changing, unbounded, and allied with scientific and military interests. As against Baconian methods, Enlightenment ideals, and market forces, Jonas advised limiting certain dangerous quests for knowledge, reigning in unbridled techno-scientific activity, and carefully decelerating modern "progress."

Critical Theory, Science, and Technology: Marcuse and Mumford

Contemporary ecological philosophy finds origins and expressions in both critical theory proper and the Frankfurt School (founded in 1923), which revised Marxism—with its focus on political economy and exploitation—through a critical encounter with scientism, technology, popular culture, art, and psychoanalysis.[44] Through the work, in particular, of Theodor Adorno, Max Horkheimer, and Herbert Marcuse (1898–1979), its theorists also advanced a new, reconciliatory understanding of nature and society. Such a reconciliation might be achieved through recovery of an "objective reason" that inheres deeply in the mind and reality as a guiding, value-laden *logos*, but which has devolved and been supplanted largely by "instrumental reason," or a technique to achieve individual, operational, and pragmatic ends. In *Dialectic of Enlightenment* (1944), Horkheimer and Adorno examined the ways in which the world has been progressively disenchanted, outer nature increasingly dominated, and inner nature repressed as society has adhered blindly to the Enlightenment goals associated with social progress, scientific knowledge, and industrial rationalization.[45] In *Eclipse of Reason* (1946), Horkheimer again advocated the reconciliation of spirit and nature, but admonished against equating the two.[46] Adorno, on the other hand, looked in *Aesthetic Theory* (1970) at the new, discontinuous relation of art to nature, the accompanying eclipse of *Naturphilosophie*, and the decelebration and destruction of natural beauty upon which autonomous aesthetics predicates itself. Elsewhere, he sought to "dialectically overcome the usual antithesis of nature and history."[47]

Through radical revisions of Marx and Freud, Marcuse formulated the first steps toward a new emancipatory science and technology, speaking at the same time of the "liberation of nature"—where external nature is treated as a subject and inner nature freed from psychic repression—and stressing the need for an aesthetic and erotic revolutionary praxis in order to create a new society. In *Counterrevolution and Revolt*, for example, he writes:

> In the established society, nature itself, ever more effectively controlled, has in turn become another dimension for the control of man: the extended arm of society and its power. Commercialized nature, polluted nature, militarized nature cut down the life environment of man, not only in an ecological but also in a very existential sense. It blocks the erotic cathexis (and transformation) of his environment; it deprives man from finding himself in nature, beyond and this side of alienation; it also prevents him from recognizing nature as a subject in its own right—a subject with which to live in a common human universe.[48]

Henry Blanke's essay explores themes related to utopia and domination in Marcuse's work and especially their relevance for ecological thought, finding in Marcuse a strong critique of technological domination, an illumination of the modes of nonrepressive reason, and a prefigurement of the insights of ecological feminism.

"All thinking worthy of the name must now be ecological," Lewis Mumford (1895–1990) remarked in *The Myth of the Machine*.[49] Toward this end, his own thought moves with vision, force, and suggestiveness, anticipating by decades many of the issues and problems with which we are now confronted. In the 1930s, for example, he spoke presciently of the need for the "restoration of the balance between man and nature," including "conservation and restoration of soils, the re-growth wherever this is expedient and possible, of the forest cover to provide shelter for wild life and to maintain man's primitive background as a source of recreation . . . reliance upon kinetic energy—sun, falling water, and wind . . . [and] conservation of minerals and metals."[50] This balance in the natural environment was to be accompanied by an equilibrium in population, industry, and agriculture, which would have the effect of undercutting much of the basis for capitalism.

Mumford advocated a science and technology based on "an earth-centered, organic, and human model" and advanced an original historical understanding and critique of technology which began with neolithic developments and extended into the rise of what he termed the megamachine age. He distinguished between a tradition of authoritarian megatechnics and one of democratic polytechnics; the former emphasizes large-scale, centralized, and hierarchical systems while the latter stresses autonomy, arts and crafts, the human scale, and direct governance by the people.[51] His proposed solutions and suggestions do not imply pessimism, neo-Luddism, or a romantic "return to nature," as they have sometimes been characterized, but rather seek to infuse the democratic and ecological practices of the past into the present, to inform mechanical and technological developments with the organic and biological, and to integrate the cities with the landscape and countryside, and the social with the natural world.

Mumford is arguably as much an ecological anarchist as an ecological socialist, a characterization proffered by Ramachandra Guha in his critical resuscitation of the neglected polymath, whose stature as an American ecological thinker, he believes, should be equal to that of Aldo Leopold and John Muir. Mumford's emphasis on the small scale, decentralization, radical democracy, regionalism, and antiauthoritarianism—in combination with his ties to Peter Kropotkin, William Morris, and Henry David Thoreau, among others—places him in the anarchist tradition, at least peripherally. Moreover, his work has been taken up by eco-anarchists like Bookchin, Roszak, and Kirkpatrick Sale. Bookchin, in particular, has developed and expanded many of Mumford's insights on technics, "second nature," decentralization, and cities in a more radical direction, addressing in his own social ecology what Guha calls that "curious silence" in Mumford's work regarding social movements.

Mumford's advocacy of regionalism, too, has found a second life in the movement for bioregionalism, which seeks to model human communities on the native biota, land, watersheds, soil, and climate of particular places, which will be "reinhabited."[52] In terms of both human values and natural entities, Mumford described regional planning in the twenties and thirties as a "New Conservation." The civic objective of this "organic planning," as he also termed it, is captured in the concept of the garden city (which Mumford borrowed from Ebenezer Howard), where regions have a basis not in "artificial boundaries, drawn on the map with the aid of the ruler" but instead coexist with a respect for the climate, soil, and vegetation as well as the needs of culture and industry.[53]

Pesticides, Pollution, and Population: Carson, Commoner, and Ehrlich

If Mumford is a forgotten but important ecological philosopher, Rachel Carson (1908–1964) was an extremely influential but unlikely environmental reformer whose work, *Silent Spring,* helped to generate and define discussion about pesticides, pollution, and environmental degradation during the 1960s and the past two decades. Yaakov Garb reexamines her work, finding in it both a deep challenge to and compromising extension of earlier perspectives on nature and the environment, and contrasting it with Bookchin's neglected but more politically astute and radical work, *Our Synthetic Environment,* which appeared a half-year earlier than Carson's landmark text. Carson's contribution as a female scientist and writer also raises issues related to ecological feminism, a movement to which she lends inspiration even if there is, as in Arendt's writing, an absent or unarticulated "woman question."

The feminist contribution to ecology and the frequent association of nature with women were not to be identified and examined fully until the mid-to-late seventies and the early eighties with the appearance of works by Carolyn Merchant, Susan Griffin, Ynestra King, Karen Warren, and others.[54] In *The Death of Nature,* for example, Merchant detailed the way in which the representation of an organic cosmos with a living, female earth at its center was replaced in the sixteenth and seventeenth centuries by a mechanistic perspective in which nature was depicted as unformed, inert matter to be controlled or *in-formed* by male society. She proceeded to explore the manner in which nature and women have been interwined historically in the male mind and how this conjunction has worked to the detriment of both.[55] In Latin and Romance languages, as in Greek, "nature" is a feminine noun, thus presented in female terms.[56] Similarly in Platonic and Neoplatonic thought, nature and matter (which are essentially passive) are feminine while the Ideas (more perfect and pure) are masculine. Though the image of nature and earth as nurturing mother has often established an ethical relationship to these realms (for example, limiting mining and deforestation during the Renaissance), the opposing image of nature as wild, disorderly, and uncontrollable has also established a pretext for "reigning her in."

In a more recent work, *Ecological Revolutions,* Merchant extends this project to look at ecological transformations, accompanied by shifts in gender relations and representations of nature, that have occurred in New England as a result of changes and contradictions in the modes of production and reproduction.[57] Theoretical efforts of this kind by ecofeminists increasingly find corresponding, concrete expressions of resistance and reconstruction by women throughout the world: protest by Native American women against the threats of uranium mining on their reservations, encampments at Greenham Common in England, opposition to the lumber industry in India, reforestation of areas in Kenya to halt desertification, work in the formation of the German Greens, and numerous other projects and actions.[58]

In addition to pollution, theorists of the 1970s addressed the population explosion and its effect on the natural world. Two figures, Barry Commoner and Paul Ehrlich, became the most visible advocates for divergent responses to population and pollution control, as Andrew Feenberg details in his essay. The contrast between their two approaches illustrated an old tension between biological and social analysis, and between the claims of science and those of philosophy and political theory. It presaged as well divisions within the environmental movement over class, capitalism, consumerism, and race, voluntarist or state solutions, and individual conscience or governmental coercion. In this respect, the spectre of Hobbes and what Feenberg terms "the politics of survival" resurfaced in an especially telling form. An understanding of this controversy and its vicissitudes will help us undoubtedly to move beyond mere survivalism and toward a more ecological, democratic culture.

Revisions of the Frankfurt School: Habermas and Bookchin

Though some important twentieth-century philosophers and thinkers have recognized and responded to the deepening crisis in our relation to the natural world, at least one major thinker on the left, Jürgen Habermas (born 1929), has balked at the notion and distanced himself from theory and action seeking to redress it. As a second-generation Frankfurt School theorist who has witnessed attempts to invoke an atavistic conception of nature, Habermas is skeptical toward *Naturphilosophie* and proposals to "resurrect nature" because of their past and potential reactionary effects. He rejects the objectivism of a natural ontology and advocates instead an instrumental relation to nature under the aegis of preserving the gains of Enlightenment thought. Joel Whitebook's essay, which appears here with a new introduction, helped to define and generate critique of Habermas's inability or reluctance to consider the political and social claims of ecology. Whitebook also sketches an alternative, more *naturfreundlich* avenue Habermas might have taken based on his theoretical commitments to communicative reason, quasi-transcendentalism, and modernity. This approach sees the ecological crisis as having social origins in the pressures of economic expansion and the attendant strains of modern science and technology.

More recently, John Dryzek has argued for an extension of Habermasian commmunicative rationality and communicative ethics to the natural world, drawing upon critical theory, the Gaia hypothesis, and an expanded understanding of agency.[59] Others, like Thomas McCarthy, have suggested different paths to Habermas's division of reason, his concept of nature-in-itself, and understanding of technology.[60] Thus, there are a growing number of attempts to revise some of the major tenets and trajectories of Habermas's work despite the fact that he himself still believes that the ecology, antinuclear, and other "new social movements" (with the exception of feminism) are primarily defensive, neoromantic, unrealistic attempts to carve out areas of liberation within the life-world.[61]

In contrast to Habermas, whom he believes is misrepresenting the best ideas and ideals of the Frankfurt School and reifying or reducing liberatory concepts, Murray Bookchin (born 1921) has been involved with radical ecological issues and direct political action since the 1940s. By reversing the direction of the theory advanced by Frankfurt School theorists and critiquing their image of nature, he has advanced one of the most developed theories of the historical relation of the idea of dominating nature to the actual domination of humans by one another. He places a new and original focus on hierarchy (as opposed to class), freedom (not simply justice), the contributions of organic cultures, and the anarchist rather than Marxist tradition. Bookchin, too, relies upon the reconstructive and integrative insights of the science of natural ecology to develop his theory of social ecology.

Social ecology stresses the graded emergence of mind from nature; the specific and complex continuities between first (or biological) and second (or cultural) nature and their transcendent synthesis into a *free* nature and ecological society; and the strength of the organismic and dialectical tradition in Western thought (especially in Aristotle and Hegel). In short, Bookchin finds in nature an evolving ground for ethics, autonomy, and freedom. In his view, nature, like culture, develops dialectically and with directionality toward ecological complexity, unity in diversity, complementarity, and spontaneity. In their essay, Andrew Light and Alan Rudy praise and appraise Bookchin's efforts, but in defending a *socialist* as opposed to *social* ecology, also raise questions regarding his ability to come to terms with social labor and global or Third World problems.

Conclusion

Philosophers, Henri Bergson once observed, "seem to philosophize as if they were sealed in the privacy of their study and did not live on a planet surrounded by the vast organic world of animals, plants, insects and protozoa, with whom their life is linked within a single history."[62] Though generally and historically accurate, the essays in this collection seek to challenge this image, offering instead ideas which emphasize our natural relations to the earth, our social creations, and each other. They explore the possibility and necessity of an ecological critique of capitalism, technology, and the history of ideas, providing a philosophical understanding of the science of ecology[63] and the politics of nature. Perhaps, then, one could add a conditional to Mephisto's famous remark in Goethe's *Faust:* "Gray . . . is all theory, green alone life's golden tree." When and if our ideas are "greened," they might help to guide and inspire our interpretations of social, political, and biological life rather than casting a hazy pall of darkness over them. At such a time, we might also move toward both democracy and ecology.

Acknowledgments

I would like to thank James O'Connor, Barbara Laurence, Peter Wissoker, and the various contributors to this volume for their assistance and encouragement in seeing this project to its completion and for responding promptly to my comments, criticisms, questions, and solicitations. I wish to thank Michael Zimmerman, John Ely, and James O'Connor for comments on this essay.

Notes

1. See J. Donald Hughes, "Theophrastus as Ecologist," in *Theophrastean Studies*, vol. 3, ed. William Fortenbraugh (New Brunswick, N.J.: Transaction Books, 1988), pp. 67–75, and Hughes, "Early Greek and Roman Environmentalists," in *Historical Ecology: Essays on Environment and Social Change*, ed. Lester J. Bilskz (Port Washington, N.Y.: Kernnikat Press, 1980), pp. 44–59.

2. It must be acknowledged, however, that conceptual hierarchies and oppositions were forged quite early. Max Horkheimer and Theodor Adorno, for example, remark:

> The categories by which Western philosophy defined its everlasting natural order marked the spots once occupied by Oncus and Persephone, Ariadne and Nereus. The pre-Socratic cosmologies preserve the moment of transition. The moist, the indivisible, air, and fire, which they hold to be the primal matter of nature, are already rationalizations of the mythic mode of apprehension. Just as the images of generation from water and earth, which came from the Nile to the Greeks, became here hylozoistic principles, or elements, so all the equivocal multitude of mythical demons were intellectualized in the pure form of ontological essences. (*Dialectic of Enlightenment*, trans. John Cumming [New York: The Seabury Press, 1972], pp. 5–6)

3. The cosmology of Empedocles (ca. 492–432 B.C.), for example, is emblematic. His egalitarian notion of *rhizomata* or "roots of all"—thematized as the four elements—is particularly significant for a philosophy of nature. The *rhizomata* rest in equipoise (*atalanton*) and are equal with respect to (1) origin and age, (2) strength and its related concept of honor, and (3) rule, which is carried out cyclically and, in turn, in a manner analogous to the democratic governance of the *polis* by citizens.

4. *Kosmos* is related in Greek to the active, transitive verb *kosmeo*, which means to arrange or set in order, so that it can be said that the physical world of nature is brought by intelligence and through a mythic cosmology into a harmonious arrangement that is necessarily both *cosmetically* beautiful and moral. That is, it is actively ordered, created,

or constituted out of a primordial disorder or chaos (at least, this is how it is characterized in Plato's work).

5. In other words, how and in what form might a new "green reason" appear?

6. See, for example, William Leiss, *The Domination of Nature* (Boston: Beacon Press, 1972); Morris Berman, *The Reenchantment of the World* (New York: Bantam, 1984); Neil Evernden, *The Natural Alien* (Toronto: University of Toronto Press, 1985); and Susan Griffin, *Woman and Nature* (New York: Harper and Row, 1978).

7. Karl Marx, *Capital*, vol. 1 (New York: International Publishers, 1967), p. 390.

8. In *The Advancement of Learning* (London: Longmans Green, 1870), Bacon argues for a natural philosophy which searches "into the bowels of nature" while at the same "shaping nature as on an anvil." *The Works of Francis Bacon*, vol. 4, ed. James Spedding et al., p. 343.

9. See in particular Bacon's *New Organon* in *Works*, vol. 4, and Descartes's *Discourse on Method*, trans. Donald A. Cress (Indianapolis, Ind.: Hackett, 1980). Descartes speaks of methods of "knowing the force and actions of fire, water, air, stars, the heavens, and all the other bodies" so as to "use these objects for the purposes for which they are appropriate, and thus make ourselves, as it were, masters and possessors of nature" (p. 33).

10. See Thomas Hobbes, *Leviathan*, ed. Michael Oakeshott (New York: Collier Books, 1962).

11. William Ophuls, *Ecology and the Politics of Scarcity* (San Francisco: W. H. Freeman, 1977), p. 151.

12. For a discussion of Catton, Hardin, and the works of others in this vein, see George Bradford, *How Deep is Deep Ecology?* (Hadely, Mass.: Times Change Press, 1989).

13. Garrett Hardin, "The Tragedy of the Commons," *Science* 162 (1968): 1243–1248 and "Living on a Lifeboat," in *Social Ethics*, ed. Thomas Mappes and Jane Zembaty (New York: McGraw-Hill, 1982), pp. 365–372.

14. Georg Lukács, *History and Class Consciousness: Studies in Marxist Dialectics*, trans. Rodney Livingstone (Cambridge, Mass.: The MIT Press, 1968), p. 234.

15. Theodor Adorno, *Aesthetic Theory*, trans. C. Lenhardt (New York: Routledge and Kegan Paul, 1984), p. 101.

16. Lynn White, "The Historical Roots of Our Ecologic Crisis," *Science* 155 (1967): 1203–1207.

17. Paul Shepard writes: "Though important to the roots of Western spiritual life, the desert for the Hebrews was not valued as a place. It was a vacuum, idealized as a state of disengagement and alienation, a symbol of the condition of the human spirit." *Nature and Madness* (San Francisco: Sierra Club Books, 1982), p. 58. See also David Miller, *The New Polytheism* (New York: Harper and Row, 1974), and George Steiner, *Bluebeard's Castle* (New Haven, Conn.: Yale University Press, 1971). For a balanced historical discussion, see Susan Power Bratton, "The Original Desert Solitaire: Early Christian Monasticism and Wilderness," *Environmental Ethics* 10, no. 1 (1988): 31–53.

18. This account by Steiner and others should not be construed as blaming the Hebrews who "created" monotheism for all its consequences or, naturally, for condoning persecution. Nor should such characterizations of the desert, which often underscore its barrenness and dualities, be taken as entirely accurate; desert ecology and literature reveal a world more complex and beautiful than mere wasteland. See, for example, Edward Abbey's *Desert Solitaire* (New York: Simon and Schuster, 1968). In his discussion of ascetic ideals, Nietzsche, too, warns: "The desert . . . where the strong, independent spirits withdraw and become lonely—oh, how different it looks from the way educated people imagine a desert!—for in some cases they themselves are this desert, these

educated people." *On the Genealogy of Morals*, trans. Walter Kaufmann (New York: Vintage Books, 1967), p. 109.

19. James Hillman, *The Dream and the Underworld* (New York: Harper and Row, 1979), p. 147.

20. See Mark Poster, ed., *Harmonian Man: Selected Writings of Charles Fourier* (New York: Doubleday and Company, 1971).

21. Murray Bookchin, *The Ecology of Freedom* (Palo Alto, Calif.: Chesire Books, 1982), p. 331.

22. Lewis Mumford, *The Story of Utopias* (New York: Viking Press, 1950), pp. 5–6.

23. Bookchin, *Ecology of Freedom*, p. 325.

24. For a study of the relationship between communitarian living and ecological thought, see David Pepper, *Communes and the Green Vision* (London: Green Print, 1991).

25. Marx and Engels criticized the approach of utopian socialists, though they credited Fourier with recognizing class antagonisms and providing a strong critical element which was valuable for working-class enlightenment. See Karl Marx and Frederick Engels, *The Communist Manifesto*, pt. 3, ed. Samuel Beer (Arlington Heights, Ill.: AMH Publishing Corporation, 1955).

26. See especially Hegel's *Phenomenology of Spirit*, trans. A. V. Miller (Oxford University Press, 1977), and his *Philosophy of Nature*, ed. M. J. Petry (London: Unwin Brothers Ltd., 1970).

27. The literature on Marx and ecology is extensive. For a sampling of views, see Alfred Schmidt, *The Concept of Nature in Marx*, trans. Ben Fowkes (London: New Left Books, 1971); John Clark, "Marx's Inorganic Body," *Environmental Ethics* 11 (1989): 243–258; Howard Parsons, *Marx and Engels on Ecology* (Westport, Conn.: Greenwood, 1978); Reiner Grundmann, "The Ecological Challenge to Marxism," *New Left Review*, 187 (May/June 1991): 103–120; John Ely, "Marxism and Green Politics in West Germany," *Thesis Eleven*, no. 13 (1986): 22–38; and the ongoing discussions in *Capitalism, Nature, Socialism*.

28. Kropotkin argues for the abolition of the distinction between city and country through a creative combination of small-scale industry and agriculture. Kropotkin advocates as well a "synthesis of human activities" which would overcome the division of society into intellectual and manual labor. He stresses the need for integration and balance—for the individual, economy, and society—and underscores the importance of decentralization and regionalism in industry. His definition of economics as "a science devoted to the study of the needs of men and of the means of satisfying them with the least possible waste of energy" is implicitly attentive to the ecological dimensions of the *oikos*. Kropotkin criticizes neglect and waste of the land and advocates a new, radical agriculture, envisaging farms on a horticultural and garden model which would support more people and be labor-intensive. See *Fields, Factories and Workshops*, ed. Colin Ward (New York: Harper and Row, 1974); Kropotkin, *Mutual Aid* (Boston: Extending Horizons Books, n.d.); and David Macauley, "Evolution and Revolution: The Ecological Anarchism of Kropotkin and Bookchin," in *Anarchism, Nature, and Society: Critical Perspectives on Social Ecology*, ed. Andrew Light (New York: Guilford Press, in press).

29. Martin Heidegger, *Introduction to Metaphysics*, trans. Ralph Mannheim (Garden City, N.Y.: Doubleday and Company, 1959), p. 3, and *Being and Time*, trans. John Macquarrie and Edward Robinson (New York: Harper and Row), 1962.

30. See Jacques Derrida, *Margins*, trans. Alan Bass (Chicago: University of Chicago Press, 1982), pp. 111–136; Theodor Adorno, *Negative Dialectics*, trans. E. B. Ashton (New York: Seabury Press, 1973), p. 62; and Jürgen Habermas, *The Philosophical Discourse of Modernity*, trans. Frederick G. Lawrence (Cambridge, Mass.: The MIT Press,

1987), pp. 131–160; Allen Megill, *Prophets of Extremity* (Berkeley: University of California Press, 1985), pp. 120–125.

31. See especially the essays collected in Martin Heidegger, *Poetry, Language, Thought*, trans. Albert Hofstadter (New York: Harper and Row, 1971).

32. Maurice Merleau-Ponty, *Phenomenology of Perception*, trans. Colin Smith (London: Routledge and Kegan Paul, 1962), pp. 52ff.

33. Maurice Merleau-Ponty, *The Visible and Invisible*, trans., Alphonso Lingis (Evanston, Ill.: Northwestern University Press, 1968), p. 130.

34. Ibid., p. 147.

35. Merleau-Ponty, "Husserl's Concept of Nature," trans. Drew Leder in *Texts and Dialogues*, ed. Hugh J. Silverman and James Barry Jr. (Atlantic Highlands, N.J.: Humanities Press, 1992); and *Themes from Lectures at the Collège de France, 1952–1960*, trans. John O'Neill (Evanston, Ill.: Northwestern University Press, 1970).

36. Quoted in Hannah Arendt, *The Life of the Mind* (New York: Harcourt Brace Jovanovich, 1978), p. 167.

37. Ibid., p. 142.

38. Friedrich Nietzsche, *Thus Spoke Zarathustra*, trans. Walter Kaufmann (New York: Penguin, 1966), p. 13.

39. Ernst Bloch, *Daz Prinzip Hoffnung*, vol. 2 (Frankfurt: Suhrkamp Verlag, 1967), p. 806. Quoted in Bookchin, *The Ecology of Freedom*, p. 34.

40. For a defense of Bloch against Jonas on points related to ecology, utopia, and community, see John Ely, "An Ecological Ethic? Left Aristotelian Marxism versus the Aristotelian Right," *Capitalism, Nature, Socialism*, no. 2 (Summer 1989): 143–154.

41. Hans Jonas, *The Phenomenon of Life: Towards a Philosophical Biology* (New York: Harper and Row, 1966), p. 283.

42. Hans Jonas, *The Imperative of Responsibility* (Chicago: University of Chicago Press, 1984), pp. 2, 8.

43. For an elaboration of these themes, see Hans Jonas, *Technik, Medizin und Ethik: Zur Praxis des Prinzips Verantwortung* (Frankfurt: Insel Verlag, 1985), and *Philosophical Essays: From Ancient Creed to Technological Man* (Englewood Cliffs, N.J.: Prentice-Hall, 1974), as well as Wolf Schäfer, "Die Büchse der Pandora: Über Hans Jonas, Technik, Ethik und die Träume der Vernunft," *Merkur* 43 (April 1989): 292–304.

44. For further discussion of these connections, see Vincent Di Norcia, "From Critical Theory to Critical Ecology," *Telos* 22 (1974–1975): 86–95; and Andrew Dobson, "Critical Theory and Green Politics," in *The Politics of Nature*, ed. Andrew Dobson (New York: Routledge, 1993); and Robyn Eckersley, *Environmentalism and Political Theory* (Albany: State University of New York Press, 1992).

45. Like Marcuse, in certain respects they seem to have anticipated ecological feminism, pointing out how women "became the embodiment of the biological function, the image of nature, the subjugation of which constituted that civilization's title to fame" and adding that "for millenia men dreamed of acquiring absolute mastery over nature, of converting the cosmos into one immense hunting ground." *Dialectic of Enlightenment*, p. 248.

46. For example, Horkheimer writes: "The equating of reason and nature, by which reason is debased and raw nature exalted, is a typical fallacy of the era of rationalization. Instrumentalized subjective reason either eulogizes nature as pure vitality or disparages it as brute force, instead of treating it like a text to be interpreted by philosophy that, if rightly read, will unfold a tale of infinite suffering. Without committing the fallacy of equating nature and reason, mankind must try to reconcile the two." *Eclipse of Reason* (New York: Oxford University Press, 1946), p. 125.

47. Theodor Adorno, "The Idea of Natural History," *Telos* 60 (Summer 1984): 111.

48. Herbert Marcuse, *Counterrevolution and Revolt* (Boston: Beacon Press, 1972), p. 60.

49. Quoted in Anne Chisholm, *Philosophers of the Earth: Conversations with Ecologists* (London: Sidgwick and Jackson, 1971), title page. In conversation with Chisholm, Mumford goes on to say, "Ecology has always been part of my thinking. . . . It hasn't been an independent subject. I think about the entire complex, the whole environment, not in terms of a fragment of it. That is ecological thinking" (p. 2).

50. Lewis Mumford, *Technics and Civilization* (New York: Harcourt, Brace and World, 1963), p. 430.

51. This distinction echoes Kropotkin's much earlier statement: "Throughout the whole history of civilization, two opposed tendencies have been in conflict; the Roman tradition and the popular tradition; the imperialist tradition and the federalist tradition; the authoritarian and the libertarian." See Peter Kropotkin, *Modern Science and Anarchism*, in *Kropotkin's Revolutionary Pamphlets*, ed. Roger Baldwin (New York: Dover Publications, 1970).

52. See, for example, Peter Berg, "What is Bioregionalism?" *The Trumpeter* 8 (1991): 6–8; and Kirkpatrick Sale, "Bioregionalism—A New Way to Treat the Land," *The Ecologist* 14 (1984): 167–173.

53. In *The Culture of Cities*, Mumford writes, "The human region . . . is a complex of geographic, economic, and cultural elements. Not found as a finished product in nature, not solely the creation of human will and fantasy, the region, like its corresponding artifact, the city, is a collective work of art" (p. 367). See also *The Lewis Mumford Reader*, ed. Donald Miller (New York: Pantheon Books, 1986), and Lewis Mumford, *The City in History* (New York: Harcourt, Brace and World, 1961), especially pp. 514ff.

54. See, for example, *Healing the Wounds: The Promise of Ecofeminism*, ed. Judith Plant (Philadelphia: New Society Publishers, 1989).

55. Carolyn Merchant, *The Death of Nature* (New York: Harper and Row, 1980).

56. One finds in Spanish, *la natura*; German, *die Natur*; French, *la nature*; and Italian, *la natura*.

57. Carolyn Merchant, *Ecological Revolutions: Nature, Gender, and Science in New England* (Chapel Hill: University of North Carolina Press, 1989).

58. Merchant discusses many of these projects in her most recent book, *Radical Ecology* (New York and London: Routledge, 1992), pp. 183ff.

59. John Dryzek, "Green Reason: Communicative Ethics for the Biosphere," *Environmental Ethics* 12 (Fall 1990): 195–210.

60. Thomas McCarthy, *The Critical Theory of Jürgen Habermas* (Cambridge, Mass.: The MIT Press, 1978), pp. 110ff. See also recent works by Henning Ottman and Mary Hesse.

61. Habermas now acknowledges an "ecological problematic," but claims that it "can be dealt with satisfactorily within the anthropocentric framework of a discourse ethic." Jürgen Habermas, "A Reply to My Critics," in *Habermas: Critical Debates*, ed. John B. Thompson and David Held (London: Macmillan, 1982).

62. William Barrett paraphrasing Henri Bergson, in Barrett, *The Illusion of Technique* (Garden City, N.Y.: Anchor Press/Doubleday, 1979), p. 365.

63. For a critique of the *science* of ecology and a defense of natural history, see Robert H. Peters, "From Natural History to Ecology," *Perspectives in Biology and Medicine* 24 (Winter 1980): 191–203. On the compatibility of ecological science and deep ecology, see Frank Golley, "Deep Ecology from the Perspective of Ecological Science," *Environmental Ethics* 9, no. 1 (1987): 45–55.

Chapter One

Nature as Artifact: Thomas Hobbes, the Bible, and Modernity

Frank Coleman

Physiologically, man in the normal use of technology (or his variously extended body) is perpetually modified. Man becomes, as it were, the sex organs of the machine world, as the bee of the plant world, enabling it to fecundate and to evolve ever new forms. The machine world reciprocates man's love by expediting his wishes and desires, namely, in providing him with wealth. One of the merits of motivation research has been the revelation of man's sex relation to the motor car.

—MARSHALL MCLUHAN, *Understanding Media*

With every tool man is perfecting his own organs, whether motor or sensory, or is removing the limits to their functioning. Motor power places gigantic forces at his disposal, which like his muscles, he can employ in any direction; thanks to ships and aircraft neither water nor air can hinder his movements; by means of spectacles he corrects defects in the lens of his own eye; by means of the telescope he sees into the far distance; and by means of the microscope he overcomes the limits of visibility set by the structure of his retina. . . . Man has, as it were, become a kind of prosthetic God.

—SIGMUND FREUD, *Civilization and Its Discontents*

For by Art is created that great Leviathan called a Common-Wealth, or State, (in latine Civitas) which is but an Artificiall Man; though of greater stature and strength than the Naturall, for whose protection and defence it was intended.

—THOMAS HOBBES, *Leviathan*

First published in *Capitalism, Nature, Socialism* 4, no. 4 (December 1993): 89–101; reprinted here with minor corrections.

Introduction

The subject of this essay is Thomas Hobbes (1588–1679), whose political thought is widely held to be pivotal to an understanding of the principles underlying the modern state.[1] Hobbes's life spanned the period of the English civil wars (1642–1646) and his major work, *Leviathan* (1651), followed the execution of Charles I at their conclusion. But it was not simply domestic upheaval that gave occasion for Hobbes's reflections on political order. Of at least equal significance were the changes occurring in paradigms of scientific explanation,[2] in the fundamental religious ethos,[3] and in social and economic relationships.[4] Hobbes's thought embraces all these changes. His method of exposition in *Leviathan* is classical in form; he proceeds according to a definition of publicly significant terms: "natural right," "covenant," "representation," "sovereignty," "law." However, the content supplied within this classical form is distinctively modern in nature, as I hope to make clear. This essay departs from conventional treatments of Hobbes's political thought by placing stress on the natural philosophy which underlies his theory of the state, bringing to view important relationships between the biblical sources of Hobbes's idea of nature, the Leviathan state, and the contemporary ecological crisis.

My thesis is that Hobbes effects a "modernity synthesis" (consisting of liberalism, capitalism, and modern science) which reassigns to man, considered universally, a new and extraordinary role, that of a "prosthetic God" who would by the artifacts of his creation duplicate and even exceed the creative accomplishments of the God of the Old Testament. Like the God of the Old Testament, this prosthetic God is the creator of a new world, one in which the deficiencies of the Creator's work are amended through the artifacts of man. I further offer that this vision has dominated the intellectual landscape of the West, particularly in the United States and that it is a principal reason that the domain of nature is presently at risk.

Let us consider the earliest expression of this vision of human purpose. Hobbes begins the *Leviathan* thus, "Nature (the Art whereby God hath made and governes the World) is by the Art of Man, as in many other things, so in this also imitated." Man can make government through pacts and covenants resembling, Hobbes says, "that Fiat, or the 'Let us make man,' pronounced by God in the Creation," and he can make "automata" or, as we would say today, machines.[5] In this passage Hobbes is directly comparing the creative powers of man with those of God, while at the same time indicating the arena in which they are to be appropriately deployed. Just as God created the world in the beginning, so will man imitate God's creative activity in an effort to remedy its evident shortcomings. Nature, God's handiwork, will be upstaged in a new order featuring the products of human artifice. And as God's artifact, the world, recedes in importance, the artifacts of man will come to occupy a position of premiere significance.

Hobbes's comparison of the creative activity of man with that of God, reinforced by a hint that man's powers may exceed those of God, may seem gratuitous (not to mention scandalous), but it is in fact essential to his purpose. So long as the world is viewed as the product of a Maker, God himself, whose power is diminished once found to originate in human agency, then to that extent the intentional structure of creation will remain hidden. Hobbes's statement that man "imitates" the artifice of God forces the reflection that man, at the threshold of the Protestant-capitalist era, should *consciously* assume some of the functions of making which were formerly the exclusive province of the Hebraic God.

In what follows, I attempt first to show that the inspiration for the manner of artifice undertaken by Hobbesian man is biblical. Hobbes borrows selectively from the Bible, overlooking or ignoring passages which are not in accord with his overall design in the course of fashioning the modernity project. Next, I shall attempt to develop the view that the artifact of existence produced by modernity has acquired the status of an ineluctable order of nature at the expense of alternative visions. Whereas in Hobbes the advance over nature represented by modernity is safeguarded through the coercive power of the sovereign, as the modernity project acquires purchase over the structures of daily life it is transformed into a social rather than a political structure. When the members of the community come to perceive themselves as prosthetic Gods, achieving self-extension through the modes of artifice which political order makes possible, then the need for a coercive sovereign to maintain the external conditions of public order is dissipated. It is at that point that modernity becomes identified with a pure order of nature, that is, a world where the artifact of modernity is camouflaged in the drapery of the "natural."

This essay views the natural world as imperiled by the modernity project. In the terms of this essay, this fear is not irrational, fantastic, or capricious. It is not irrational because nature, in Hobbes as in the Scriptures from which he draws his inspiration, is nothing other than an artifact produced by man as he strives to overcome the tensions between pain and imagination in his encounter with the world. Thus defined, there are no limits in nature which man can transgress in the process of remaking the world closer to his heart's desire. As in the Old Testament, there truly is no such phenomenon as nature. There is only creation. As the Old Testament God created a world and found it good, so the modernity project—fashioned by the prosthetic gods of our age—has given us a world associated with an invincible order of things. Those who question the wisdom of this invincible order may well encounter the rebuke of Isaiah, "Shall the potter be regarded as the clay; that the thing made should say of its maker, 'He did not make me'; or the thing formed say of him who formed it, 'He has no understanding'?" (Isa. 29:16).

Biblical Creation

The Bible begins with a project of world creation. It records that a solitary, powerful, procreative will—God—creates the world in the beginning through fiat (Gen. 1, 2). As the creation epic makes clear, the world has no independent physical existence save as the product of God's creative will. God's activity as a powerful, procreative force is duplicated on many subsequent occasions. He makes Adam, a word properly translated from the Hebrew as mankind. He makes wombs, formerly barren, to become fertile (Gen. 17, 25:21). He covenants with Abraham, promising that his descendants will be a mighty nation as numerous as the sands of the seashore or the stars of heaven (Gen. 15:5, 22:17), a promise inherited by and fulfilled in the people of Israel.

Although the world produced in the beginning is pronounced "good," it is so only because it is an artifact, and its role thereafter is confined to serving as a backdrop against which numerous additional acts of artifice may be displayed. One scholar notes that for the Old Testament mind, there truly is no such thing as nature (*natura*); there is only creation (*creatura*).[6] Another notes that there is no word in Hebrew for "nature" in its current meaning as "the totality of processes and powers that make up the universe."[7] God's original act, the creation of the world, conceives the entire cosmos as the proper domain for acts of artifice. All forms of material, cultural, and political creation originate with God, are substantiated by Israel and the world, and are unhindered in the manner of their expression by the natural realm.[8]

This biblical outlook is evident in many passages where the object of celebration is God's power over creation, not the existence of natural process. God, the psalmist says, transforms the desert landscape. "He turns a desert into pools of water, a parched land into springs of water. And there he lets the hungry dwell, and they establish a city to live in; they sow fields and plant vineyards, and get a further yield. By his blessing they multiply greatly; and he does not let their cattle decrease" (Ps. 107). It is not hyperbole for the prophet, Isaiah, to suggest that through God's project of world re-creation, "Every valley shall be lifted up, and every mountain and hill laid low; the uneven ground shall become level, and the rough places plain, and the glory of the Lord shall be revealed" (Isa. 40:4). One searches the Bible in vain for evidence that there might exist natural processes independent of God's agency.

An important object of God's project of world creation and many subsequent acts of re-creation is to enable Israel to overcome the physical limitations of a desert landscape. Speculatively, we may note that to the extent that God succeeds in his project of world creation and re-creation (fertilizing wombs formerly barren, transforming the desert ecology, multiplying the children of Israel, making the land prosper) to that extent, man will become more like God. Liberation from the features of aversive experience—the project of

God—will liberate man from awareness of his body, and liberation from his body will enable him to share in the sublimated activities of the procreative God through whom the world is transformed. We may push these speculations further by noting that belief in a disembodied, powerful, procreative God serves an important need of a desert people. The harsh, forbidding landscape of Israel could be overcome through a God whose project of world creation surpassed its limiting conditions.

Let us go further still by remarking on the intuition concerning the nature of creation that is implicit in the biblical record. There, the essential attribute of Israel is as body. As Pedersen suggests, Israel is to be considered as one moral, collective, *physical* person.[9] When the Israelites swear to the Mosaic covenant, it is "with one voice" (Exod. 24:3). When they go up to do battle with the Midianites, they smite them "as one man" (Judg. 6:16).

In contrast, the distinguishing characteristic of God is that he is without body. He is a disembodied, procreative spirit engaged in a project of world transformation, but lacking a body, he cannot engage in his project of world transformation independently. He must do so vicariously through Israel. Israel is therefore the sole means by which God's existence is substantiated and through which his project of world transformation may be executed. By the same token, God is the sole means through which Israel can achieve self-extension through acts of artifice.

The relationships between God and Israel duplicate the relationships between pain and imagination, body and spirit, in the creation process. Let us define pain as an intentional state consisting of acute, aversive embodiment, unrelieved by objectification. To be in great pain, Scarry suggests, is to be "de-worlded," that is, to be divested of the objects which make up one's world or even the capacity to imagine them.[10] Conversely, the imagination is an intentional state distinguished by objectification without embodiment.[11] To put this another way, we may think of the imagination as restoring the world, but solely in the form of the objects of the imagination. Sentience is the field of awareness framed by the polar opposition of pain and imagination, and it acts continually to modify perceptual, emotional, and somatic states of consciousness. For example, the manner in which the desert is perceived, as merely scenic or as hostile, is a function of whether one is traveling along the desert floor in an air-conditioned Scenicruiser, or whether one is on foot and solitary, lost in the great wastes of Death Valley. How an approaching thunderstorm is perceived turns upon whether you are hastily seeking the safety of a doorstep or are seated before the evening fire.

Scarry theorizes that creation springs from our nature as sentient beings. More specifically, creation is born from the tension between pain and imagination as each struggles to become the other's missing, intentional counterpart.[12] Creation offers a means for resolving this tension because it supplies a freestanding artifact, thus relieving the imagination of the burden of holding an object before the mind. Simultaneously, it places an object in the world, thus relieving the body of the burdens of solitary, aversive experience. Pain and the imagination, while opposed intentional states (the latter a form of extension without embodiment and the former a manner of embodiment without extension), are nevertheless complementary because each seeks to resolve the deficiencies of its intentional state in the direction indicated by its opposite. Through creation we escape the interior, self-enclosed field of sentient awareness framed by pain and the imagination, putting in its place a freestanding artifact.[13]

The biblical account of God and Israel enacts the above intuition concerning creation. Israel, the body in pain, seeks relief through objectification; God, the imagination totally extended into its objects, seeks relief from the burden of objectification through embodiment. Through the creation process the tension between pain and the imagination, as each seeks to become the other's missing intentional counterpart, is resolved. Israel, God's creation, is relieved of the burdens of solitary, aversive experience by becoming a freestanding artifact in history. God, Israel's creation, is liberated from the burdens of objectification because he now has a freestanding artifact, Israel, through whom he achieves embodiment.

Of these two artifacts, God and Israel, only the latter is an acknowledged creation. Although God and Israel occupy interdependent roles in the process by which the world is to be transformed, God's role as Maker is preeminent. God's distinctive position is recognized by numerous prohibitions on unauthorized forms of making by humans. For example, the building of the Tower of Babel and the creation of a native language are viewed as forms of human making which lack sanction. Thus, the Tower of Babel must be and is dealt with in a horrific manner (Gen. 11:1–9). A greater offense is the making of graven images which are intended to offer a tangible substitute for the disembodied and unrepresentable God of Israel. Such activity, also regarded as audacious and wicked insolence, meets with a devastating reply (Exod. 32:20; Lev. 26:1–2). Finally, the Ten Commandments make clear that God's making alone is to be honored, as in the keeping of the Sabbath, the day on which he rested from his project of world creation. Man participates in God's project of world transformation by *abstaining* from labor on this hallowed day (Deut. 5:12).

From these and like instances, one gathers that there is strong presumptive evidence against the propriety of any form of human making by Israel which does not have the express sanction of God. This precautionary stance is warranted by the danger that Israel, having reappropriated its role in the creation process, will come to "de-construct" its relationship with the Maker; that having de-constructed this relationship, it will re-capture a sense of its own agency, that having re-captured a sense of their own agency, the whole project of transforming the world to reflect the attributes of human sentience will stand imperiled. In some cases, the hazard of de-construction is directly addressed: "Woe to him," warns Isaiah, "who strives with his Maker, an earthen vessel with the potter!" (Isa. 45:9). Israel did not make God. God made Israel. "It is He that hath made us and we are his." (Ps. 100).

The great contradiction of the Old Testament is that while Israel is at least an equal partner in the creation process (after all, it bears the brunt of the effort through which imagination achieves embodiment), the Old Testament mind requires that all the values which arise from this process be credited to God's account.[14] Aversive experience is overcome through Israel, an artifact in history, but the reciprocal and often beneficial effects of artifice upon consciousness—particularly its capacity for remaking human sentience—is truncated or denied.

To illustrate, consider that the Hebraic faith has no literature apart from the Bible, the record of God's procreative power. It has no sculpture, painting, or theater because as mimetic arts, they might be put to use representing God, who is the pure, disembodied principle of creative will.[15] These limitations arise from the emphasis placed by the Old Testament on God's overwhelming power as our Maker. Not only does he possess the authority to make us, but he possesses the correlative power of destroying us (Deut. 4:25–32; Jer. 4:23–26). Indeed, instances of the latter capacity are used as evidence in support of the former.

Modernity

Let us return to the introduction to *Leviathan*. Hobbes assigns "man"—formally "universal man," but in substance the new prosthetic God stepped forward onto history's stage—the role formerly occupied by the great Artificer of the Bible. Nature, the art by which God has made and governed the world, will by the art of man be imitated (perhaps exceeded?) in the making of objects which, unlike those of God's, have a recognizable and recoverable source in man himself. As a result, they may even have more utility in overcoming aversive experience than the objects authored by God.

Further, as the prosthetic God assumes the epic proportions of the solitary, male artificer of the Old Testament, he will spread the facts of human sentience across the realm of material, cultural, and political creation, becoming, like God, both author of a new world and divested of the attributes of embodiment. That is, like the God of the Old Testament, he will become a disembodied spirit who is author of its own world. By virtue of the reciprocal effects of artifice upon consciousness, the prosthetic God will remake himself and, in so doing, will become like God who in the biblical account made man. This, I believe, is the significance of Hobbes's enigmatic statement that the making of the state is like that "Fiat, or the 'Let us make man,' pronounced by God in the Creation."

The making of the state, a "Mortal God," compares with the making of Israel in the Old Testament narratives. It supplies the condition under which the modernity project can commence. As in the Old Testament narratives, the project is to transform the world to reflect the attributes of human sentience. But there is a difference. In Hobbes's case, the state is an unrecognized but recoverable artifact because it is the witting product (by contract) of human agency. In the Old Testament, God is and must remain an unrecognized and unrecoverable artifact because his power as Maker is extinguished once the role of human agency in his own creation is perceived.

Let us consider this distinction between an unrecognized but recoverable artifact, the state, and an unrecognized and unrecoverable artifact, the Hebraic God. Scarry proposes that artifacts be distinguished on the basis of whether observation of the human component in their creation will interfere with their reciprocating effects. For example, affixed to artifacts such as poems, films, and paintings is a highly personal signature. This is so much the case that pointing to two objects in a room a person may say, "This is a Millet and that one is a Van Gogh." About to lower a needle to a record, a person will intone "Handel."[16] In such a case, identification of the human component in the creation of the artifact not only does not interfere but it assists in discharging the reciprocating function of artifice. We may say that because this class of artifacts bears a personal signature, its nature as an artifact is not only recognized and recoverable, but self-announcing.

A second class of artifacts, Scarry suggests, bears a general rather than personal signature, with the consequence that the human component in its creation is recoverable, if not recognizable. As one maneuvers through the dense sea of artifacts that sustain daily life—tablecloths, dishes, lamps, city parks, streets, language, streetlights, armchairs, and so forth—one does not actively perceive these objects as humanly made. However, if one stops for any reason to think about their origins, one can with varying degrees of success recover the fact that they all have human makers; moreover, this recognition will not diminish their usefulness.[17] The Leviathan state is an instance of such a class of artifacts.

A third class of artifacts is distinguished by the circumstance that observation of their human origin will interfere with the reciprocating task for which they were created. In order for the artifact to discharge its function, the earlier, human arc of its creation must remain undisclosed. Marx alleges that capital occupies this role in Western society since, though created by labor, once created it transforms labor into a "commodity," a thing made by capital.[18]

The Hebrew God is an illuminating example of this class of artifacts.

Scarry observes, "There would be no point in inventing a god if it did not in turn reinvent its makers: all that is untrue is that the power of recreation originates in the object, a mistake that occurs by attending only to the second half of the total arc of action."[19] The stress placed by the Old Testament on God's role as "maker" (Ps. 100) is only matched, as we have seen above, by his exclusive role in the creation process. Isaiah rebukes those who see a role for human agency in the creation process (Isa. 29:16). The Old Testament God belongs to that class of artifacts which cannot discharge its reciprocating function once the role of human agency in its own creation is perceived.

Hobbes is plainly a great deal more interested in producing the circumstances where the creation of artifacts falling into classifications one and two (recognized and recoverable or unrecognized but recoverable) can occur, unhindered by the animosities of the state of nature, than he is in sponsoring the conditions under which the production of artifacts falling into classification three (unrecognized and unrecoverable) will occur. In one of the most remembered passages of *Leviathan*, he makes clear that failure to create the state means forgoing the benefits of artifice.

> In such condition (of Warre), there is no place for Industry; because the fruit thereof is uncertain; and consequently no Culture of the Earth; no Navigation, nor use of commodities that may be imported by Sea; no commodious Building; no Instruments of moving, and removing such things that require much force; no Knowledge of the face of the Earth; no Account of Time; no Arts; no Letters; no Society; and which is worst of all, continuall feare and danger of violent death: And the life of man, solitary, poore, nasty, brutish, and short.[20]

Nevertheless, Hobbes is quite mindful that the circumstances under which the production of artifacts falling within classifications one and two can occur, requires the creation of the state. He reverts to the language of classification three to illuminate the genesis of the state, referring to the state as a "Mortall God."[21] At the same time, he reassigns to man the role occupied by God in the Bible. Unlike Israel, whose maker is God and whose matter is man, the state is derived from man who, as Hobbes emphasizes, is both "the Matter thereof, and the Artificer."[22]

The presumption in Hobbes is that while the Bible is of indispensable value in understanding the sources of human creation, its teaching must be applied in a new way. Whereas in the biblical account Israel is an artifact whose maker is God and whose matter is man, in the Hobbesian account, man is both the matter and the maker of the Leviathan state. The artificial character of the state is revealed through the deconstructionist exercise of imagining away the presence of constituted authority, while retaining in the mind's eye the dominant elements of the political culture of which one is a member. Once in the "state of nature," absent the authority of kings and bishops yet forcefully aware of one's partnership in a biblical culture, one can begin to see oneself acting in the paradigmatic role of God in the creation epic.

Man's assumption of a new role, that of a prosthetic God whose capacities rival those of the Old Testament God, presents Hobbes with a new problem. Chief among these is that the objectifying powers of the Old Testament God can no longer be called upon in the service of maintaining political order. Hobbes indicates awareness of this problem when he places a sword in the hand of the sovereign, "that by the terror thereof, he (the sovereign) is inabled to forme the wills of them all, to Peace at home, and mutual ayd against their enemies abroad."[23]

The sword has the object of securing the internal unity of will which characterized the life of Israel under the sovereignty of God, but even so, the sword does not solve Hobbes's problem. As soon as the sovereign deploys it, the subject, under the terms of the covenant, *rightfully* recovers his original right of self-rule, as Hobbes acknowledges.[24] For example, Hobbes says that the subject who throws down his weapon and flees from the field of battle commits "Cowardice" but not "Injustice."[25] It is not injustice because the covenant no longer serves the objective of "commodious living"[26] which led us to undertake it. Should the sovereign use the sword to prick us on into battle, what began as defection may lead, justifiably, to insurrection. Hobbes misleadingly proposes to secure the advantages of "commodious living" through the coercive power of the state. This approach fails because the covenant safeguards the prior and absolutely superior right of the subject to disobey the state when brought into collision with either his life or, more broadly conceived, the "means of so preserving life as not to be weary of it."[27]

The problem of securing the advantages of artifice begins to clear when it is realized that consent to the state is more than consent to a determinate person (or persons) occupying the office of the sovereign. It is consent to a political order consisting of capitalism, modern science, and liberalism which sovereignty makes possible. The elements of this political order provide unparalleled advantages to the subject in obtaining self-extension through artifice. As the subject takes on the new powers of a prosthetic God, exerting dominion over the natural world, he becomes wedded to the political order through which his enhanced powers are assured.

Put somewhat differently, modern order requires the creation of an artifact of existence which will remedy the imperfections of the "natural state" in a variety of dimensions, and it requires the installation of a civil sovereign who will secure the same. The final project of this artifact of existence is the replacement of "nature" by "artifice" in every dimension of existence—economic, social, civil, and scientific/technical. Ultimately, the sovereign is sustained not so much by the coercive power of the state as by the opportunities for self-extension through artifice which modern order secures.

Hobbes's "modernity synthesis" embraces three elements: classical liberalism (not to be confused with welfare-state perspectives), capitalism, and modern science. These are the constituent elements of an order within which the powers of the prosthetic God are released.

Hobbes founds *classical liberalism* on the basis of a private need for public order. Yet, public order is an artifact, created and sustained only insofar as it intersects with the private needs of a whole society of independently situated actors. Each is differently constituted in terms of intelligence, temperament, and personal possessions to perceive the good for themselves in radically different ways, and each is armed with an "inalienable right"[28] to pursue their private vision of the good life unhindered by public authority in all matters delimited as nonpublic, or private regarding.[29] In Hobbes's view, it is even consistent with the conditions under which public order is created to prefer private need to public order in cases of conflict.[30] Later formulations of classical liberalism, as in the version put forward by Locke and the framers of the U.S. Constitution, lay stress on separation of powers, indirect election, and a written constitution as a means of avoiding collision between private needs (rights) and public order.

Hobbes also defends incipient *capitalism*, here defined as a political economy which elevates the rights of possessive individualism over a prescriptive allocation of work and resources.[31] The doctrine of possessive individualism holds that the individual is sole owner of his talents and energies, owes nothing to society for them, and that whatever he sets aside through the exercise of these talents is his to dispose of as he likes. This doctrine is of indispensable importance in the creation of market relationships where men "freely" contract for the sale of their labor-power. In a manner congenial to incipient capitalism, Hobbes reconceives the person as not only a prosthetic God but also a commodity who will sell his talents and energies to the highest bidder in exchange relationships. The value of a man is no more, Hobbes announces, than his "Price," and that is a value determined not by the "seller" but by the "buyer."[32] While sweeping away a prescriptive allocation of resources and labor, Hobbes is firm about the binding nature of contracts. His notion that it is the function of the sovereign to enforce valid contracts assures capital of a means of securing agreements reached in the market.

A third element of the modernity synthesis effected by Hobbes consists in the potential of *modern science*. Hobbes invites us to suspend belief in the phenomenal characteristics of the natural world (such as color, taste, or sound) and to postulate a world of matter moving, "(unless something els hinder it) eternally."[33] We cannot rely upon memory for assistance in reconstituting the world we formerly knew for memory is but "decaying sense"[34]; we cannot rely upon reason for assistance in reconstituting the world we knew for there is "no right Reason constituted by Nature"[35]; we cannot rely upon faith for assistance in reconstituting the world we knew because faith is merely "fear arising from ignorance of causes."[36] What we must rely upon in reconstituting the world is science, particularly the reported accounts of the behavior of the world supplied by science in terms of the mobility, number, extension, and impenetrability of body. Science supplies us with knowledge of the true causes of things, such knowledge consisting of causal explanations in terms of the forces sufficient to produce intended effects.[37] Science also provides us with a means of dominating the world. When we have knowledge of the forces sufficient to produce intended effects, we can operate upon the world to produce the results we intend.

So defined, the modernity synthesis supplies an avenue through which man can undertake a task of world transformation, unimpeded by nature, community, or public authority, in a manner that can be rivaled only by the God of the Old Testament. Overarching each of the above conceptions of man—capitalist, scientific, and liberated (classical liberal) man, is the conception of man as a prosthetic God who, in imitation of the God of the Old Testament, will achieve self-extension through the products of artifice. Since the main outlines of the modernity project forged by Hobbes have obtained an unshakeable hold on the public imagination, it is useful to review the main arguments of *Leviathan* in its behalf.

Defect in Nature

Hobbes's argument for the modernity project is based on the perception of defect in nature. The crucial task confronting the new, prosthetic God is that of overcoming defect in nature. This task arises from the circumstance that while a man's desire is infinite ("Felicity," Hobbes says,"is a continuall progresse of the desire, from one object to another; the attaining of the former being still but the way to the later"),[38] the power of a man to obtain the objects of his future, apparent desire is finite. It is finite because it is limited by other men, each of whom is similarly motivated and each of whom therefore imposes restrictions upon the power of all others.[39] A man might be content with "moderate power," Hobbes contends, but the presence of others forces the consideration that "he cannot assure the power and means to live well, which he hath present, without the acquisition of more."[40]

Hobbes universalizes the "Right of Nature," an absolute right which extends "even to anothers body."[41] Thus, by the Right of Nature, all men have an equal right to pursue the objects of their desire, using whatever powers they may be able to deploy and in whatever manner they choose to exercise them on behalf of their privately defined vision of "felicity."

However, while it may be the case that the Right of Nature is uniform, it is reasonably apparent that the distribution of power is not. The defect read by Hobbes into "nature," arising from the incommensurability of man's infinite desire and his circumscribed power, is one which, if not created, is exacerbated by capitalist forms of economic organization. In the latter the whole object of economic activity is to achieve a net transfer of labor-power away from those who lack access to land, materials, and machinery and to those who own the means of production.[42]

The condition of man in the state of nature, Hobbes says, is one of "warre."[43] The condition of war, assimilated by Hobbes to "nature," in truth arises from the circumstance that the "Right of Nature," which Hobbes universalizes, belongs effectively only to those who are the owners of land, labor, machinery, and materials. These are the men empowered to set a price for labor in market relationships. Capital reconceives of the person as a bundle of natural and instrumental powers "exchangeable for benefit,"[44] then purchases that labor-power with "moveables . . . gold, silver, and money . . . a commodious measure of the value of all things between Nations."[45] The stage is thus set among men for a contest over "Power after power"[46] or war. The form the contest assumes is dictated by competitive market relationships between buyers (of labor-power) and sellers (of labor-power), in short, between capital and labor. The conditions which give rise to war suggest that it is not simply the product of defect in nature, as Hobbes proposes, but a consequence of capitalist forms of economic activity.

If capitalism is what generates the perception of defect in nature, it is also capitalism, assisted by modern science, that shows Hobbes the way out. The manner in which capitalism extricates us from the presence of defect in nature is through the biblically derived project of dominion over the earth. The powers of the Old Testament God are reassigned to capital and modern science which thereafter proceed to produce an artifact of civil society paralleling and reinforcing the artifact of the state. For present purposes, it is not so much the political role assigned to capitalism and science in sustaining the modern state that is of interest, but the concept of nature *as artifact* underlying all three.

To summarize, Hobbes proposes the creation of the state, a "Mortal God," in order that the conditions may exist through which further acts of artifice become possible. Just as the covenant between God and Israel is crucial to the prospects for self-extension through material, cultural, and political creation, so the creation of the state through covenant relationships is a necessary condition for self-extension through the artifacts of civil society. Hobbes's reference to the state as an "Artificiall Man" and a "Mortal God" calls attention to that other artifact in history, Israel, and its creator. Just as Israel creates a political sovereign who supplies it with the conditions of extension through artifice, so Hobbes calls upon the members of the "state of nature," each of whom effectively claims the same rights and powers as the Old Testament God, to create a "Mortal God." The sovereign (in turn) creates his creators as a people[47] and provides the conditions through which the lesser, commercial and mechanical, forms of artifice become possible.[48]

To continue, Israel created an unrecognized and unrecoverable artifact (because removed from human agency), God, through whom they were themselves created and enabled to escape privation and bondage. Analogously yet in a different way, Hobbes proposes the creation of an unrecognized yet recoverable artifact (because the product of human agency), the state, through whom the conditions are created for a variety of other forms of material and cultural production and for escape from the brutish conditions of natural existence. For Hobbes, the state is a prosthetic device designed to overcome the presence of perceived defect in nature. Its political stability depends on the activity of modern science and capitalism in assuring that the subject's dominion over the natural world, originating in the act of state-making itself, is enlarged.

Discontents of Modernity

In reviewing the modernity project, it is important to acknowledge that it has realized its principal endeavor. This is the biblically derived goal of self-extension through artifice. Freud pays tribute to this triumph in his description of the modern, prosthetic god. As McLuhan well knows, the transformation of sex organs into motorcars is a less respectable achievement. Both McLuhan and Freud wrote when the costs imposed by the advances of modernity were becoming apparent. Patently, there is considerable irony and paradox in the prosthetic God. Man, the prosthetic God who creates the state, in the end becomes captive to his own creation; he is dependent on the artifact of modernity for his survival no less than Israel was dependent on its Old Testament God for its continued survival.

If one doubts the truth of this proposition, consider the effect of divesting modern man of the privatized extensions of self to which he has become accustomed: the ubiquitous TVs, ATVs, PCs, RVs, CPs (cellular phones), CBs, CDs, ATMs, and SDs (satellite dishes) and, of course, the ubiquitous automobile. These extensions of the self are of a piece with Hobbes's biblically derived emphasis upon the rights of the hermetic, hypertrophied self and, what is more, they are of a piece with the modern state (also delineated by Hobbes) which secures and promotes them. No less than ancient Israel, modern man succumbs to objectification by the products of his own creation.

The descendants of Hobbes, chiefly located in the U.S. political setting, view the advantages of modernity as best secured through emphasis on its liberal and capitalist components. These components safeguard, above all, the *site of artifice*, the prosthetic God of Thomas Hobbes,[49] and they bring in their train the attendant ideas of nature as mere artifact, of political community as constituted by independently situated artificers, and of the good life as affording opportunity for privatized extensions of the self. In the version in which

Hobbesian theory is transmitted to these shores through Locke, there is, if anything, a radicalization and extension of these basic views.

Consider Locke's connecting role. First, Locke hedges the site of artifice with additional safeguards. We transfer our executive right of nature to the civil sovereign, but never our property right.[50] By contrast, Hobbes holds that property is an effect of sovereignty and that the sovereign may redistribute property if, motivated by prudential reasons, he considers it wise to do so.[51] Second, nature is further marginalized by Locke who views it in its unappropriated state as "waste."[52] Locke saw the great unappropriated commons of America of that time as waste and straightforwardly proposed its rapid appropriation and commodification as a way to defuse conflicts at home generated by land enclosure.[53] Third, following the introduction of money as a speculative instrument no less than as a means of exchange, Locke shifts the locus of artifice from generic man (in the conditions of a subsistence economy)[54] to capital.[55] Thus, Lockean political arrangements empower a particular class of artificers, the "Industrious and the Rational,"[56] who are knowledgeable about the use of capital instruments as a way to further their dominion over man and the natural world.

Finally, Locke makes the Hobbesian modernity synthesis more palatable by vesting final authority and power in a group of institutions, a representative and hereditary assembly and monarch, rather than in a single institution, monarchy, as preferred by Hobbes.[57] This alteration in the distribution of power, as Hobbes was well aware and which Locke understood as well,[58] does nothing to change the principles on which authority in modern political community is constituted; Locke's redistribution of power does not affect the legal principle of sovereignty. More simply, Locke popularizes Hobbesian views, dressing them to suit the prejudices of their intended audience and thereby enabling them to gain acceptance.

If the natural world is to be safeguarded from the consequences of the modernity project, Hobbes himself will have to be exposed to the terrors of deconstruction in a manner similar to his own recovery of human agency through deconstruction of the role occupied by Israel's Maker.

Hobbes erroneously considered the modernity project in need of safeguarding by the coercive powers of the state. We now know that it is not the coercive power of the state that sustains the modernity project, but the circumstance that we have all become, as Hobbes anticipated in other passages, prosthetic Gods who are accustomed to achieving self-extension through artifice. The difference between Hobbes and our present circumstance, therefore, is that the modernity project visualized by Hobbes as novel and controversial is viewed by our eyes as identical with pure nature. If this view is to be disturbed, the place to begin is with a deconstruction of the artifact of consciousness which modernity has produced.

An important way of replying to Hobbes is by contesting the concept of nature on which his vision of modernity is premised. No more than a sketch can be offered here, but it ought at least to include the observation that the view of nature as mere by-product of human artifice violates ordinary language usage. In ordinary language usage, the term nature is commonly employed to signify "that which exists and functions independently of human contrivance or interference, that which is other than the human, over against the human, sharply distinct from culture."[59] Such a definition emphasizes that nature is other than artifice, even not-artifice, precisely the opposite of the Hobbesian definition. The Wilderness Act of 1964 recognizes the importance of conceiving nature as Other when it speaks of wilderness as "an area where the earth and its community of life are untrammeled by man; where man, himself, is a visitor who does not remain" and where there exist opportunities "for solitude or a primitive and unconfined type of recreation."

The value of preserving the otherness of nature becomes evident when the distinction between nature and artifice is erased by significant actors. Consider one of the products of such erasure: the Disney empire where Bambi, Thumper, and the Lion King cavort as aery poltergeists and where artificial trees display the effects of electronically simulated seasons. Disneyland is ludicrous, freakish. An aberration of this kind is only possible in a culture where the Hobbesian perspective on the natural world has become an artifact of consciousness.

The transformation of *consciousness* into an artifact opens a second line of contestation to Hobbesian premises. If nature is mere artifact as Hobbes says, then man, who is a part of nature, must be inescapably an artifact as well. Hobbes says as much in the famed introduction when he speaks of the "Heart, (as) but a Spring, and the Nerves, but so many Strings; and the Joynts but so many Wheeles, giving motion to the whole Body, such as was intended by the Artificer."[60] If man is mere artifact, corresponding to the artifact of nature, then despite Hobbes's profession of liberal political principles, it is no violation of nature for the sovereign to create an artifact of consciousness parallel to and reinforcing that of the state. At the end of Book 2, we find the sovereign and his ministers busily engaged in imposing a new version of the Decalogue on the members of the political community.[61] The members of the political community are to be taught to avoid speaking "irreverently" of the sovereign, that it is "evil" to hold the sovereigns of other nations above their own, that "reforme" of the Common-Wealth is likely to bring on its destruction, and that the desire for political change is truly a lusting after strange gods.[62]

Here Hobbes has surely overstepped his argument. As students of Rousseau, Hegel, and Marx, we know that man is indisputably a reflexive being, distinguished by a capacity for making his own consciousness an object of reflection and hence of reordering his own nature and political community in keeping with an alternative vision. Further, turning the tables on Hobbes, we may say that since man is a reflexive being, so must nature (of which man is a part) be reflexive. Nature, therefore, is more than simply Other, an artifact, or unconscious process; it is itself a reflexive being which, like man, takes up modes of self-reflexiveness within itself.[63] Such a view belongs neither to a naturalistic perspective—in which man is subsumed within the order of nature as one more function of the powers and processes that make up the natural world—nor to that advanced by Hobbes, in which nature is the by-product of human artifice. Rather, it belongs to what may be termed as "historico-natural" view in which man occupies an intermediate position between nature and culture in the course of history, objectively determined by neither.

If there is merit to the historico-natural outlook, then the view of nature as Other (taken by many wilderness advocates and environmental groups) is inadequate as a line of defense against the modernity project articulated by Hobbes. The reason it is inadequate is that the Hobbesian project includes within itself the intention of creating a political culture which reinforces and extends the aims of the Leviathan state. The dominion spoken of by Hobbes is not simply over the earth, but over man as well. Thus, the battle must be joined in the realm of political culture no less than in observable conflicts over the management of natural resources. It will have much to do with the idea of nature held by the parties to the conflict, no less than the political resources at their disposal. It is a conflict which pertains to the universities and institutions of higher learning no less than to the formal institutions of government.

Notes

1. C. B. Macpherson,*The Political Theory of Possessive Individualism* (New York: Oxford University Press, 1962); William Connolly, *Political Theory and Modernity* (Oxford: Basil Blackwell, 1989); Leo Strauss, *The Political Philosophy of Hobbes* (Chicago: University of Chicago, Press, 1965).

2. Steven Shapin and Simon Schaffer, *Leviathan and the Air Pump* (Princeton, N.J.: Princeton University Press, 1979); David Johnston, *The Rhetoric of Leviathan* (Princeton, N.J.: Princeton University Press, 1986).

3. Michael Walzer, *Exodus and Revolution* (New York: Basic Books, 1985); Christopher Hill, *Puritanism and Revolution* (New York: Schocken, 1958); R. H. Tawney, *Religion and the Rise of Capitalism* (New York: Mentor, 1954).

4. Karl Polanyi, *The Great Transformation* (Boston: Beacon Press, 1944); Macpherson, *The Political Theory of Possessive Individualism*.

5. Thomas Hobbes, *Leviathan*, edited and with an introduction by C. B. Macpherson (Baltimore: Penguin Books, 1968).

6. Gerhard von Rad, *Genesis* (Philadelphia: Westminister Press, 1961).

7. Gordon D. Kaufman, "A Problem for Theology: The Concept of Nature," *Harvard Theological Review*, 65 (1972): 337–366.

8. The clear presumption for artifice at the expense of an independently existing, natural realm has elicited an argument that the roots of the environmental crisis lie in biblical modes of consciousness. This argument was presented first in the path-breaking essay by Lynn White, "The Roots of the Environmental Crisis," *Science*, 155 (1967):

1203–1207. A review of the literature surrounding this debate appears in Roderick Nash, *The Rights of Nature* (Madison: University of Wisconsin Press, 1989). The present essay owes a debt to White's article, but it attempts to deepen the argument by focusing on the biblical contribution to modernity. God's role as sovereign over nature is emphasized at the expense of the more familiar stress on his role as sovereign over history, deliverer of his people, and lawgiver. To put this another way, the emphasis is on the book of Genesis at the cost of giving due attention to the books of Exodus, Judges, Numbers, or Deuteronomy.

9. Johannes Pedersen, *Israel: Its Life and Culture* (London: Oxford University Press, 1926).

10. Elaine Scarry, *The Body in Pain* (New York: Oxford University Press, 1985) p. 200.

11. Ibid., pp. 164–165.

12. Ibid., pp. 161–180.

13. Ibid., p. 170

14. Ibid.

15. Hiram Caton, *The Politics of Progress* (Gainesville: University of Florida Press, 1988).

16. Scarry, *The Body in Pain*, p. 314.

17. Ibid., pp. 312–313.

18. Karl Marx, *Economic and Philosophic Manuscripts*, trans. Tom Bottomore (New York: Ungar, 1988).

19. Scarry, *The Body in Pain*, p. 311.

20. Hobbes, *Leviathan*, p. 186.

21. Ibid., p. 227.

22. Ibid., p. 82.

23. Ibid., pp. 227–228.

24. Ibid., p.192.

25. Ibid., p. 270.

26. Ibid., p. 186.

27. Ibid., p. 192.

28. Ibid.

29. The notion that Hobbes both originates and is master of a distinctive tradition within political thought, natural right theory (here called "classical liberalism"), is shared by eminent scholars. See Strauss, *The Political Philosophy of Hobbes*; Ernest Barker, *Greek Political Theory* (New York: Barnes and Noble, 1961); and Macpherson, *The Political Theory of Possessive Individualism*.

30. See Hobbes's discussion of the "recalcitrant soldier" above.

31. Macpherson, *The Political Theory of Possessive Individualism*.

32. Hobbes, *Leviathan*, p. 132.

33. Ibid., p. 88.

34. Ibid.

35. Ibid., p. 111.

36. Ibid., p. 170.

37. For an account of Hobbes as the progenitor of a science of politics, see Tom Sorel, *Hobbes* (New York: Routledge and Kegan Paul, 1986); and Johnston, *The Rhetoric of Leviathan*.

38. Hobbes, *Leviathan*, p. 160.

39. Ibid., p. 161.

40. Ibid.

41. Ibid., pp. 189–190.

42. Macpherson, *The Political Theory of Possessive Individualism*.

43. Hobbes, *Leviathan*, p. 186.

44. Ibid., pp. 150–151.

45. Ibid., p. 300.

46. Ibid., p. 161.

47. Ibid., p. 228.

48. Ibid., pp. 299, 387.

49. It is worth remembering that the only right secured to the members of the U.S. political community by the Constitution (1787) as it was initially drafted and presented to the states for ratification was that of "authors and inventors . . . to their respective writings and discoveries" (Article 1, Section 8, Clause 8).

50. John Locke, *Second Treatise*, in *Two Treatises of Government*, edited and with an introduction by Peter Laslett (New York: Mentor, 1965), sects. 87–89.

51. Hobbes, *Leviathan*, pp. 202, 295–296.

52. Locke, *Second Treatise*, sects. 37, 42.

53. Ibid., sects. 37, 48. See Macpherson, *The Political Theory of Possessive Individualism*.

54. Ibid., sects. 27–34.

55. Ibid., sect. 50.

56. Ibid., sect. 34.

57. Ibid., sect. 213. Cf. Hobbes, *Leviathan*, p. 241.

58. Ibid., sect. 151, and see comment by Laslett.

59. Kaufman, "A Problem for Theology," p. 345. It may be asked how it has come to pass that Hobbes's view of nature has prevailed given the presence of a conflicting idea embedded in ordinary language. A full reply would be most interesting. One course which might be pursued is to consider the role of the advertising industry. Taking a cue from Hobbes, it employs nature as a "referent system" and redirects its value away from its standing as an independent, self-subsistent entity toward an endorsement of privatized, adventitious consumption (e.g., Marlboro country). Thus, ordinary language values are overriden by the hegemonic power of a system of signs derived (ultimately) from Hobbes and Locke. See the excellent introductory work on the semiotics of advertising by Judith Williamson, *Decoding Advertisements* (New York: Marion Boyars, 1988).

60. Hobbes, *Leviathan*, p. 81.

61. Ibid., pp. 379–381.

62. Ibid.

63. Kaufman, "A Problem for Theology."

Chapter Two

Charles Fourier: Proto-Red-Green

Joan Roelofs

Introduction

Each era finds in Charles Fourier (1772–1837) a source for its own ideas. To his immediate disciples, he was the prophet of communitarian socialism. His greatest influence in the United States was during the 1840s, when his ideas were popularized by Albert Brisbane in the *New York Tribune*. Twenty-nine Fourierist phalanxes were created in the United States; the best known was Brook Farm in Roxbury, Massachusetts.[1] Even Christian ministers applauded the greatly desalinated version of Fourier which crossed the Atlantic.[2]

Marx and Engels acknowledged his devastating critique of capitalist civilization and his greatness as a satirist.[3] Nineteenth-century Russian revolutionaries saw Fourier as the Westernizer, modernizer, and feminizer of the peasant collective.[4] In the twentieth century, surrealist poets appreciated his wildest reveries and, along with the rebels of 1968, hailed his pleas for instinctual liberation and his radical anarchism, which dethroned conventional wisdom and scholarly authority as well as churches, governments, and banks.[5] Semiotician Roland Barthes lauded his utopian vision, but even more his role as logothete, founder of a language.[6]

Today, we can see Fourier as a source and inspiration for red-green theory. His writings brilliantly illuminate basic green principles: small is beautiful, steady-state economy, quality of life, not accumulating consumer durables, work as play, postpatriarchal values, abolition of hierarchy, and respect for

First published in *Capitalism, Nature, Socialism* 4, no. 3 (September 1993): 69–88; reprinted here with minor corrections.

nature. He also reminds us of what we have forgotten, oppressions and dysfunctions that current green theory and practice virtually ignore. These include isolation and loneliness, especially of the elderly, the unattractive, and the outcast; the unhealthy, immoral, and boring nature of most work; and the failure of the family to achieve its supposed purposes while excelling as a locus of wasteful consumption. His proposed solution, entirely relevant to the contemporary world, was collective living and consumption, which he regarded as economically, environmentally, and socially superior to the individual family system.

In addition to Fourier's substantive concerns, he provides the valuable addition of humor. Reading Fourier is recreational and therapeutic as well as educational. It reminds us that hilarity can be safely overproduced and overconsumed in a steady-state economy, and even that puns should be valued because from scant resources, they provide a double or triple product. Fourier's "world war of meat pies" is not only a future possibility; it can enliven our dark days right now.[7]

Fourier was born in Besançon, France, to a family of silk merchants. He received a classical seminary education, then became a traveling silk salesman, continuing his education through newspapers, periodicals, and penetrating observation. He was attuned to the voices of the time: the Enlightenment and its *Encyclopedia;* the considerable utopian literature of the eighteenth century; travelers' tales, especially those of Tahiti; and Rousseau, who also was impressed with "savages" but who came to different conclusions. He may well have been influenced by the rituals of "androgynous Masonry," such as the Parisian "Knights and Nymphs of the Rose"; he may even have known of the German order of "Harmony."[8]

Fourier's experience of postrevolutionary France and the developing capitalist industrial order led to his sharp critique of "civilization," as he called contemporary society. Enlightenment ideals and his own idiosyncratic concept of humankind had been equally betrayed

> Liberty is illusory if the common people lack wealth. When the wage-earning classes are poor, their independence is as fragile as a house without foundations. . . . *Equality of rights* is another chimera. . . . The first right of men is the right to work and the right to a *minimum* . . . and the right to pleasure. Fraternity . . . can only be realized if four conditions are satisfied: Comfort for the people and the assurance of a splendid minimum; The education and instruction of the lower classes; General truthfulness in work relations; The rendering of reciprocal services by unequal classes.[9]

Engels's tribute is worth quoting.

> We find in Fourier a criticism of the existing conditions of society, genuinely French and witty, but not upon that account any the less thorough. Fourier

> takes the bourgeoisie, their inspired prophets before the revolution, and their interested eulogists after it, at their own word. He lays bare remorselessly the material and moral misery of the bourgeois world. . . .
>
> Fourier is not only a critic; his imperturbably serene nature makes him a satirist, and assuredly one of the greatest satirists of all time. He depicts, with equal power and charm, the swindling speculations that blossomed out upon the downfall of the revolution, the shopkeeping spirit prevalent in, and characteristic of, French commerce at that time. Still more masterly is his criticism of the bourgeois form of the relations between the sexes, and the position of women in bourgeois society. He was the first to declare that in any given society the degree of woman's emancipation is the natural measure of the general emancipation. But Fourier is at his greatest in his conception of the history of society. . . . Civilisation moves in "a vicious circle," in contradictions which it constantly reproduces without being able to solve them; hence it constantly arrives at the very opposite to that which it wants to attain, or pretends to want to attain, so that, for example, "under civilisation poverty is born of superabundance itself." Fourier, as we see, uses the dialectic method in the same masterly way as his contemporary, Hegel.[10]

In contrast to "civilization," Fourier designed a society (or community) which not only allowed for great abundance and luxury with minimal resource use, but also permitted the full expression of all human passions. Complete harmony was possible without the need for repressing human desires or reforming humankind. Indeed, Fourier called his ideal society Harmony.

Fourier on Work and Social Psychology

Fourier's doctrine began with psychology. He believed that people were born with certain personality types based on their dominant passions. He posited twelve basic passions: five sensual; four of the soul (friendship, love, family, and ambition); and three distributive (the cabalist, lover of intrigues; the butterfly, lover of change and contrast; and the composite, desiring to combine pleasures of sense and soul). The superior individuals were those of the greatest complexity and the largest number of dominant passions. Fourier believed that all passions, manias, and desires were good; otherwise, God would not have created them. He attributed crime, all social pathology, and dysfunction to repression; with the proper organization of society, all tastes would become socially useful or at least innocuous.

Fourier's phalansteries (also translated as phalanxes or phalanges), communities of 1,620 people, were designed to include male and female representatives of all the basic personality types (a requirement for the proper arrangements of work and love). They would be rich and poor, young and old, and people of all persuasions. Because of his selective tolerance for some inequality as well as the pleasures of sharp dealing, gambling, conspicuous consumption, and ostentatious philanthropy, some have declared that Fourier was not a socialist. However, he believed that these inequalities had considerable benefits and very minor costs. In the phalanstery, everyone was guaranteed a generous minimum of food, lodging, clothing, entertainment, education, medical and dental care, and sex. While there were to be three classes of dining rooms, catering to those with more or less refined taste (and with the possibility of mobility), even the third-class dining room would provide better food than the rich now eat.

An affluent standard of living for all would be possible because of the savings permitted by "uniting into combined households," the avoidance of waste, the labor-intensive production of necessities and luxuries, the extremely

Figure 2.1. "Human Happiness—food for the asking—in the Fourieriest utopia." Illustration by Jean Grandville, *Un autre monde*, 1844.

high productivity of Harmony's ecstatic workers, and the elimination of "12 classes of parasites. Unlike Robert Owen and Karl Marx, Fourier did not foresee abundance deriving from mass production. His objective was to achieve "1) The greatest possible consumption of different kinds of food; 2) the smallest possible consumption of different kinds of clothing and furniture; . . ." Because of the odiousness of such work, all manufactured goods would have to be nearly indestructible: "furniture and clothing will last an extremely long time. They will become *eternal*."[11]

No one would be laboring to support capitalists, middlemen, idlers, priests, economists, bureaucrats, armies and navies, or various other parasites. Wives of the rich as well as all children would become workers. Fourier's discovery of the "theory of passionate attraction," a breakthrough he compared to Newton's discoveries, meant that people would voluntarily enroll in all those (and only those) types of work which satisfied their particular combination of passions. In addition, work would be spiced with competition, intrigue, sex, and pageantry. Fourier believed that no occupation should be pursued for more than two hours at a time. Necessary dangerous work, as in chemical plants and glassworks, would be rotated so that one person might spend only two or three hours per week in those places, but no matter how enticing, all work would be done in short sessions. This would have the additional advantage of promoting equality and solidarity, as those who were leaders in one field would be novices in another. The major productive work in Harmony would be horticulture which, along with animal husbandry, gamekeeping, and fish farming, would supply a large part of the diet. Rather than bread, legumes would be the staple food, with high consumption of fruits and vegetables.

A wide variety of occupations would exist in Harmony and Fourier imagined that they would be developed to the highest standards. For example, "The doctors of the phalange will be specialists in preventive medicine: their interest is to see that no one falls ill. In Harmony, doctors (and dentists) will always work as a team in a group. They will be collectively remunerated in proportion to the general health of the phalange, and not according to the number of ailments or number of patients treated."[12] Dirty work would be joyfully pursued by the Little Hordes, teams of children who (according to Fourier) had a penchant for filth, noise, and "disgusting" tasks such as removing reptiles from the roads. In contrast, the Little Bands, those children with a taste for elegance, would have responsibility for maintaining the decorative side of the phalanstery and correcting the grammar of their elders.

There would be trade, partying, and joint enterprises with the rest of the world, which also would be organized into phalansteries. Most notable were the "industrial armies," mustered for ecological projects such as reclamation of deserts, reforestation, and building canals (no "industrial" example was given by Fourier). The expenditure, he pointed out with the logic of a commercial traveler, would be much smaller for a productive army. Besides the savings in slaughtered men, burnt cities, and devastated fields, there would be savings in the cost of equipment plus the benefit of the work accomplished.[13]

Children would be educated in Harmony by following their instincts, imitating older children, finding mentors, and participating in the work of the community. Miniature workshops with tiny tools would be irresistible. Opera—whose educational value derived from the great variety of skills required, including complex planning—would serve as a prime educational tool as well as a phalanstery-integrating activity. (For Fourier, there was nothing more enticing than the orchestration of vast diversity.) Adults, children, and members of all classes would participate; a prince might well be a member of the chorus line.[14] Although some people might prefer painting sets to performing, Fourier expected that children would be trained in singing and playing instruments from an early age. Indeed, he suggested that all the working groups of the phalanstery should sing distinctive anthems at the beginning and end of each session. While classes were to be offered in geography, history and politics, children would attend only when they felt the urge, at whatever age this might occur.

Fourier proposed a radical re-creation of the "amorous world." He opposed monogamy and the nuclear family because they were uneconomic, but even more, because they did not fulfill any of their supposed purposes. Like most other aspects of "civilization," marriage encouraged corruption and harbored misery for almost everyone—most wives, husbands, and children as well as the unmarried. It was obvious to him that monogamy did not satisfy sexual needs. On the contrary, the widespread debauchery which he observed indicated to Fourier that marriage was an unnatural institution. "The philosophers," he wrote, "should take account of these observations; they should watch how people actually behave and try to make some use of the conduct which they are unable to prevent. . . . Without exaggerating one can estimate the number of forbidden love affairs as seven times that of sanctioned conjugal relationships."[15]

Fourier also indicted the family because of its oppressiveness for women. Neither their sexual nor intellectual needs could be properly fulfilled in it. Fourier, an early feminist, believed in the inherent superiority of women in matters intellectual and political. "In the combined order, education shall have restored woman to the use of her faculties, [now] smothered by a social system which engrosses her in the complicated functions of our isolated households."[16] Women's personalities had been warped by long years of training in duplicity for the purpose of snaring a husband. This energy was in any case wasted for once snared, the merry-go-round began.

However, Fourier did not devalue "traditional women's work." On the contrary, the marital arts, especially cooking, gardening, child rearing, and lovemaking, were to become the most important activities in the future. However, the isolated household did not permit the complexity, refinement, and pleasure that could be attained in establishments of 1,620 persons. The family was not a guarantor of security; even the "normal" family was constantly threatened by death or departure of spouse, children, or parents. Falling out of love, boredom, or "internal migration" were constant risks even where technical fidelity prevailed. Sterility was another possible disaster: "Children come in torrents to people who are unable to feed them, but rich families seem particularly subject to sterility."[17] The institution of the family created special hardships for those excluded: the single for whatever reason, including unattractiveness. A particular concern of Fourier's was the elderly, whom he saw as socially, vocationally, and sexually isolated.

Was all the sacrifice worth it because the family was a wonderful nest for raising children? Fourier thought not. "In the family system children spend all their time crying, quarreling, breaking things and refusing to work."[18] Children were oppressed by child rearing which concentrated on breaking their wills and fitting them to society. He believed that a better method would encourage children's instincts for imitation and play. Society must respect nature and provide for the harmless release of all desires and passions; otherwise, the repression would result in a "countermarch" of evil and violence. Finally, Fourier saw the nuclear family as the enemy of community. The "wondrous inventions" of science and industry needed to be matched by a "social order which will assure our happiness," that is, a communal combination of skills and passions. Fourier sketched in elaborate detail his "new amorous order" in which marriage would be abolished, housework and childcare collectivized, and a sexual minimum made the right of all.

How was this sexual minimum to be assured? Fourier's solution was not the anarchistic one of "free love" which, like the market, tends to reward the strong, the quick, and the beautiful. First, organization and planning would make it easy for everyone (including shy people) to get into the act. Each evening, "Every individual must go to the Exchange to arrange his work and pleasure sessions for the following days. It is there that he makes plans concerning his gastronomic and amorous meetings and, especially, for his work sessions in the shops and fields. Everyone has at least twenty sessions to arrange, since he makes definite plans for the following day and tentative ones for the day after."[19] These love meetings were to be consensual (as the work sessions were voluntary) so that some people might be left out. However, help was on the way from "courts of love" to enforce rules for the different groups in which people voluntarily enrolled. (Fourier excluded children under fifteen from the amorous arrangements.)

Some of the voluntary "orders" were to be dedicated to sexual service. For example, industrial armies also provided the occasion when virgins (males and females of the "Vestalate") were to choose their first lovers. These armies were to be accompanied by the order of Bacchants and Bacchantes, whose function was to "recover the wounded" (i.e., to take care of those virgins whose choices were not reciprocated).

Other devices included "sympathetic matching," performed by the venerable order of Confessors and Confessoresses, and the sexual philanthropy of Angelic Couples. These apparently salacious fantasies were not for the furtherance of what Woody Allen calls "meaningless sex." Fourier believed that only when one had assurance that one's urgent physical needs would be met, could more complex amorous ties based on spiritual, intellectual, and affectionate affinities flourish. In fact, his revolution in love would have to be postponed. "It will first be necessary to purge the globe of syphilis and other skin diseases. Until this is accomplished, Harmony will be more circumspect about love than civilization now is."[20]

In Fourier's cosmology, the planets were bisexual creatures which reproduced and played various educational roles in the solar system. The earth would not emit the proper aroma for universal equilibrium until Harmony was an actuality. Then the seas would turn into lemonade, more helpful animals would appear, and humans would be seven feet tall, live 144 years, and grow tails to assist in navigation. Much of this material has been warmly received by poets.

His attitude toward nature was friendly, but certainly anthropocentric. He allotted to one of his work groups the function of ensuring humane treatment of domestic animals, but another group was to dispose of repulsive reptiles. In Harmony, even more animals would be drawn into the human domain. "Zebras and quaggas will have been domesticated like the horses and donkeys of today; beavers will be building their dams and establishing their communities in the midst of the most heavily populated areas; herds of vicuna will be as commonplace a sight in the mountains as are sheep today; and a multitude of other animals like the ostrich, the deer, the jerboa, etc., will come and join forces with man as soon as his company becomes attractive to them, which it can never be in the civilized order!"[21]

Fourier believed that world population should stabilize at three billion, when both human fertility and infant mortality would decline: "It would be much better to have fewer children but keep them."[22] People would sleep little and eat enormous quantities, with sugared fruit replacing bread as a staple food—a clear prediction of the Fruit Loop breakfast. Another prophecy was that the Northern Crown would "heat the arctic glacial regions with reflected light," melting the polar icecap and making Siberia warmer than Florence.[23]

Aside from its therapeutic value, the full appreciation of Fourier's work awaits a spiritual red-green.[24] Nevertheless, his attitude toward the natural world could add some spice (nutmeg) to the continuing debate about the role of nature in relation to human existence.

Fourier's Relevance for Red-Green Theory

How might Fourier's doctrines assist in the creation of red-green theory today? Can we adopt a forefather? We hardly need to; Fourierism has always been present as green in the red. In any case, Fourierism has simmered along as part of the major socialist traditions, erupting into full boil on occasion as in nineteenth-century Russia or in Edward Bellamy's *Looking Backward*, which was so important to the development of the (U.S.) Socialist Party. Such detailed descriptions of socialism were precisely why Marx called Fourier utopian. Nevertheless, utopias may well be effective recruiters. Even Lenin was influenced by the Fourierist-inspired novel by Nikolai Chernyshevsky, *What is to Be Done?*

After the Russian Revolution, Lenin dismissed as "infantilism" the abolition of the family proposed by Kollontai and other Fourierist revolutionaries. The road not taken might have resulted in a very different fate for communism which, in addition to the burdens imposed by capitalist encirclement, was stressed by the wastefulness of individual consumption and the dysfunctions of Soviet marriage, nepotism, shopping, and poor nutrition and health, deriving from the attempted maintenance of the traditional family.

Fourier is of value to red-greens today both for the charming scenarios of life in a steady-state economy and for his serious consideration of many issues which Greens, though seeking to be a comprehensive movement, have ignored. Foremost is a basic matter of human ecology: the family. Other issues of importance include feminism, agriculture, work, and the liberation of instinct. Fourier's indictment of the traditional conception of the family is conventional wisdom today, yet few are willing to take the next step and design institutions which are more in accordance with human needs and desires. Greens should be particularly interested in replacing this dead horse for several reasons.

First, the nuclear family is the locus of consumerism. Fourier was concerned with the waste of energy for heat and light in individual households. Today, polluting, nonrenewable resources keep our large, empty houses warm all day. Further, each family must have not merely a cider press and goose roaster, but cars, VCRs, crockpots, stockpots, candlepins, recycling bins, wine racks, Nordic Tracks, mowers, rowers, and blowers. When Greens urge that we live more simply, they are accused of elitism. We are told that growth is needed so that the poor will benefit from "development." Yet development, even where it has actually increased the GNP per capita, has not always increased well-being. The use of infant formula and other canned, surplus, Western, agricultural products is just one example. Besides, the path of development is a no-win game for nowhere in the world will people be satisfied until each family has the "normal stock" of Western consumer goods. Even if that were possible, the environmental consequences of this level of production would be staggering. Greens in the affluent countries could provide a model for the poorer parts of the world by living better with less. That is essentially the argument of Rudolf Bahro, who shares many of Fourier's enthusiasms.[25]

All could enjoy an opulent standard of living through communal consumption. Not only could manufactured goods (such as vehicles and grand pianos) be shared, but many of our current "needs" are merely the consequence of our isolated lives (e.g, transportation and communication equipment) or compensations for monotonous or sedentary work (e.g., hobby and sports gear). Our propensity to overconsume is often based on the poverty of our emotional lives, and it is precisely this type of enrichment which Fourier expected the phalanstery to provide. Community (or updated tribalism) may be a basic principle of human ecology, which must be consciously re-created with attention this time around to human liberation and environmental protection.

Second, it is clear that throughout the world the family is in decline as a supportive institution; in many places, it hardly ever existed as such. Disintegration becomes more obvious under the spotlight of activism for human, women's, and children's rights. Today we do not have to be an observant traveling salesman of silk to know that armies of children live alone, work, and die on the streets of major cities; that even in "good homes" children and women are physically and sexually abused; and that whole nations and "respectable" families balance their budgets through prostitution and drug sales.

Additional stress on contemporary families stems from several sources. First, there is women's desire for equality in society, the workplace, and the home. As Fourier clearly saw, the institution of marriage does not sit well with an equal couple. Second, bisexuality adds complications to the fragile traditional arrangement. Third, today there is a continuing alteration in the sex ratio; perhaps one reason for this is the decline in maternal mortality. There are more women than men in almost all countries (except China, India, Abu Dhabi, Albania, and a few others). This imbalance is even greater if we consider only those men and women interested in and capable of forming stable families. The conditions for heterosexual monogamy for many women who desire to have children do not now exist. However, for married men, the abundance of single (including divorced) women means that apparently better deals are constantly available, and this hardly strengthens the family.[26]

Fourier's solution of security within a community, for both adults and children, would remove much of the stress and resignation to inhuman terms in marriage when no alternative is seen. With lost, or no love, all would not be lost. Even for "happy" people, if they are single parents, homosexual parents, or couples in which both work outside the home or are politically active, a communal family would have obvious advantages. Parenting could be a shared function. Monogamous couples might still exist, but there would be no social or economic pressure for monogamy per se. Groups with unequal numbers of men and women could provide for needs now so imperfectly satisfied under familism.

Another major problem with the monogamous family is that people are living much longer, often in poor condition, and neither traditional nor welfare-state approaches to the care of the elderly are satisfactory. The traditional way was for women to care for elderly relatives, with the rich employing servants to assist in the task. Today, domestic service or filial servitude is not compatible with the freedom women seek. Further, women may be fully employed single parents, which can make caring for the elderly an enormous burden. The same problems apply in the case of children and adults who are disabled or, for that matter, any of our rugged individualists who become temporarily ill or injured. The welfare-state approach is to institutionalize people in public or private nursing homes or halfway houses, with care provided by paid workers. Some of these institutions may be splendid and some of the workers are saints, but isolation still results. Even in the model welfare states of the Nordic countries, dissatisfaction with the bureaucratic solution to caregiving is widespread.

The communal approach would be preferable. Care becomes part of the work of the community and can be shared by many. Those who enjoy such work might spend more time at it, but they would be less isolated and more likely to find others willing to relieve them. Aristotle's notion of ruling and being ruled in turn, which he considered the essence of a genuine *polis*, could be expanded to include the idea of caring and being cared for in turn. Reports of nineteenth-century communal societies in the United States indicate that this function was performed successfully. Before the days of Social Security, older people without relatives, especially the poor, had few options for a decent life; the poorhouse was a dreaded institution. Communities such as Oneida and the many Shaker communes provided care in a familiar setting, and there is some indication that communists were healthier and lived longer than those in "normal" families.

A fourth benefit of communal organization relates to the quality of life, particularly the important matter of the diet. Today tradition, modified by commercial advertising, determines most domestic decisions. Yet even those who are aware of a scientifically proper diet (which relies heavily on organic fruits, vegetables, grains, and legumes with an occasional small animal such as a goose or frog) find preparing it on a daily basis difficult and time-consuming. In a communal setting, food could be grown locally and prepared elaborately to the highest health and aesthetic standards. As the work would be done in company, with intrigues, assignations, and puns flying about, it would hardly be laborious. In addition to the health benefits, there would be great savings of energy and other resources. The kind of food we eat, the amount, and the distribution methods are among the most wasteful aspects of civilization (and we would have to add the diet-food industry and much of medical practice to this waste).

A second aspect of Fourier's doctrine which deserves attention today is his feminism, which was amazingly radical for a male even by today's standards. In the name of equality and liberation of women, he does not turn women into men, as in Plato's *Republic*, although he likewise proposes to abolish marriage and the family. Fourier believed women to be intellectually superior to men, but their minds were dragged into the dust by the need to focus on petty household chores.[27] He also advocated women's sexual liberation, which he saw as their substantive right (regardless of age or cellulite) to have lovers and to choose partners of either or both sexes. Hence the guaranteed sexual minimum. This is different from the demand for "free love" of most leftists (e.g., Engels), which is mostly a right to be chosen, depending on one's marketability.

Fourier's feminism can be justly accused of going beyond equality. The phalanstery not only ends the separation between "productive" and "reproductive" work, but values most highly "traditional" women's work. Thus, Fourier is seen as wanting to "feminize" the world so that peace, cooperation, environmental protection, nurturance, comfort, pleasure, decoration, the arts, relationships, and food occupy the highest rungs.[28] (That sounds suspiciously like the Green project). Fourier does not want to reform men or repress their natures. Let them have their tournaments, intrigues, mock wars, and crusades; just make sure no one gets hurt and that rewards also go to the fastest grape peeler and the fluffiest omelette makers so that all egos get massaged.[29] The phalanstery then, answers not only the Third World criticism that reduced consumption will doom them to drudgery and spartan living, but also the feminist complaint that only that which is marketed is valued. Even by market standards, there is growing recognition that human capital is most important for productivity.

Other modern problems brilliantly addressed by Fourier include the increasing disdain for both agricultural and manufacturing work. By all Green standards, our contemporary methods of producing and distributing food are destructive. Banana Republics, the Green Revolution, pesticides, cattle raising, monoculture, agribusiness, suicide-prone family farmers, hard tomatoes, frozen carrots, canned spinach, long shopping trips, child labor, huge private refrigerators, fast food, cholesterol, chips, soft-drink machines, microwaves, cans, bottles, boxes, wrappers—all must go into the dumpster for the sake of the planet and our own health and sanity. The preferred alternative, endorsed by energy economics, is labor-intensive, diverse, organic farming in proximity to consumers.[30] As a voluntary undertaking, organic farming can be a great delight, but farm life, albeit cowless, has its long stretches of "rural idiocy." Family farms have depended on the exploitation of family members who are increasingly unwilling today to accept such a role. Women do not want to be farmers' wives and children want to go to the disco, not the silo. The solution is clearly communal farming with the ambiance of cruise ships, which also would provide varied nonfarm work, permitting the economic use of capital equipment and enough labor and jolly work parties to keep everything repaired and mended shipshape. Fourier's plan is bioregional, precluding international or even long-distance trade in basic subsistence.

Greens need to confront Fourier's insistence on the liberation of instinct. We seem to have forgotten that the Green movement in Europe and the United

States is an offspring of the 1960s, whose theorists, Herbert Marcuse and Norman Brown, indicted the repressiveness of Western society. The demands for wholeness and happiness are precisely the point of a radical movement. Breton's *Ode to Fourier* and its introduction by translator Kenneth White evoke the New Left by bringing together Fourier, the Native American culture of New Mexico, Lewis Mumford, Taoism, and surrealism. "What unites Fourierism and surrealism is the notion of a communal life lived under the pleasure-principle rather than, as pessimistic Freud believed it must be, the reality principle, with its consequent suppression of the less controllable and libidinous levels of human being."[31]

Now everyone wants to liberate the magical child within. That deserves some place, too, in red-green theory and practice. Greens claim to oppose the domination of nature, including human nature, but they rarely discuss what that might entail. We do not have to accept Fourier's psychology to see how much repression there is in our lives or how our natural rhythms are forced to fit the needs of capitalism and bureaucracy. The resolution-packed, siestaless Green conclaves frequently imitate these same unnatural patterns.

Finally, however bizarre parts of Fourier's proposal are, he reminds us that all benefits and burdens need to be distributed equally or fairly. Although Fourier assumed some financial inequalities in Harmony, he indicated how insignificant these would be. All would enjoy an enhanced quality of life (food, lodging, clothing, medical care, entertainment); autonomy and pleasure in work; and opportunities for leadership. Mental and physical labor would not be separated. All classes, sexes, and ages would be integrated in all activities, except that the children would not participate in sex. He also realized the importance of nonmaterial benefits such as sex, social life, and honors and awards (e.g., ego goods). These need to be distributed more equally not only for the sake of justice, but also because they take the pressure off hard goods made of scarce resources. Change of fashion will not be necessary when neither status nor love is gained by it.

Criticisms and Conclusion

There is plenty more inspiration in the treasure trove of Fourier; perhaps the readers will dip into this hilarious and instructive body of writings. What are some of his shortcomings as a red-green visionary? Most obvious is his lack of "political correctness." He was delighted with every type of person, had a special love for lesbians, and attributed lower-class crudeness and violence to socialization. However, he also had some strong ethnic prejudices; these contradicted his own doctrine that basic personality types were formed from various combinations of inherent passions, not shaped by ethnic or other socialization.

His whole theory rides on the assumption that harmony will be automatic, that "the free development of each will be the condition for the free development of all." Since all necessary work will be accomplished because of "passionate attraction," no coercive institutions will be needed, and because he sees no serious conflict, he does not provide for any democratic means of decision making. Thus, Fourier may well be accused of "reckless optimism."[32] Fellow Frenchman André Gorz, who shares much of Fourier's vision, sees the persistence of a "realm of necessity" in the best of future societies.[33] This would still allow most time for play or voluntary work, but social discipline and coercion, hence democratic institutions, would remain necessary.

As for Fourier's new amorous world, it is not certain that all his devices would get everyone into the act. Further, would one be thrilled spending the weekend with a swain when one is merely a punishment, a merit-badge requirement, or an object of philanthropy? Perhaps it would be satisfying, for people pay prostitutes and still believe that the deal is worthwhile. We acquire sexual partners for many reasons other than lust (e.g., money, brains, power, security). Rousseau thought that in a "state of nature" anyone would do, and perhaps Fourier is following this approach. (Or maybe in a state of desperation, anyone would do.) He regarded relationships based on spritual and intellectual affinities as the best, but when the best was not available, Woody Allen's "meaningless sex" would fill the bill.

We may also doubt his predictions of loving relations among all one's current and past lovers, a basic postulate of Harmony where they continue living together. Jealousy may be mitigated in a situation where there is no fear of material abandonment and all enjoy high self-esteem in a supportive community, but can it really disappear? A related problem in both love and work is that although Fourier was confident, we cannot really say what human nature is, or what it would mean to live without dominating or repressing nature. Furthermore, we may not want to as it seems that part of "nature's plan" for women is pregnancy, childbirth, and lactation. Is birth control unnatural? Are some means natural (e.g., herbal) and others not? What about abortion?

Fourier's new world has been deemed excessively rural. He ignores the benefits of industrialism which Owen, Saint Simon, Marx, and Engels saw as the major means for abolishing poverty. Instead, reduction of need and wasteful consumption, and elimination of exploitation by parasites (including armies and navies), would yield abundance. He posits enormous productivity from his joyful workers without mass-production techniques. Nevertheless, he was not an advocate of primitivism or peasant culture and did not share Rousseau's enthusiasm for Sparta. Harmony would include the best of city life (without the noise, crime, or traffic jams) and artistic creativity would surpass that of "civilization." His vision heralded the Garden City concept, which remains a worthy Green ideal.[34]

Fourier instinctively opposed factories. Perhaps the benefits of industrialism have never exceeded the costs: so much unseen exploitation and unmeasured social, health, and environmental damage. It may permit a larger population, but is this a clear benefit? Is the industrialized diet clearly superior to the preindustrial which, especially for the poor, included a larger proportion of wild and unrefined food? Even primitive people could satisfy their needs with only a few hours of work. Industrialization creates needs and it does not satisfy them very well. While we might concede medical and dental progress, it is not clear that they depend on the factory system, coal burning, or extensive agriculture. Appropriate technology can be developed even in rural areas.

Despite reservations about details and even some basic premises of Fourierism, it can nevertheless supply inspiration for red-greens in search of theory. His radical solutions to the problems of consumerism, the elderly, the lonely, the family, agriculture, the nature of work, the full expression of human potentialities, and the waste and corruption of capitalist civilization could be quite practical. The time for communitarianism may have come as pressures mount on individuals, families, and the environment, and corporations desperately make erotic pitches to sell exotic goods to neurotic consumers.

Notes

1. Carl Guarneri, *The Utopian Alternative: Fourierism in Nineteenth Century America* (Ithaca, N.Y.: Cornell University Press, 1991).
2. H. H. Van Amringe, *Association and Christianity* (Pittsburgh: J. W. Cook, 1845).
3. Frederick Engels, *Socialism: Utopian and Scientific* (New York: International Publishers, 1989), p. 39.
4. See Alexandra Kollontai, "Soon," in *Selected Writings of Alexandra Kollontai*, translated with an introduction and commentaries by Alix Holt (New York: W. W. Norton and Company, 1977), p. 232.
5. André Breton, *Ode to Fourier*, translated with an introduction by Kenneth White (London: Cape Goliard, 1970); M. C. Spencer, *Charles Fourier* (Boston: Twayne, 1981), p. 132.
6. Roland Barthes, *Sade/Fourier/Loyola*, trans. Richard Miller (Berkeley: University of California Press, 1989), p. 3.
7. Charles Fourier, *Le nouveau monde amoureux*, in *Oeuvres complètes*, vol. 7 (Paris: Anthropos, 1967), pp. 339–386.
8. C. W. Heckethorn, *The Secret Societies of All Ages and Countries* (New Hyde Park: University Books, 1965), pp. 84–90.
9. Jonathan Beecher and Richard Bienvenu, eds. and trans., *The Utopian Vision of Charles Fourier: Selected Texts on Work, Love, and Passionate Attraction* (Boston: Beacon Press, 1971), p. 161. Henceforth cited as *UV*.
10. Engels, *Socialism: Utopian and Scientific*, pp. 38–39.
11. *UV*, p. 288.
12. David Zeldin, *The Educational Ideas of Charles Fourier* (New York: Augustus M. Kelley, 1969), p. 72.
13. Ibid., p. 109.
14. Ibid., p. 92.

15. *UV*, p. 172.

16. Mark Poster, ed., *Harmonian Man: Selected Writings of Charles Fourier* (Garden City, N.Y.: Doubleday, 1971), p. 210.

17. *UV*, p. 182.

18. Ibid., p. 99.

19. Ibid., p. 253.

20. Ibid., p. 395. Fourier assumed that birth control was available. The traditional French method (coitus interruptus) did not prevent disease.

21. Ibid., p. 404.

22. Ibid., p. 505.

23. Ibid., p. 405.

24. The hieroglyphs evoke Tantric Yoga, in which chakras (or lotuses) are related to colors, sounds (mantras), organs of the body, and emotions. Perhaps Fourier knew of Tantra, as he was well versed in geography. Another idea that wafted in is the newly respectable "aromatherapy." As for algae on ponds, it is not exactly lemonade, but it is the new food according to some New Agers. Parallels with some of Fourier's ideas can be found in the works of Rudolf Steiner (though I have not found evidence of direct influence), whose disciples are prominent among the German Greens. See Anna Bramwell, *Ecology in the Twentieth Century: A History* (New Haven, Conn.: Yale University Press, 1989).

25. Rudolf Bahro, *Building the Green Movement*, trans. Mary Tyler (Philadelphia: New Society Publishers, 1986).

26. See Marcia Guttentag and Paul Secord, *Too Many Women: The Sex Ratio Question* (Beverly Hills, Calif.: Sage Publications, 1983).

27. As Alexandra Kollontai so perceptively noted, even if all chores are equally divided, men still demand an unequal share of the ego-maintenance work. See her *Autobiography of a Sexually Emancipated Communist Woman*, ed. Iring Fetscher, trans. Salvator Attanasio (New York: Schocken Books, 1975), p. 7.

28. Diana Coole, *Women in Political Theory* (Boulder, Colo.: Rienner, 1988), p. 178.

29. Of course, all these activities are open to women. Unfortunately, there is no evidence that sublimation of aggression works; the most sports-minded nation is also the most violent in internal and external relations.

30. Juan Martinez-Alier, *Ecological Economics* (Oxford: Blackwell, 1987).

31. Kenneth White, Introduction to Breton, *Ode to Fourier*.

32. Nicholas Riasanovsky, *The Teaching of Charles Fourier* (Berkeley: University of California Press, 1969), p. 242.

33. See André Gorz, *Paths to Paradise: On the Liberation from Work*, trans. Malcolm Imrie (Boston: South End Press, 1985).

34. See Ebenezer Howard, *Garden Cities of Tomorrow* (1898; reprint, Cambridge, Mass.: The MIT Press, 1965).

Chapter Three

Martin Heidegger: Antinaturalistic Critic of Technological Modernity

Michael E. Zimmerman

In the early 1970s, during an era of growing environmental crisis, social decay, and Third World challenges to triumphal First World narratives, a number of critics began asserting that neither modernity's industrial systems (whether in their liberal capitalist or state socialist guises) nor its legitimating political ideologies could be reformed, but rather would have to be fundamentally changed if modernity's noble goals of universal political emancipation and material well-being were to be realized without causing ecological calamity and human alienation. In the following two decades, these critics helped to alter the political landscape dramatically, so that today most politicians depict themselves as greatly concerned about environmental affairs. I shall call the critics "radical environmentalists," who include not only participants in Green politics and environmental activism but also deep ecologists, social ecologists, and ecofeminists. Most radical environmentalists agree that state socialism does not provide a genuine alternative to capitalism, for both ideologies share modernity's anthropocentric view that nature is nothing but raw material for human ends. Many radical ecologists asserted that modernity's aim of conquering nature to enhance the human estate is not only suicidal but inconsistent with some of humanity's deepest yearnings, including the need to establish an ecologically sustainable and spiritually and psychologically sound relationship with the nonhuman world.

Having renounced modern political heroes ranging from John Locke to

Karl Marx, radical environmentalists sought alternative theoretical justification for their claim that humankind must learn to live with nature instead of trying to conquer it. Some members of the deep-ecology branch of radical environmentalism (including the present author) once believed that the thought of Martin Heidegger (1989–1976) could help to articulate that movement.[1] After all, Heidegger was one of the only leading twentieth-century European thinkers to have condemned the destruction of nature at the hands of modern economies. His later writings sought to redefine humankind and nature so as to envision an alternative to industrial ideologies that justify treating humans and nonhumans alike as commodities.[2] Because modern activism generated the political and economic institutions responsible for destroying nature and diminishing humankind, Heidegger concluded that more such activism would only worsen the situation. He believed that only the gift of a "new beginning," involving a dramatic change in human self-understanding, could provide an alternative to modernity's industrial-technological imperative. Although in what follows I shall focus on efforts to use Heidegger's thought to articulate themes in deep ecology, I believe that all radical environmentalists have something to learn about the political drawbacks that have emerged in connection with those well-intended efforts.

There are several points of agreement between Heidegger's thought and deep ecology, in particular the notion that humankind's highest possibility and obligation is not to dominate nature through technological means, but rather to "let things be" in the twofold sense of allowing them to manifest themselves according to their own possibilities, and of allowing them to pursue their own destinies with as little interference as possible. However, there also are some important differences. First, Heidegger's thought was intertwined with reactionary political views, a fact that compromises efforts to read him as a theoretical antecedent of deep ecology since deep ecologists oppose fascism. Nevertheless, Heidegger's political experience provides deep ecologists and other radical environmentalists with food for thought. Second, unlike most deep ecologists, Heidegger did not appeal to the science of ecology either to justify his critique of industrial modernity or to articulate his vision of a postmodern world. Though admitting that scientific findings are valid in the appropriate domain, he insisted that as positivism, science arrogantly overreaches itself.

Heidegger abjured all forms of "naturalism," according to which humans can be adequately defined as intelligent animals that have arisen by virtue of blind, material, evolutionary processes. This antinaturalistic stance led some deep ecologists to suspect that despite many indications that his thought had a green dimension, Heidegger adhered to the humanity-nature dualism and anthropocentricism that have helped to generate the ecological crisis. As we shall see, however, Heidegger's antinaturalism had the virtue of leading him to reject Nazism's biological racism, despite his attraction to certain other elements of its program.

Political and conceptual problems notwithstanding, I shall argue that radical ecologists can learn from Heidegger's philosophy. Challenging modernity's notions that nature is merely raw material for enhancing human purposes and that humans are nothing more than clever animals struggling to survive and to prosper, Heidegger maintained that we must understand human existence in terms of its relation to a transcendent dimension that confers on humankind the responsibility not only for preserving itself, but also for caring for all entities. His own misdirected political views and actions, however, remind us of the dangers involved in any antihumanist critique of modernity's universalistic, emancipatory aims.

In what follows, I first offer further comments on the complex relationships among Heidegger, deep ecologists, and more traditional social radicals. Then, after discussing the major features of Heidegger's ontology, I describe his complex relationship with National Socialism. Finally, I suggest what deep ecologists and other radical environmentalists may learn from Heidegger's thought and political misadventures.

Heidegger, Deep Ecology, and Reactionary Politics

Deep ecologists assert that leftists, despite purporting to pose a radical alternative to capitalism, in fact share capitalism's anthropocentric humanism, which promotes a wholly instrumentalist attitude toward nature. As deep ecologist George Sessions says, "An ecologically harmonious social paradigm shift is going to require a *total* reorientation of the [anthropocentric] thrust of Western culture."[3] In effect, deep ecologists take seriously Marx's own claim that after capitalism has gained control of nature to such an extent that material scarcity can be eliminated, communist revolution will dismantle the class structure standing in the way of human self-actualization, which will require the technological control of many natural processes. Far from being diametrically opposed to one another, then, Marxism and liberal capitalism share modernity's anthropocentric belief that humankind can and should use science, technology, and industry to master nature in order to further humanity's material and political interests. Despite efforts by leftist theorists to read Marx as an environmental thinker, deep ecologists and many other radical environmentalists conclude that Marx's anthropocentrism inevitably leads to an instrumentalist view of nature.[4] Certainly what he had in mind by the "self-actualization" of *humanity's* potential cannot be reconciled with deep ecology's ideal of "self-realization" for *all* beings.

Like Heidegger, deep ecologists hold that modernity's quest to emancipate humanity by gaining power over nature has become the quest for power for its own sake. The industrial system is no longer a means to human ends; rather, humans and everything else are means to the irrational and self-destructive end of growth and control for their own sakes.[5] Heidegger would have agreed with Sessions's remarks that "the diminishment of man and the diminishment of the planet and its nonhuman inhabitants [are] essentially one and the same problem."[6] Deep ecologists correctly assert that an anthropocentric, cost-benefit approach to dealing with "natural resources" predominates in industrialized countries. Further, mainstream environmental protection has focused largely on curbing industrial pollution and extending the life span of resources. Although such measures are an important first step in the right direction, deep ecologists (like other radical environmentalists) maintain that "reform environmentalism" merely addresses the *symptoms* of the environmental crisis. Admitting that the origins of this crisis are highly complex, deep ecologists maintain that two crucial factors are anthropocentrism and humanity-nature dualism which, both in state socialism and liberal capitalism, promote an exclusively instrumental attitude toward nature. Deep ecology is "deep," then, because it asks deeper questions about the origins of our ecological problems than do theorists for the predominant political ideologies of modernity.[7]

Unfortunately, deep ecologists have not always asked deep enough questions about the political antecedents of positions similar to their own. Deep ecologists were not the first to offer a sweeping condemnation of communism and capitalism, to assert that the apparently opposed systems of capitalism and communism shared a destructive underlying ideology, or to call for a revolutionary spiritual transformation that would lead to a new age of ecological stability and social harmony. Years ago, discovering that Heidegger offered a similar diagnosis of and alternative to modernity's various ills, I proposed that he be regarded as an intellectual antecedent of deep ecology. Only later did I fully understand that Heidegger's interpretation of modernity was in many (though not all) respects consistent with National Socialism, which condemned capitalism and communism alike for causing the destruction of "blood and soil" (*Blut und Boden*), "homeland and people" (*Heimat und Volk*).[8] Having become aware of the reactionary aspect of Heidegger's thought, most deep ecologists have stopped citing him as an intellectual predecessor. Moreover, they have become much more cognizant of the fact that a totalizing critique of modernity could inadvertently lend support to incipient forms of ecofascism.

In view of the resurgence of neofascist politics in Europe and the Americas, all radical environmentalists have an obligation to inform themselves about how Nazi Germany sought to fuse its racist ideology with many laudable ecological concepts. Denying that either capitalism or communism could possibly be reformed, Nazi ideologues championed Hitler's revolutionary "third way" beyond modernity's rootless, cosmopolitan, urban, materialistic, and rationalistic (i.e., according to the Nazis, altogether *Jewish*) ideologies, attitudes, and institutions. Rejecting the social atomism of French, English, and U.S. capitalism, as well as the materialism of Soviet Marxism, the Nazis maintained that the German *Volk* could be saved only by a leader who could show the *Volk* how to live according to nature's eternal laws. In instituting many of the world's first serious environmental-protection laws, the Third Reich revealed its green side, though lamentably this nature-protection impulse was not only based on racism, but was also soon contradicted by the Nazi decision to expand its industrial economy in order to support territorial expansion and military conquest.

Though many young people who became involved in radical environmentalism during the 1970s were unfamiliar with the perverse green side of Nazism, progressive thinkers did *not* forget that the Nazis promoted racial purity and environmental protection while condemning the universal emancipatory aims of modernity. This helps explain why the Green movement could not arise in West Germany until the 1970s, that is, after the emergence of the postwar generation.[9] Radical environmentalists are ill-informed if they believe that Marxists and liberal theorists criticize radical environmentalism solely because it threatens their growth-oriented economies. Instead, such criticism is also advanced because progressive thinkers (here I mean the best socialist and liberal theorists) recall National Socialism's celebration of the link between a healthy race and an unpolluted landscape. In view of that link, socialist and liberal theorists are understandably suspicious of radical environmentalists who either condemn modernity in toto, or who claim that they are neither left nor right, but "out in front." If environmentalists are correct in saying that modernity gives rise to totalizing systems with ecologically destructive tendencies, supporters of modernity are also right in replying that radical ecology may inadvertently help to generate ecofascism by promoting an antiuniversalistic antihumanism that undermines modernity's emancipatory aims.[10]

Radical environmentalists are faced with the daunting task of envisioning new social forms and modes of awareness that retain the beneficial aspects of modernity, while at least minimizing its ecologically destructive dark side. For radical ecology to succeed, then, it must help to transform the emancipatory aims of modernity as well as the means by which they are to be achieved. Ideally, the achievement of human rights and self-realization should not have to come at the cost of exterminating other species and drastically altering whole ecosystems.[11] Herein lies the kernel of truth both in Heidegger's contention that authentic humanism involves learning to let things be, and in deep ecology's claim that human self-realization cannot be achieved apart from the self-realization of all beings.

Heidegger's Thought in Its Historical Context

As a German, Heidegger was born among people with a great appreciation for the natural world. The German people have had a long-standing interest in natural environments and human cultures that have been relatively untouched by modern civilization. During the late nineteenth and early twentieth centuries, this interest prompted a number of German theorists, scientists, and politicians to criticize the ecologically and socially devastating consequences of the rapid industrialization being pushed by the newly unified and nationalistic Germany, which was competing with its economically and politically powerful French and British rivals. During the period before and after World War I, German geographers and biologists, some of whose arguments are very similar to those used by today's ecological activists, were prominent participants in international conferences devoted to preserving wild species and native peoples from destruction at the hands of modern economic systems and predatory colonial practices. Using the example of U.S. success with establishing national parks, leaders of Germany's popular *Naturschutz* (nature protection) movement demanded that similar measures be taken to preserve Germany's own natural monuments, and to protect the landscape from threats ranging from rampant industrialization to billboards.[12]

Not surprisingly, the *Naturschutz* movement proved particularly attractive to people with conservative or reactionary political views. Most progressive Germans (socialist and liberal alike) shared in modernity's instrumental view of nature, and thus had little interest in a movement inspired in part by romantic ideals of a harmonious relation between humankind and the land. Because the *Naturschutz* movement was supported primarily by people who believed that environmental destruction and social alienation were inevitable products of industrial modernity, Nazis spouting antimodernist rhetoric were more easily able to capture that movement for their own purposes. Even though pre-1933 *Naturschutz* members were generally conservative, however, they did not always share Nazism's militaristic, jingoistic, and racist outlook.

Germany's defeat in World War I, the disastrous Versailles Treaty, the terrible inflation, and the economic collapse of 1929 all helped to swell the ranks of those who considered modernity to be an insidious disease spread by greedy English, French, and American capitalists and by evil Soviet communists. Liberalism lost many followers to various conservative parties, even though socialism managed to retain a strong core of working-class supporters. The Weimar Republic had little chance of success, not only because many Germans regarded liberal democracy as a foreign imposition of the victorious French and British, but also because left- and right-wing forces wanted it out of the way so that they could finally meet in a more direct contest. In 1933, Hitler came to power in part because he promised to restore Germany's pride, dignity, and economic stability, not only by ridding the country of the foreign ideologies that were destroying the German homeland, but also by controlling the "vermin" that were polluting German blood.

Heidegger, arguably the most influential philosopher of this century, grew up in the context of Germany's agonizing attempt to cope with the social, political, economic, cultural, and environmental pressures of modernity. Because he was raised in a small country town in southwest Germany, whose landscape he loved to explore on skis and on foot, he might well have been drawn to the *Naturschutz* movement, though there is no evidence that he joined it. His appreciation for natural phenomena, and his later conviction that authentic human existence involved a close relationship with local landscape and dialect, help to explain his distrust not only of the culturally homogenizing effects of large cities, but also of the ideologies that supported their land-destroying industries. Heidegger's critical comments about industrial modernity, which have proved appealing to many who are not deep ecologists, can be traced at least in part to his boyhood love for nature. As we shall see, however, what he meant by "nature" was a far cry from the "ecosystems" studied by modern scientists and referred to by deep ecologists. Instead, Heidegger defined "nature" as akin to what he meant by the "being" of entities.

Heidegger maintained that his effort to resurrect the question of being had world-historical significance. Though influenced by Oswald Spengler's best-selling book, *The Decline of the West*, Heidegger rejected the author's explanation for decline. In Heidegger's view, it was forgetfulness of being that had initiated Western humankind's long post-Platonic decline, which eventually culminated in the technological nihilism of modernity. Recovery could only take the form of a new encounter with what had been forgotten: the being of entities.

Before discussing any further Heidegger's account of the decline of the West, we need to review what he meant by being (*Sein*), human existence (*Dasein*), and the relationship between them. According to Heidegger, "being" refers not to a superior entity (e.g., Platonic forms, Aquinas's God, Hegel's Absolute, or Nietzsche's will to power) that constitutes the ground, source, or foundation for all other entities, but rather to the "self-manifesting" of entities. For something "to be" (*sein*), then, means for it to show up, to present itself, to stand forth as an entity having this or that meaning, this or that worth.

The German word *sein* is an infinitive, translated as "to be." In German, a gerund is formed simply by capitalizing the first letter of the infinitive; hence, *sein* becomes *Sein*. To form a gerund in English, however, one takes the infinitive form of a verb, drops the "to," and adds "-ing" to the verb stem. Thus, "to be" becomes "being." In part because the gerund "being" is often used as the equivalent of static nouns such as "entity" or "thing," the active verbal dimension of "be-ing" may be overshadowed by the substantive or thinglike connotations of "being." In conceiving of being as a superior kind of entity, we overlook what Heidegger called the "ontological difference" between being (*Sein*) and the entity (*das Seiende*). (*Das Seiende*, a participial form of *sein*, means something close to the substantive usage of the English gerund, "being.") Being, he insisted, *cannot* be understood as any sort of entity or thing. For something "to be," then, does not mean for a thing merely to subsist as a material object, but rather for a thing to disclose itself in such a way that it can be encountered, described, evaluated, understood, and appreciated *as* a thing.

Only one sort of entity, however, human *Dasein*, can encounter other entities in such a way as to notice that they "are." We know that animals encounter things because animals alter their behavior as a result of such encounters, for example, a hungry lion sees prey or a lizard comes upon a rock warmed by the sun. What sharply differentiates animals from humans, however, is that *only* the latter exists in a historical-linguistic world in which entities can show up *as* entities.[13] Supposedly, only humans are endowed with a usually unthematized "understanding of being" that enables them to differentiate between being and entities. Human *Dasein* constitutes the appropriately receptive site, clearing, or "absencing" (*Abwesen*) necessary for the being or "presencing" (*Anwesen*) of an entity to occur. *Dasein* is composed of two terms: *Da*, meaning "there," and *sein*, meaning "to be." *Dasein* means, then, to be the "there" (clearing, opening, absencing) necessary for the "being" (showing up, self-manifesting, presencing) of entities.

Thus conceived in terms of its ontologically disclosive existence, *Dasein* can scarcely be called an "entity" in any ordinary sense. Hence, just as being utterly transcends all ontic or entity-related categories, so too human *Dasein* cannot be defined in terms of categories applicable to nonhumans. Moreover, *Dasein* cannot be understood as "consciousness" or "subjectivity," since these terms refer to ontic phenomena that show themselves and thus "are" *within* the clearing constituted by *Dasein*. In other words, subjects and objects alike are *entities* that can "be" only insofar as *Dasein* exists understandingly as the temporal opening for being. While entities "are," only Dasein "exists." To exist means to execute the continual act of transcendence that constitutes the temporal-historical clearing in which entities manifest themselves as entities. Just as for Kant time constituted the pure form of intuition necessary for encountering objects, so for Heidegger temporal transcendence constitutes the receptive horizon necessary for entities to manifest themselves as entities.

Despite early Heidegger's debt to Kant and German idealism, he later concluded that the terminology of transcendental philosophy was tainted by the very subjectivism that he sought to overcome. After the "turn" in his thought, he tried to surpass all subjectivism and anthropocentrism by emphasizing that the "essence" of human existence, *Dasein*, is not a "property" or "possession" of humanity; instead, humanity itself is *appropriated* as the temporal-historical clearing in which things can "be." In this sense, the "clearing" is a cosmo-ontological event, not merely a human one. Hence, in his later writings, Heidegger spoke of *Dasein*, not of "human" *Dasein*.

According to later Heidegger, the clearing is articulated by language, which constitutes the "house of being."[14] Language is a gift that imposes on human *Dasein* the "supreme obligation" of bearing witness to entities, in the sense of "letting them be." Letting-be can mean an active engagement with things for the sake of allowing them to reveal themselves in ways consistent with their own possibilities, rather than as objects that are disclosed solely according to the demands of the power-oriented human subject. Letting-be can also mean letting things alone so that they can pursue their own course without undue human interference. By allowing things to manifest themselves, human *Dasein* is not doing them a favor, as if *Dasein* were deigning to give voice to its inferiors. Instead, human existence is fulfilled only to the extent to which it *serves* the ontologically disclosive process that transcends such existence. Though often existing dualistically as a subject trying to control objects, human *Dasein* can also exist nondualistically when it stops clinging to any form of subjectivity. Once subjectivity has receded from the clearing, entities can be manifest *without a separate subject* to apprehend them.[15] In this nondualist condition, which has many similarities with Buddhism, there would seem to be no room for domination or violence.[16]

According to early Heidegger, human *Dasein*—as the peculiar temporal transcendence necessary for understanding or disclosing the being of entities—exists in an "uncanny" (*unheimlich*) condition as "no-thingness." This condition is uncanny because for an entity to exist as nothingness means for it *to be aware of itself as* intrinsically finite, dependent, and mortal. Recoiling from awareness of this mortal condition, *Dasein* tries to conceive of itself as a kind of thing that can protect itself by controlling its environment. In denying its own mortality, human *Dasein* flees from the awesome responsibility and opportunity of existing resolutely as the clearing in which entities can manifest themselves in ways appropriate to their own possibilities.

In early Heidegger, then, *Dasein* can exist either inauthentically (*uneigentlich*, in an unowned way) or authentically (*eigentlich*, in an owned way). One is faced with this decision when, in the midst of everyday or routine existence, the mood of *Angst* threatens to remind us that we are mortal nothingness, not substantial things. To exist inauthentically means to repress *Angst* by becoming wholly absorbed in or distracted by entities (such as work, sex, drugs, food, or power) so that one forgets about their *being*, and in particular about one's own uncanny mode of being—existence. To exist authentically, in contrast, means to let *Angst* disclose one's mortal existence, thereby enabling one resolutely to "own" or to affirm it instead of disowning or fleeing from it. Such resoluteness helps make possible the generation of a new mode of temporality (later, a new poetic utterance) within the horizons of which entities can manifest themselves in novel ways, especially in ways that are more consistent with the innermost possibilities of things themselves. In the routines of everyday life, however, such possibilities are generally concealed.

Despite later Heidegger's concern that modern technology had reduced nature to industrial resources or "standing reserve" (*Bestand*), the early Heidegger—apart from some suggestive remarks—usually did not criticize the instrumental disclosure of entities. Indeed, early Heidegger may have contributed to the technological understanding of being by saying that *Dasein* discloses things authentically as resources for human purposes.[17] In *Being and Time*, we read that everyday *Dasein* exists according to the identity-constituting social roles and possibilities assigned to it at birth, when *Dasein* is "thrown" into the world. Here, "world" means not the natural cosmos, but rather the interrelated network of meaningful reference relationships that constitute the pathways of one's existence. These purposive reference relations stand in the service of that for the sake of which human activity is ultimately undertaken, namely, human *Dasein*. In the farmer's world, then, soil reveals itself as a rich or poor resource for growing crops.

Soon after *Being and Time*, however, Heidegger began to seek an alternative to a world in which entities can only manifest themselves as either scientific objects or industrial resources. In the 1930s, he suggested that a new world could arise only by virtue of an original encounter with *physis*, usually translated as "nature" but rendered as "being" by Heidegger. "*Physis* is being itself, by virtue of which entities become and remain observable."[18] He maintained that cosmic *logos* works through great poets and statesmen to "bring to a stand" the overpowering surge of *physis*/being. "Bringing being to a stand" means founding a delimited historical world in which being or *physis* can manifest itself intelligibly, that is, within *limits*, so that entities can be encountered and understood as entities. Language articulately discloses the being of entities. In regard to its ontologically disclosive capacity, however, language is not a human possession. Rather, language is cosmic *logos* acting through human existence so as to reveal the "jointing" or articulation that *logos* grants to *physis*. Not being a linguistic idealist, Heidegger asserted that authentic language lets *physis* reveal its own inherent structure.

By emphasizing that *logos* transcends humankind, Heidegger defended himself against the charge of anthropocentrism. Human existence is needed as the historical clearing through which *physis* can manifest itself as articulated by *logos*, but human existence is not the source of *physis* or *logos*. The "humanity-nature" relationship can never be healed until human *Dasein* recognizes that humans are not merely animals who use language as a tool, and that "nature" is not the totality of material particles in human service. Ultimately, neither human existence nor *physis*/being refer to "things" at all, but rather to the ontological conditions necessary for things to emerge into presence and thus "to be."

Heidegger's later concept of nature was strikingly different from his early concept, according to which "nature" was simply an entity that manifests itself within a given human "world." In 1936, he began distinguishing between "earth" and "world," so as to take into account the fact that earth (nature) involves an extrahistorical dimension that can never be fully disclosed in any human world.[19] "World" refers primarily to the historical-linguistic clearing in which entities can show themselves, while "earth" refers primarily to the tendency of entities to conceal themselves even as they show themselves within a given historical world. World and earth contend with each other; world wants to compel earth to become completely accessible within the historical-linguistic domain, but earth can never become totally accessible.

Heidegger's view of "nature" was influenced by Hölderlin's idea that nature as the All-One (*en panta*) was akin to Greek *physis*, the being of entities.[20] Such being, in Heidegger's opinion, could not be explained in terms of causal processes studied by science; rather, natural things could be interpreted in causal terms only because they first *manifested* themselves. When Hölderlin spoke of "homecoming," then, he meant humankind's return to its initial proximity to the origin that "gave" to the early Greeks their encounter with the sheer presencing of entities. For Heidegger, a homecoming was needed to prepare the German *Volk* for a new encounter with being, or nature, or Hölderlin's "holy wildness." Only such an encounter could free human *Dasein* from its currently constricted existence, which compels entities to "be" solely in instrumental terms. If entities began showing up in terms of possibilities proper to themselves, then—presumably—the ontologically expanded human *Dasein* would spontaneously begin treating entities with greater respect and care. In the one-dimensional technological world, however, such respect is impossible for there are no limits to human activity. Since only humans seem out of harmony with *physis* and its governing *logos*, only humans overstep the appropriate limits. Heidegger states that "the unnoticeable law of the earth" prevents the bee and birch tree from overstepping their limits. Unfortunately, in the technological era *Dasein* lacks such limits and thus seeks to devour the whole planet.[21]

In his later metaphor of humanity as a shepherd, Heidegger developed *Being and Time's* notion that to be human means to care for oneself and others, as well as about things. Later Heidegger stressed that caring was primarily a matter of letting things be manifest in terms of their most appropriate possibilities, rather than in terms of the expectations of the objectifying subject. Abandoning his earlier existential tendencies, Heidegger ceased describing inauthenticity in terms of an individual's flight from anxiety, but rather as a historical-cultural phenomenon resulting from the increasing self-concealment of being. Cut off from being, human *Dasein* gradually became blind to its own obligation to exist as the mortal openness in which things can "be." Conceiving of itself as a particular kind of thing, the "clever animal," *Dasein* set out to conquer all other entities in order to insure its security and to contribute to its ever-growing power.

During the 1930s, inflamed by political passions, Heidegger used violent terms to describe the world-founding activity needed to establish an alternative to the technological world. Later on, however, he began developing a less voluntaristic concept of a new *ethos*, "dwelling in the fourfold," which he hoped could provide such an alternative. For Heidegger, an *ethos* is a genuine world, one that is constituted by the limits necessary for cherishing, caring for, and saving things. Saving things, however, is not to be understood as an action of the human subject. Rather, "To save really means to set something free into its own presencing."[22] Seeking to remove the subjective aspect of the idea that things can manifest themselves only through the clearing established by human *Dasein*, later Heidegger suggested that the clearing is in fact constituted by the "fourfold" of earth and sky, gods and mortals. Further, this fourfold is itself gathered together by "things." Any thing—footbridge or bench, mountain or stream, ring or poem—can perform the world-founding function formerly restricted to human works.[23] World and thing are reciprocally related: "The world grants to things their presence. Things bear world. World grants things."[24] World, then, is the articulated unity of earth and sky, gods and mortals that are assembled in and through the thing.

Heidegger uses the example of a jug to show how thing and world grant and sustain one another.[25] A jug draws together *earth*, which provides water for the grapes in wine; *sky*, which provides sunshine for the ripening fruit; *gods*, to whom we offer a libation for the gift of wine and life; and *mortals*, who by partaking of the wine celebrate the mystery of thing and world. The jug draws together the differing ways in which things can be manifest: as natural things, gods, mortals. These things are inherently interrelated manifestations of being: "The united four are already strangled in their essential nature when we think of them only as separate realities, which are to be grounded in and explained by one another."[26] World is not constituted by a transcendental subject, but instead constitutes a luminous realm drawn together by the things manifesting themselves within it. Heidegger used the term *Ereignis* to describe how things join together and manifest themselves in the fourfold.

Critics have often charged that Heidegger's otherwise charming account of the fourfold fails to confront the harsh realities of modern political economies. By the 1940s, however, he had—for good reason—recused himself from further political involvement. Since politics was part of the problem, continuing to engage in it would only make matters worse. The idea of the fourfold (*Geviert*), then, is not to be seen as a set of political recommendations or as a call to action, but rather as an antipode to the "enframing" (*Gestell*) that compels modern humankind to undertake the brazen task of the technological conquest of nature. Given the extent to which yesterday's technological "solutions" (e.g., nuclear power) have given rise to today's problems (e.g., how to dispose of hideously poisonous nuclear wastes), one might well appreciate Heidegger's counsel that we need to step back from our incessant action. Meditative "thinking" may help to reveal the *historical* character of the technological disclosure of the being of entities. Realizing that our current mode of disclosure is not absolute or eternal, but instead is only temporary, constitutes a necessary step toward the advent of an alternative disclosure.

Heidegger's Involvement with National Socialism: Antinaturalistic Ontology versus Racist Naturalism

In 1933, after joining the Nazi Party in an elaborate public ceremony, Heidegger was installed as Rektor of Freiburg University.[27] Proclaiming his allegiance to Hitler, he became one of many German philosophers seeking to influence the direction of the "movement."[28] After the war, he explained that he briefly supported Nazism because he saw it as an effort to counter the devastating, technological disclosure of being. Contrary to such self-justifying postwar statements, however, his "engagement" with Nazism was momentous and enduring. Moreover, there are conceptual links between Heidegger's thought and reactionary politics. During the 1930s, for example, his account of the decline of the West coincided with the Nazi view that modernity was the culmination of a long degeneration of the West, and especially the German *Volk*, since the glorious age of the ancient Greeks.

According to Heidegger's controversial account of the "history of being," the technological era is the final stage in the long decline of the West from its great beginning in the age of the ancient Greeks. The early Greeks had a primordial encounter with being, which they called *physis*. For Heidegger, *physis* named not only the power that generated entities, but also the ontological power that appropriated Greek humanity as the measure-giving clearing required for entities to present themselves in delimited ways. While early Greek humanity existed in the service of *physis*, Plato supposedly initiated the metaphysical tradition which became increasingly blind to the ontological difference between being and entities. Projecting the activity of handcraft production onto the cosmos, Plato interpreted being as the "forms" (ontological blueprints) necessary for producing spatiotemporal things.

The productionist metaphysics of Plato and Aristotle degenerated into Rome's causal-material understanding of what it means for things to be. For medieval humanity, "to be" meant to be produced as a creature of the Creator, the Supreme Being. For Descartes and subsequently for modern humanity, "to be" meant to be an object for the self-certain, calculating subject. For technological humanity, "to be" means to be a raw material in the planetary production process in which all things—including people—are revealed as raw material to be transformed, switched about, stored up, and used in accordance solely with the ever-growing power demands of the technological system.[29] Under the sway of productionist metaphysics, humanity has forgotten (1) that there is an ontologically disclosive dimension that is *prior to* the causal-material dimension, and (2) that human existence is not the master of entities, but rather is in the service of the self-disclosure of entities. Seeking to decenter the voracious, self-assertive, clever animal of modernity, Heidegger regarded both the "rights" of the ego-subject and the "prerogatives" of nation-states as manifestations of an arrogant subjectivity that lacks insight into humankind's higher possibility.

Only a beginning equal in power to the one initiated by the Greeks, so Heidegger believed in the 1930s, could restore ontological depth to things that had been flattened in the one-dimensional technological world. Influenced by Ernst Jünger's claim that the technological age is governed by the *Gestalt* of the worker, Heidegger wanted to move the National Socialist German *Workers* Party toward a postproductionist, post-technological age in which work would cease being the laboring animal's indiscriminate commodification of all things, but instead would be attuned to humankind's ontological gift of disclosing things according to their own possibilities. Though eventually concluding that Nazism was merely another expression of the demented productionist metaphysics, he never explicitly renounced his belief in National Socialism's "inner truth and greatness."

Despite his support for the Nazi regime, Heidegger took some risks in combating the naturalism that formed the basis for Nazi racism. In his view, such naturalism was linked to modernity's scientific-technological understanding of being. In promoting naturalism, then, Nazism was unwisely promoting a crucial element of the very modernity that it claimed to attack. Heidegger called for an entirely different account of nature than that offered by modern science, important elements of which were accepted by Nazi ideologues. Indeed, Heidegger proclaimed that humans are *not* animals, that an abyss lies between the hand of the ape and that of the human.[30] Heidegger's antinaturalistic attitude was so pronounced that one of his former students, Karl Löwith, accused him of perpetuating the anthropocentricism and dualism so characteristic of the metaphysical and theological traditions which he purported to overcome![31] Another former student, Hans Jonas, charged that Heidegger held the Gnostic view that humanity is radically different from the natural world.[32] Heidegger's antinaturalistic attitude would seem to disqualify him as a theoretical forerunner of ecology movements which hold that humans are intelligent animals.

Heidegger used his accounts of "homecoming" and "nature" in conjunction with his own version of Nazism, but his understanding of nature was radically different from that justified Nazi racism. Unlike his Nazi colleagues, for whom getting "back in touch" with nature meant purifying the blood, reactivating the instincts, and achieving contact with occult cosmic forces, for Heidegger it meant encountering *physis* or being in a new way. Despite basic differences between his sophisticated view of nature and the Nazi view (or views, for there were several), he found it politically expedient to speak in a way that was consistent with the politically important *völkisch* theme that the *Volk* should live in accord with "nature."

The *völkisch* movement, which presaged the rise of National Socialism, insisted that the issue of the nature-*Volk* relationship was crucial for guiding

Germany in the face of the pressures of industrial modernity. According to George Mosse, the *völkisch* movement encouraged people to live "in accordance with nature" rather than conquering it. "Not within the city, but in the landscape, the countryside native to him, was man fated to merge with and become rooted in nature and the Volk."[33] Historian Robert Pois has argued that National Socialism, influenced by *völkisch* thinking, was a "religion of nature." According to Pois, Hitler renounced the evils of modernity, proclaimed the need to renew contact with the elemental natural forces, and stressed the importance of restoring ancient folk-customs, traditions, and attitudes.[34] These ideas appealed not only to the alienated bourgeoisie of Germany, but to other Europeans as well, representing the "overcoming of alienation, not through some hideous form of class war, *but rather through a revolution of consciousness*, the result of which would be a new sense of rootedness and belonging. . . ."[35]

Hitler's movement sought to replace the degenerate, other-worldly, transcendental belief systems of Judaism and Christianity with an immanent, this-worldly, scientifically grounded, nature-revering religion. Hence, one Nazi ideologue attacked man's "hubris and guilt" for trying to "master" nature, for such an attempt could only destroy the "natural foundations" of life.[36] According to such ideologues, anthropocentrists ignorantly assumed that nature was made for humanity. One author asserted, "According to our conception of nature, man is a link in the chain of living nature just as any other organism."[37] In 1939, Alfred Baeumler praised the view that man "must be understood *as a part of Nature*."[38] As a part of nature, people had to follow its laws in order to survive. For the Nazis, such "laws" included the necessity of racial "purity." As is well known, the Nazis practiced infanticide, euthanasia, genocide, and similar practices to in order to purge "racial parasites," degenerates, and others who posed a threat to the health of the *Volk*.

Curiously, in certain respects Heidegger's early thought has more in common with Judaism, especially its idea that the divine "source" of things radically transcends the world, than it does with Nazi doctrines about the immanence of nature in the "blood" of the *Volk*. For early Heidegger, being utterly transcends all entities; nature does not give rise to *Dasein*, but instead becomes manifest within the world opened up *by Dasein*. Even when trying to accommodate his thought to political trends, however, Heidegger never abandoned the idea that human beings are not merely "natural" entities. Rather, they are charged with taking upon themselves the often unwelcome responsibility of letting things be manifest. Regrettably, as John D. Caputo has observed, Heidegger neglected to take seriously enough the Jewish and Christian claim that human existence, by dint of its association with and dependence on the transcendent, also must take upon itself the enormous *moral* responsibility of living according to a law that runs counter to natural human inclinations.[39] Enthralled by the Greek fascination with the aesthetic, that is, with encouraging and encountering the *manifestness* of things, Heidegger neglected the Jewish and Christian preoccupation with the moral, that is, with being responsible for the ontical well-being of the *Other*. Resolving to do the "hard" and "dangerous" things needed to let being reveal itself anew, the Nazi Heidegger turned a blind eye to the suffering of those who were crushed in this brutal process.

What Radical Environmentalists Can Learn from Heidegger

We may approach the issue of what Heidegger may teach today's radical environmentalists by examining an issue about which they and Heidegger would profoundly disagree. Heidegger claimed that there is a greater danger than the destruction of all life on earth by nuclear war.[40] For radical environmentalists, it is hard to imagine anything more dangerous than the total destruction of the biosphere! Heidegger argued, however, that worse than such annihilation would be the totally technologized world in which material "happiness" for everyone is achieved, but in which humanity would be left with a radically constricted capacity for encountering the being of entities. This apparently exorbitant claim may be partially mitigated by the following consideration. If human existence lost all relationship to transcendent being, entities could no longer show themselves at all, and in this sense would no longer "be." Who needs nuclear war, Heidegger asked rhetorically, if entities have already ceased to be?[41] For many environmentalists, such a question reveals the extent to which Heidegger remained part of the human-centered tradition that he wanted to overcome. By estimating so highly human *Dasein's* contribution to the manifesting of things, Heidegger may well have underestimated the contribution made by many other forms of life, for which the extinction of humankind's ontological awareness would be far preferable to their own extinction in nuclear war!

Even though celebrating the natural world, and appreciating the extent to which humankind is dependent on and related to it, are preconditions for most forms of environmentalism, Heidegger is right that certain kinds of naturalism are dangerous. If a naturalism asserts that humans are simply animals just like any other animals with no special rights or standing, that life is a struggle for survival, and that genetic and ecological purity are crucial factors in that struggle, such naturalism can readily be used again as it has been in the past: to justify horrendous social practices. Hence, while deep ecologists rightly criticize those theories that portray humankind as having no relation to nature other than a domineering one, they should also be wary of the possible implications of theories that say humankind is no more than a clever animal. Despite Heidegger's ascription of such a reductive naturalism to both Marxism and liberalism, a more charitable interpretation of both ideologies arrives at a different conclusion.

Marxism and liberalism alike share the Enlightenment conviction that human history introduces upon this planet phenomena that cannot be understood in purely natural terms. The ecologically destructive consequences of these modern ideologies can be attributed to the fact that they tend to *dissociate* humanity from nature rather than carefully distinguishing between humanity's organic-naturalistic dimensions, on the one hand, and its historical-spiritual dimensions, on the other. Dualism between humanity and nature leads to serious ecological (and social) problems. Yet, conflation of humanity and nature may lead to significant problems of a different sort. Portraying humans as nothing but intelligent animals may prove appealing when compared with world-views which attribute to humankind a relation with a transcendent dimension that imposes an onerous moral and historical responsibility. Because Heidegger resisted naturalistic reductionism and affirmed the transcendent dimension of human existence, he eschewed a central feature of Nazi ideology. Rather than accepting facile answers to what is meant by "nature" and "humankind," deep ecologists and other radical environmentalists must recall that the most crucial political question concerns the relationship between these two terms.

Jeffrey Herf has convincingly argued that National Socialism proved so popular with many Germans because it provided a satisfying answer to the question of humanity's relation to nature. Nazi Germany went on to acquire such enormous military power, however, by conflating its mythical, nature-revering, antimodernist *Volksgemeinschaft* with the industrial technology made possible by modernity.[42] In a technological and nationalistic epoch, then, deep ecologists must be wary of calling for a "future primitive" mentality. As Tim Luke has pointed out:

> A reenchantment of Nature in Nordic myth and new Aryan ritual produced V-2s, Auschwitz, Me-262s, and nuclear fission, while covering itself in fables of Teutonic warriors true to tribal *Blut und Boden*. Ideologists of industrial fascism openly proclaimed it to be *anti-modern* and *future primitive*. Nazism also condemned industrialism and the overpopulation of most other societies as it propounded a very peculiar vision of reinhabiting its self-proclaimed and historically denied *Lebensraum*. One should not assume that deep ecology will lead necessarily to a fascist outcome; yet, the deep ecologists must demonstrate why their philosophy would not conclude in a similarly deformed fusion of modernity with premodernity.[43]

In moments of social stress, the temptation is great to impose one's views on others. At the peak of Heidegger's political zeal, he proclaimed that in the new Nazi Germany, "The individual by himself counts for nothing."[44] Thus, he was silent while thousands of German socialists, communists, liberals, and other "un-German" types were rounded up into concentration camps near Freiburg. After the War, moreover, he refused to comment on the Nazis' attempted genocide perpetrated against the Jews. During difficult times, he apparently concluded, difficult things have to be done. What can radical environmentalists learn from Heidegger's political involvement with a dark political movement that had an undeniably green aspect? As ecological problems increasingly give rise to social and political unrest, we can expect the emergence of a revolutionary vanguard whose aim will be to compel "selfish" people to "do the right thing" ecologically, for example, by agreeing to follow Draconian measures to limit human reproduction. During times of "ecological scarcity," this vanguard will want to round up the ecologically unenlightened so that the Earth may recover from the "human cancer" afflicting it. We may readily see potential parallels between talk of "human cancer" (a concept that is scarcely new to environmentalists) and Nazi talk of the "human vermin" who were a threat not only to the purity of the German *Volk*, but to the very future of human life on Earth. Heidegger's experience teaches us that we should be wary of revolutionaries prepared to use dangerous means, including the elimination of political freedoms, to achieve even noble ends, including protection of the Earth from the predations of industrial capitalism and communism.

The connections that I once forged between Heidegger's thought and deep ecology have provided grist for the mill for deep ecology's critics, including Murray Bookchin.[45] Bookchin maintains that Heidegger and deep ecologists alike fail to see that ecological problems stem not from some generalized anthropocentrism, but rather from hierarchical and authoritarian power structures that allow *some* humans to dominate others while simultaneously misusing the natural world. According to Bookchin, by failing to engage in close ideological critique of contemporary economic forces, deep ecologists move toward a dangerously misanthropic antihumanism.

In reply to similar charges, Heidegger once asserted that he favored a "higher humanism," one that appreciated humanity's true ontological calling more than did modernity's anthropocentric humanism. Likewise, deep ecologist Warwick Fox warns against the fallacy of misplaced misanthropy: criticizing an arrogant anthropocentrism does not mean that one is misanthropic.[46] Like Fox, Arne Naess also maintains that deep ecology is consistent with the progressive aim of enhancing genuine human interests.[47] Moreover, in his inaugural essay on deep ecology in 1973, Naess emphasized its "anti-class posture," based on the conviction that oppressive elites exploit other humans and nature as well. Just as the progressive tradition insists that humanity can liberate itself from ignorance, enslavement, and suffering, deep ecologists can be read as saying that humanity can not only free itself from political domination, but also can liberate nature from the consequences of arrogant anthropocentrism.[48] Lacking the metaphysical basis for a progressive interpretation of history, however, deep ecologists have difficulty in explaining why we should *expect* that humanity will move away from such anthropocentrism.

Some people associated with deep ecology, such as Fritjof Capra, speak of an incipient turning point or paradigm shift that will move humankind beyond the dualistic and anthropocentric basis of communism and capitalism. The call for a total, extrapolitical paradigm change is appealing, as Robert Pois has noted, "when one bears in mind the rape of planet Earth that has occurred as the most significant result of predatory capitalism [and, I might add, state socialism—MEZ]."[49] Though I agree that neither capitalism nor communism can solve the ecological crisis that they have helped to generate, and though I join in the search for alternatives, I am unwilling to engage in a wholesale condemnation of modernity. Needed is a more nuanced critique that would seek both to expand and to transform modernity's emancipatory goals, while questioning its arrogant anthropocentrism.

Even though in some ways human existence may transcend the organic dimensions of life, such transcendence provides no excuse for mistreating and degrading the life forms with which humanity has evolved over thousands of years. Many Enlightenment thinkers, despite sympathizing with a naturalistic account of "man and world," retained dissociative attitudes toward nature, the body, emotions, and the female. Such attitudes led these thinkers not only to ignore the needs and worth of nonhuman life, but also to conflate at times the noble goal of liberating humankind from external compulsion and material depravation with the hubristic goal of freeing "man" from all reminders of finitude and dependence. In modernity, the fantasy of turning the finite ego-subject into an eternal god is strong, indeed. Declining to go along with "ecocentric egalitarianism," which obscures humankind's perhaps exceptional mode of historical awareness, I also disavow an arrogant anthropocentrism, which pretends that the human being *as such* is somehow the source of all meaning, value, and purpose. Whatever special qualities with which humanity may be endowed, they are best understood, in my view, as having cosmological significance: they are manifestations of the universe's capacity for bringing itself to awareness. Because other life forms on this planet may play their own version of this role, their extinction may be an even greater loss than we now comprehend.

If we are successfully to develop a "third way" beyond capitalism and communism, and beyond arrogant anthropocentrism and antihumanistic biocentrism, we must do so without regressing psychologically and socially to collective states that erase the gains won by the difficult process of human individuation. Because of the masculinist, dissociative slant of such individuation, however, it must be transformed in such a way that some day, a truly free person will be recognized as one who has integrated the corporeal and the spiritual, the immanent and the transcendent. For a community of such persons, the practice of wantonly destroying nonhuman life and ecosystems in the name of "human freedom" and "progress" would make as little sense as enslaving some people for the "liberation" of others.

Acknowledgment

Thanks to David Macauley for critical suggestions that improved the present essay.

Notes

1. Concerning deep ecology, see Warwick Fox, *Toward a Transpersonal Ecology* (Boston: Shambhala, 1990); Bill Devall and George Sessions, *Deep Ecology* (Layton, Utah: Peregrine Smith Press, 1985); Alan Drengson, *Beyond Environmental Crisis* (New York: Peter Lang, 1989); Dolores LaChapelle, *Sacred Land, Sacred Sex, Rapture of the Deep* (Silverton, Colo.: Finn Hill Arts, 1988); and Michael E. Zimmerman, *Contesting Earth's Future* (Berkeley and Los Angeles: University of California Press, 1994).

On Heidegger and deep ecology, see Michael E. Zimmerman, "Toward a Heideggerean *Ethos* for Radical Environmentalism," *Environmental Ethics* 5, no. 2 (1983): 99–131; Zimmerman, "Implications of Heidegger's Thought for Deep Ecology," *The Modern Schoolman* 64 (November 1986): 19–43; Zimmerman, "Rethinking the Heidegger-Deep Ecology Relation," *Environmental Ethics* 15, no. 3 (1993): 195–224.

A sampling of other essays depicting Heidegger's thought as compatible with contemporary radical environmentalism includes Herbert Marcuse, *One-Dimensional Man* (Boston: Beacon Press, 1964); Hwa Yol Jung, "The Ecological Crisis: A Philosophical Perspective," *The Centennial Review* 17 (Winter 1974): 1–28; Hwa Yol Jung and Petee Jung, "To Save the Earth," *Philosophy Today* 19, no. 2 (1975): 108–117; Michael E. Zimmerman, "Heidegger on Nihilism and Technique," *Man and World* 9, no. 4 (1975): 394–414; Zimmerman, "Beyond Humanism: Heidegger's Understanding of Technology," *Listening* 12 (1977): 74–83; Zimmerman, "Heidegger and Marcuse: Technology as Ideology," *Research in Philosophy and Technology* 2 (1977): 245–261; Zimmerman, "Marx and Heidegger on the Technological Domination of Nature," *Philosophy Today* 23 (Summer 1979): 99–112; George P. Cave, "Animals, Heidegger, and the Right to Life," *Environmental Ethics* 4 (Fall 1982): 249–254; Bruce V. Foltz, "On Heidegger and the Interpretation of Environmental Crisis," *Environmental Ethics* 6, no. 4 (1984): 323–338; Laura Westra, "Let It Be: Heidegger and Future Generations," *Environmental Ethics* 7, no. 7 (1985): 341–350; Paul Shepard, "Homage to Heidegger," in *Deep Ecology*, ed. Michael Tobias (San Diego, Calif.: Avant Books, 1985); Neil Evernden, *The Natural Alien: Humankind and Environment* (Toronto: University of Toronto Press, 1985), pp. 60–72, 199–121; Joseph Grange, "Being, Feeling, and Environment," *Environmental Ethics* 7 (Winter 1985): 351–364; and Charles Taylor, "Heidegger, Language and Ecology," in *Heidegger: A Critical Reader*, ed. Hubert L. Dreyfus and Harrison Hall (Cambridge, Mass.: Blackwell, 1992).

2. In what follows, "early Heidegger" refers to work written during the era of *Being and Time* (1927), while "later Heidegger" refers to work that he wrote beginning in the early 1930s.

3. George Sessions, "Ecological Consciousness and Paradigm Change," in Tobias, *Deep Ecology*, p. 30.

4. For an insightful critique of Marx's anthropocentrism, see John Clark, "Marx's Inorganic Body," *Environmental Ethics* 11, no. 3 (1989): 243–258. Literature on this topic has become extensive, especially as leftist theorists have attempted to show that socialism is more concerned about ecological matters than is capitalism. See Alfred Schmidt, *The Concept of Nature in Marx*, trans. Ben Fowkes (London: New Left Books, 1971); William Leiss, *The Domination of Nature* (Boston: Beacon Press, 1974); Kostas

Axelos, *Alienation, Praxis, and Techne in the Thought of Karl Marx,* trans. Ronald Bruzina (Austin: University of Texas Press, 1976); Howard L. Parsons, *Marx and Engels on Ecology* (Westport, Conn.: Greenwood Press, 1977); Murray Bookchin, "Marxism as Bourgeois Sociology," in *Toward an Ecological Society* (Montreal: Black Rose Books, 1980); Donald C. Lee, "On the Marxian View of the Relation Between Man and Nature," *Environmental Ethics* 2 (Spring 1980): 3–16; Charles Tolman, "Karl Marx, Alienation, and the Mastery of Nature," *Environmental Ethics* 3 (Spring 1981): 63–74; Val Routley, "On Karl Marx as an Environmental Hero," *Environmental Ethics* 3 (Fall 1981): 237–244; Hwa Yol Jung, "Marxism, Ecology, and Technology," *Environmental Ethics* 5 (Summer 1983): 169–172; Rudolf Bahro, *Building the Green Movement,* trans. Mary Tyler (Philadelphia: New Society Publishers, 1986); Andrew McLaughlin, "Ecology, Capitalism, and Socialism," *Socialism and Democracy* 10 (Spring/Summer 1990): 69–102; and many essays in *New Left Review; Capitalism, Nature, Socialism; New German Critique,* and elsewhere.

5. See Martin Heidegger, *The Question Concerning Technology,* trans. William Lovitt (New York: Harper and Row, 1977); and Herbert Marcuse, *One-Dimensional Man* (Boston: Beacon Press, 1964). See also the massive two-volume work by William Lovitt and Harriet Brundage Lovitt, *Modern Technology in the Heideggerian Perspective* (Lewiston, N.Y.: The Mellen Press, 1994).

6. George Sessions, "Ecological Consciousness and Paradigm Change," p. 28.

7. For a critical discussion of the meaning of "deep" in "deep ecology," see Fox, *Toward a Transpersonal Ecology.*

8. See Anna Bramwell, *Blood and Soil: R. Walther Darré and Hitler's "Green Party"* (Bourne End, England: The Kensal Press, 1985); and Bramwell, *Ecology in the 20th Century: A History* (New Haven, Conn.: Yale University Press, 1989).

9. See Kim R. Holmes, "The Origins, Development, and Composition of the Green Movement," in *The Greens of West Germany,* ed. Robert L. Pfaltgraff Jr. et al. (Cambridge, Mass. and Washington, D.C.: Institute for Foreign Policy Analysis, Inc., 1983), pp. 30ff.

10. See Robert C. Paehlke, *Environmentalism and the Future of Progressive Politics* (New Haven, Conn.: Yale University Press, 1989).

11. See James W. Nickel and Eduardo Viola, "Integrating Environmentalism and Human Rights," *Environmental Ethics* 16, no. 3 (1994): 265–273.

12. See Raymond H. Dominick III, *The Environmental Movement in Germany* (Bloomington: Indiana University Press, 1992).

13. Regarding Heidegger's controversial views on animals, see his lectures from 1929–1930, *Die Grundbegriffe der Metaphysik,* ed. Friedrich-Wilhelm von Herrmann, *Gesamtausgabe* vol. 29/30 (Frankfurt am Main, Germany: Vittorio Klostermann, 1983); David Farrell Krell, *Daimon Life: Heidegger and Life Philosophy* (Bloomington: Indiana University Press, 1992); Jacques Derrida, "*Geschlecht* II: Heidegger's Hand," in *Deconstruction and Philosophy,* ed. John Sallis (Chicago: University of Chicago Press, 1987); and Derrida, *De l'esprit: Heidegger et la question,* chap. 6 (Paris: Editions Galilée, 1987).

14. Martin Heidegger, "Letter on Humanism," trans. Frank A. Capuzzi with J. Glenn Gray, in *Basic Writings,* ed. David Farrell Krell (New York: Harper and Row, 1977), p. 164.

15. See Michael E. Zimmerman, *Eclipse of the Self: The Development of Heidegger's Concept of Authenticity* (Athens: Ohio University Press, 1986).

16. See Michael E. Zimmerman, "Heidegger, Buddhism, and Deep Ecology," in *The Cambridge Companion to Heidegger,* ed. Charles Guignon (New York: Cambridge University Press, 1993).

17. Hubert L. Dreyfus, "Between Techne and Technology: The Ambiguous Place of Technology in *Being and Time,*" *Tulane Studies in Philosophy,* ed. Michael E. Zimmer-

man, 32 (1984): 23–36. For a discussion of Dreyfus's essay, see Zimmerman, *Heidegger's Confrontation with Modernity: Technology, Politics, and Art* (Bloomington: Indiana University Press, 1990).

18. Martin Heidegger, *An Introduction to Metaphysics*, trans. Ralph Mannheim (Garden City, N.Y.: Anchor Doubleday, 1961), p. 12. I have replaced Mannheim's somewhat awkward "essents" with "entities."

19. On earth and world, see Martin Heidegger, "The Origin of the Work of Art," in *Poetry, Language, Thought*, trans. Albert Hofstadter (New York: Harper and Row, 1971). See also Michel Haar, *The Song of the Earth*, trans. Reginald Lilly (Bloomington: Indiana University Press, 1993).

20. See John Llewelyn, "Ontological Responsibility and the Poetics of Nature," *Research in Phenomenology* 19 (1989): 3–26; Llewelyn, *The Middle Voice of Ecological Conscience* (New York: St. Martin's Press, 1991).

21. Martin Heidegger, *The End of Philosophy*, trans. Joan Stambaugh (New York: Harper and Row, 1973), 109.

22. Heidegger, *Poetry, Language, Thought*, p. 150.

23. Ibid., p. 182.

24. Ibid.

25. See ibid., pp. 167ff.

26. Ibid., p. 180.

27. Among the many works on Heidegger's politics, see Victor Farias, *Heidegger and Nazism* (Philadelphia: Temple University Press, 1989); Hugo Ott, *Martin Heidegger* (Frankfurt: Campus Verlag, 1988); Tom Rockmore, *Heidegger's Nazism and Philosophy* (Berkeley: University of California Press, 1992); Richard Wolin, *The Politics of Being* (New York: Columbia University Press, 1990); and Zimmerman, *Heidegger's Confrontation with Modernity*.

28. See Hans Sluga, *Heidegger's Crisis: Philosophy and Politics in Nazi Germany* (Cambridge, Mass.: Harvard University Press, 1994).

29. Heidegger, *The Question Concerning Technology*, p. 16.

30. For examples of Heidegger's rejection of the animality of humankind, see his "Letter on Humanism."

31. See Karl Löwith, "Zu Heideggers Seinsfrage: Die Natur des Menschen und die Welt der Natur," in *Aufsätze und Vorträge. 1930–1970* (Stuttgart, Germany: W. Kohlhammer Verlag, 1971).

32. See Hans Jonas, *The Gnostic Religion* (Boston: Beacon Press, 1972). Elsewhere, however, Jonas criticized Heidegger's thought for having a "profoundly pagan character" and for involving an "essential immanentism." See Jonas, *The Phenomenon of Life* (Chicago: University of Chicago Press, 1966), pp. 248, 249.

33. George L. Mosse, *The Crisis of German Ideology* (New York: Grosset and Dunlap, 1964), p. 15.

34. Robert A. Pois, *National Socialism and the Religion of Nature* (London: Croom Helm, 1986).

35. Ibid., p. 30, emphasis added.

36. Ibid., 40.

37. As quoted in ibid., pp. 42–43.

38. As quoted in ibid., p. 62.

39. John D. Caputo, *Demythologizing Heidegger* (Bloomington: Indiana University Press, 1993).

40. Martin Heidegger, *Discourse on Thinking*, trans. John M. Anderson and E. Hans Freund (New York: Harper and Row, 1966), pp. 55–56.

41. See Michael E. Zimmerman, "Humanism, Ontology, and the Nuclear Arms Race," *Research in Philosophy and Technology* 6 (1983): 151–172; and Zimmerman, "Anthropocentric Humanism and the Arms Race," in *Nuclear War: Philosophical Perspectives*, ed. Michael Fox and Leo Groarke (New York: Peter Lang Publishers, 1985).

42. Jeffrey Herf, *Reactionary Modernism* (New York: Cambridge University Press, 1984).

43. Tim Luke, "The Dreams of Deep Ecology," *Telos* 76 (Summer 1988): 65–92. Quotation is from p. 78.

44. Cited in Ott, *Martin Heidegger*, p. 231.

45. See Murray Bookchin, "Social Ecology versus 'Deep Ecology,' " *Green Perspectives*, nos. 4–5 (1987): 1–23.

46. Fox, *Toward a Transpersonal Ecology*, pp. 19–20.

47. Arne Naess, "The Arrogance of Anti-Humanism?" *Ecophilosophy* 6 (1984): 9.

48. See Robyn Eckersley, *Environmentalism and Political Theory* (Albany: State University of New York Press, 1992); and Andrew McClaughlin, *Regarding Nature: Industrialism and Deep Ecology* (Albany: State University of New York Press, 1993).

49. Pois, *National Socialism and the Religion of Nature*, p. 151.

Chapter Four

Merleau-Ponty and the Voice of the Earth

David Abram

Introduction

Slowly, inexorably, members of our species are beginning to catch sight of a world that exists beyond the confines of our specific culture—beginning to recognize, that is, that our own personal, social, and political crises reflect a growing crisis in the biological matrix of life on the planet. The ecological crisis may be the result of a recent and collective perceptual disorder in our species, a unique form of myopia which it now forces us to correct. For many who have regained a genuine depth perception—recognizing their own embodiment as entirely internal to, and thus wholly dependent upon, the vaster body of the Earth—the only possible course of action is to begin planning and working on behalf of the ecological world which they now discern.

And yet ecological thinking is having a great deal of trouble taking root in the human world—it is still viewed by most as just another ideology; meanwhile, ecological science remains a highly specialized discipline circumscribed within a mostly mechanistic biology. Without the concerted attention of philosophers, ecology lacks a coherent and common language adequate to its aims; it thus remains little more than a growing bundle of disparate facts, resentments, and incommunicable visions.

It is my belief that the phenomenological investigations of Maurice Merleau-Ponty provide the seeds of a new and radical philosophy of nature that remains true to the diversity of experience within the biosphere of this planet.

First published in *Environmental Ethics* 10, no. 2 (Summer 1988): 101–120; reprinted by permission, with minor corrections.

In this paper I show why a phenomenology that takes seriously the primacy of perception is destined to culminate in a renewed awareness of our responsibility to the Earth, and why the movements toward an ecological awareness on this continent and elsewhere have much to gain from a careful consideration of Merleau-Ponty's discoveries.

Merleau-Ponty was born on the west coast of France in 1908. He studied at the Ecole Normale Supérieure in Paris, where he began teaching upon receiving his certificate in philosophy in 1931. A careful student of developments in psychology and the natural sciences, he was a powerful innovator within the phenomenological tradition inaugurated by Edmund Husserl, and he won wide recognition after the publication of *Phenomenology of Perception* in 1945. Influenced by Marx's profound sense of the material relations that underlie our ideas and ideals, Merleau-Ponty's version of phenomenology was far more embodied than that of the the other phenomenologists, and more attentive to the nuances of political engagement. With Jean-Paul Sartre and Simone de Beauvoir, he founded *Les Temps Modernes*—a journal of cultural and political critique for which he was editor-in-chief from 1945 to 1952. In 1952 he was named to fill the prestigious chair of philosophy at the Collège de France, a position which he held until his sudden death in 1961. Merleau-Ponty was also a prose stylist of stunning originality. The work of such theorists as Michel Foucault and Jacques Derrida owes much to certain formulations in the late work of Merleau-Ponty; nevertheless, many of his most radical insights have yet to be discovered by later thinkers and activists.

I will be, of necessity, simplifying his work. I am not, moreover, interested in merely repeating his ideas twenty-five years after his death; I wish rather to accomplish a creative reading of his writings in order to indicate, not necessarily what Merleau-Ponty knew he was saying, but rather what was gradually saying itself through him. Where this interpretation moves beyond the exact content of Merleau-Ponty's texts, it is nevertheless informed by a close and long-standing acquaintance with those texts. Since I am here interested less in the past than in the future of his project, I have organized this paper in accordance with the plan that Merleau-Ponty himself proposed in the final working note that he wrote down in March 1961, shortly before his unexpected death, in which he projected three major sections for the new book on which he was working: first "The Visible," then "Nature," and finally "Logos."[1]

The Visible

> The visible about us seems to rest in itself. It is as though our vision were formed at the heart of the visible, or as though there were between it and us an intimacy as close as that between the sea and the strand.[2]

> There must be depth since there is a point whence I see—since the world surrounds me. . . .[3]

The great achievement of Merleau-Ponty's major completed work, *Phenomenology of Perception*,[4] was that it showed that the fluid creativity we commonly associate with the human intellect is an elaboration or recapitulation of a deep creativity already underway at the most immediate level of bodily perception. Phenomenological philosophy had, since its inception, aimed at a rigorous description of things as they appear to an experiencing consciousness. Yet the body had remained curiously external to this "transcendental" consciousness. Merleau-Ponty was the first phenomenologist to identify the body, itself, as the conscious subject of experience. Transcendence, no longer a special property of the abstract intellect, becomes in his *Phenomenology* a capacity of the physiological body itself—its power of responding to other bodies, of touching, hearing, and seeing things, resonating with things. Perception *is* this ongoing transcendence, the ecstatic nature of the living body.

By thus shifting the prime focus of subjectivity from the human intellect to what he called the "body-subject" or the "lived body," Merleau-Ponty uncovered the radical extent to which all subjectivity, or awareness, presupposes our inherence in a sensuous, corporeal world. And this presupposed world is not entirely undefined, it is not just *any* world, for it has a specific structure—that is, it exists in both proximity and distance, and it has a horizon. More specifically, this always-already-existing world is characterized by a distant horizon that surrounds me wherever I move, holding my body in a distant embrace while provoking my perceptual exploration. It is a world that is structured in *depth*, and from *Phenomenology of Perception* on, *depth*—the dimensional spread from the near to the far—becomes the paradigm phenomenon in Merleau-Ponty's writings.

The depth of a landscape or a thing can often be construed by the body-subject in a number of different ways: that cloud that I see can be a small cloud close overhead or a huge cloud far above; meanwhile what I had thought was a bird turns out to be a speck of dust on my glasses. Depth is always the dimension of ambiguity and confusion. The experience of depth is the experience of a world that both includes one's own body and yet spreads into the distance, a world where things hide themselves not just beyond the horizon but behind other things, a world where indeed no thing can be seen all at once, in which objects offer themselves to the gaze only by withholding some aspect of themselves—their other side, or their interior depths—for further exploration. Depth, this mysterious dimension, which every schoolchild knows as the "third" dimension (after height and breadth), Merleau-Ponty asserts is the *first*, most primordial dimension, from which all others are abstracted.[5] To the student of perception, the phenomenon of depth is the original ambiguity: it is depth that provides the slack or play in the immediately perceived world, the instability that already calls upon the freedom of the body to engage, to choose, to *focus* the world long before any verbal reflection comes to thematize and appropriate that freedom as its own. And so the experience of depth runs like a stream throughout the course of Merleau-Ponty's philosophizing—from the many analyses of visual depth and the incredible discussion of visual focusing in *Phenomenology of Perception*[6] to the extended meditation on depth in his last complete essay, "Eye and Mind"[7]—a subterranean stream which surfaces only here and there, but which ceaselessly provides the texture of his descriptions, the source of his metaphors. As he himself asserts in a late note:

> The structure of the visual field, with its near-bys, its far-offs, its horizon, is indispensable for there to be transcendence, the model of every transcendence.[8]

It is no accident that the crucial chapter of his final, unfinished work is entitled "The Chiasm,"[9] a term commonly used by neuroscientists to designate the "optic chiasm," that place in the brain where the two focusing eyes intertwine. Yet Merleau-Ponty always maintained a critical distance from the sciences that he studied, acknowledging specific discoveries while criticizing the standard, Cartesian interpretations of those findings. Merleau-Ponty was one of the first to demonstrate, contrary to the assertions of a dualistic psychology, that the experience of depth is not *created* in the brain any more than it is *posited* by the mind. He showed that we can discover depth, can focus it or change our focus within it, only because it is *already there*, because perception unfolds *into* depth—because my brain, like the rest of my body, is already enveloped in a world that stretches out beyond my grasp. Depth, which we cannot consider to be merely one perceptual phenomenon among others, since it is that which *engenders* perception, is the announcement of our immersion in a world that not only preexists our vision, but prolongs itself beyond our vision, behind that curved horizon.

Indeed, if I attend to my direct sensory experience, I must admit that that horizon I see *is* curved around me, as surely as the sky is arched overhead, like a dome, like a vault. Examining the contours of this world not as an immaterial mind but as a sentient body, I come to recognize my thorough inclusion within this world in a far more profound manner than our current language usually allows. Our civilized distrust of the senses and of the body engenders a metaphysical detachment from the sensible world—it fosters the illusion that we ourselves are not a part of the world that we study, that we can objectively stand apart from that world, as spectators, and can thus determine its workings from outside. A renewed attentiveness to bodily experience, however, enables us to recognize and affirm our inevitable involvement in that which we observe, our corporeal immersion in the depths of a breathing Body much larger than our own.

Often it takes a slightly unusual circumstance to disturb our metaphysical distance from the corporeal world. On certain days, for instance, when the sky is massed with clouds, I may notice a dense topography that extends overhead as well as underfoot, enclosing me within its layers, and so come to feel myself entirely *inside*. In general, if I pay close attention to bodily perceptions over a period of time, I may notice that the primordial experience of depth is always the experience of a sort of interiority of the external world, such that each thing I perceive seems to implicate everything else, so that things, landscapes, faces all have a coherence, all suggest a secret familiarity and mutual implication in an anonymous presence that subtends and overarches my own. I may discern, if I attend closely, that there is a certain closure which is suggested by the horizon and its vicissitudes, a sort of promise, in the distance, of a secret kinship between the ground and the sky, a fundamental nonopposition, a suggestion that ground and sky are not two distinct entities but two layers or leaves of one single power, two leaves that open as I move toward that horizon and that close up, behind me, back there.[10]

The importance of the visible horizon for all of Merleau-Ponty's interrogations can lead us to realize that the "world" to which he so often refers is none other than the Earth, that the coherent unity of the "visible" which slowly emerges through Merleau-Ponty's analysis of perception—the "field of all fields," or the "totality wherein all the sensibles are cut out"[11]—is not the abstract totality of the conceivable universe but the experienced unity of this enveloping but local world which we call Earth.

By *Earth*, then, I mean to indicate an intermediate and mediating existence between ourselves and "the universe," or, more concretely, between humankind and the Sun, toward whom our "pure" ideas seem to aspire directly, forgetful that it is not we, but rather the Earth that dwells in the field of the Sun, as we live within the biosphere of the Earth. Much of Merleau-Ponty's work implies a growing recognition of this enveloping existence, which is only local by current scientific standards, but which is truly total for our perception. Hence, in his later writings, he begins to speak not just of the "world" but of "this world" or "our world":

> Universality of our world . . . according to its configuration, its ontological structure which envelops every possible and which every possible leads back to.[12]

Indeed *all* of phenomenology, with its reliance upon the Husserlian notion of "horizon," is tacitly dependent upon the actual planetary horizon that we perceive whenever we step outside our doors or leave behind the city. As Merleau-Ponty has written, "it is by borrowing from the world structure that the universe of truth and of thought is constructed for us."[13] His thesis of the primacy of perception suggests that *all* of our thoughts and our theories are secretly sustained by the structures of the perceptual world. It is precisely in this sense that philosophies reliant upon the *concept* of "horizon" have long been under the influence of the actual visible horizon that lies beyond the walls of our office or lecture hall, that structural enigma which we commonly take for granted, but which ceaselessly reminds us of our embodied situation on the surface of this huge and spherical body we call the Earth.

Yet we should not even say "on" the Earth, for we now know that we live *within* the Earth. Our scientists with their instruments have rediscovered what the ancients knew simply by following the indications of their senses: that we live within a sphere, or within a series of concentric spheres. We now call those spheres by such names as the "hydrosphere," the "troposphere," the "stratosphere," and the "ionosphere," and no longer view them as encompassing the whole universe. We have discovered that the myriad stars exist quite far beyond these, and now recognize these spheres to be layers or regions of our own local universe, the Earth. Collectively these spheres make up the atmosphere, the low-viscosity fluid membrane within which all our perceiving takes place.

While science gains access to this knowledge from the outside, philosophy has approached it from within. For once again, the entire phenomenological endeavor has taken place within a region of enquiry circumscribed by a tacit awareness of Earth as the ground and horizon of all our reflections, and the hidden thrust of the phenomenological movement is the reflective rediscovery of our inherence in the body of Earth. We can glimpse this trajectory most readily in certain essays by Husserl such as his investigations of the "Phenomenological Origin of the Spatiality of Nature," in which Husserl refers again and again to "Earth, the original ark," and speaks enigmatically of Earth as that which precedes all constitution,[14] as well as in the later essays of Heidegger which are a direct invocation of "earth" and "sky" along with "mortals" and "gods" in "the fourfold."[15] Nevertheless, it is in Merleau-Ponty's work that the full and encompassing enigma of Earth, in all its dense, fluid, and atmospheric unity, begins to emerge and to speak.

This new sense of Earth contrasts with Heidegger's notion of "earth" as that which remains concealed in all revelation, the dark closedness of our ground which he counterposes to the elemental openness of "sky." The Merleau-Pontian sense of Earth names a more diverse phenomenon, at once both visible and invisible, incorporating both the deep ground that supports our bodies and the fluid atmosphere in which we breathe. In discovering the body, or in discovering a new way of thinking the body and finally experiencing the body, Merleau-Ponty was also disclosing a new way of perceiving the Earth of which that body is a part. To assert, as he did throughout the course of his life, that the human intellect is a recapitulation or prolongation of a transcendence already underway at the most immediate level of bodily sensation—to assert, that is, that the "mind" or the "soul" has a carnal genesis—is to suggest, by a strange analogy of elements that stretches back to the very beginnings of philosophy, that the sky is a part of the Earth, to imply that the sky and the Earth need no longer be seen in opposition, that this sky, this space in which we live and breathe, is not opposed to the Earth but is a prolongation, even an organ, of this planet. If the soul is not contrary to the body, then human beings are no longer suspended between a dense inert Earth and a spiritual sky, no more than they are suspended between Being and Nothingness. For the first time in modern philosophy, human beings with all of their thoughts and their ideas are enveloped within the atmosphere of this planet, an atmosphere that circulates both inside and outside of their bodies: "There really is inspiration and expiration of Being, respiration in Being."[16] Although Merleau-Ponty never quite gives the name Earth to this unity, he does write of "the indestructible, the Barbaric Principle,"[17] of "one sole sensible world, open to participation by all, which is given to each,"[18] of a "global voluminosity" and a "primordial topology,"[19] and of the anonymous unity of this visible (and invisible) world.[20] He writes of "a nexus of history and *transcendental geology*, this very time that is space, this very space that is time which I will have rediscovered by my analysis of the visible and the flesh,"[21] but without calling it by name. In another luminous passage he writes of "the prepossession of a totality which is there before one knows how and why, whose realizations are never what we would have imagined them to be, and which nonetheless fulfills a secret expectation within us, since we believe in it tirelessly.[22] But again, this totality remains anonymous.

I suspect that Merleau-Ponty had to write in this way because what was anonymous then did not finally lose its perceptual anonymity until a decade after his death, when the first clear images of the Earth viewed from space were developed, and our eyes caught sight of something so beautiful and so fragile that it has been known to bring a slight reordering of the senses. It is a picture of something midway between matter and spirit, an image for what Merleau-Ponty had written of as the "existential eternity—the eternal body."[23] Of course, in one sense such images of the Earth present the ultimate *pensée de survol*, that nonsituated "high-altitude thought" of which Merleau-Ponty was so critical (and indeed these images have become the worst kind of platitude in recent years, used in globalizing advertisements for everything from automobiles to detergents). But we should not be tricked into thinking that he would have brushed them aside on that account. For this philosopher of perception, such photographs (and their proliferation in the world) would undoubtedly have been disturbing indeed, but decisive, like catching sight of oneself in the mirror for the first time.

In any case, it is enough here to recognize (1) that Merleau-Ponty sensed that there was a unity to the visible-invisible world that had not yet been described in philosophy, that there was a unique ontological structure, a topology of Being that was waiting to be realized, and (2) that whatever this unrealized Being is, we are in its depths, and of it, like a fish in the sea, and that therefore it must be disclosed from *inside*. These points are clear from his published notes, where, for example, in a note from February 1960, he writes of his project as "an ontology from within."[24]

Nature

> It suffices for us for the moment to note that he who sees cannot possess the visible unless he is possessed by it, unless he is of it, unless . . . he is one of the visibles, capable, by a singular reversal, of seeing them—he who is one of them.[25]

> Do a psychoanalysis of Nature: it is the flesh, the mother.[26]

In the book on which he was working at the time of his death, published posthumously, with working notes, as *The Visible and the Invisible*, Merleau-Ponty makes a significant terminological shift. He refers much less often to the body—whether to the "lived-body," upon which he had previously focused, or to the "objective body," from which it had been distinguished—and begins to speak more in terms of "the Flesh." Indeed he no longer seems to maintain the previously useful separation of the "lived-body" from the "objective body"; rather, he is now intent on disclosing, beneath these two perspectives, the mystery of their nondistinction for truly primordial perception. The singular "objective body" had lingered quietly in Merleau-Ponty's writings—a residual concept, and a minor concession to the natural sciences, that was necessary as long as the rest of sensible or "objective" nature remained unattended to in his work, as long as nonhuman nature remained the mute and inert background for our human experience. However, with the shift from the "lived-body" to the "Flesh"—which is both "my flesh" and "the Flesh of the world"—Merleau-Ponty inaugurates a sweeping resuscitation of nature, both human and nonhuman.

As a number of commentators have suggested, it is likely that Merleau-Ponty's move from the lived-body to the Flesh constitutes less a break than a logical continuation of his earlier stylistic move to de-intellectualize transcendence in *Phenomenology of Perception*.[27] In the language and argumentation of that earlier work, Merleau-Ponty managed to shift the locus of subjectivity from the human intellect to the body-subject. In *The Visible and the Invisible*, Merleau-Ponty follows through on that first shift by dislodging transcendence as a particular attribute of the human body, and returning it to the whole of the sensuous world of which this body is but a single expression. Merleau-Ponty accomplishes this by describing the intertwining of the invisible with the visible—by demonstrating that the invisible universe of thought and reflection is both provoked and supported by the enigmatic depth of the visible, sensible environment:

> The visible is pregnant with the invisible, . . . to comprehend fully the visible relations one must go unto the relation of the visible with the invisible.[28]

Thus, the invisible, the region of thought and ideality, is always inspired by invisibles that are there from the first perception—the hidden presence of the distances, the secret life of the Wind which we can feel and breathe but cannot see, the interior depths of things, and, in general, all the invisible lines of force that constantly influence our perceptions. The invisible shape of smells, rhythms of cricketsong, or the movement of shadows all, in a sense, provide the subtle body of our thoughts. For Merleau-Ponty our own reflections are supported by the play of light and *its* reflections; the mind, the whole life of thought and reason is a prolongation and expansion, through us, of the shifting, polymorphic, invisible natures of the perceptual world. In the words of Paul Elouard, "there *is* another world, but it is *in* this one."[29] Or as Merleau-Ponty himself writes in one note, all the "invisibles," including that of thought, are "necessarily enveloped in the Visible and are but modalities of the same transcendence."[30] The "flesh" is the name Merleau-Ponty gives to this sensible-in-transcendence, this inherence of the sentient in the sensible and the sensible in the sentient, to this ubiquitous element which is not the objective matter we assign to the physicists nor the immaterial mind we entrust to the psychologists because it is older than they, the source of those abstractions:

> There is a body of the mind, and a mind of the body. . . . The essential notion for such a philosophy is that of the flesh, which is not the objective body, nor the body thought by the soul as its own (Descartes), [but] which is the sensible in the twofold sense of [that which is sensed and that which senses].[31]

The "flesh" is the animate element which Merleau-Ponty has discovered, through his exploration of pre-objective perception, to be the common tissue between himself and the world:

> The visible can thus fill me and occupy me because I who see it do not see it from the depths of nothingness, but from the midst of itself; I the seer am also visible. What makes the weight, the thickness, the flesh of each color, of each sound, of each tactile texture, of the present, and of the world is the fact that he who grasps them feels himself emerge from them by a sort of coiling up or redoubling, fundamentally homogenous with them; he feels that he is the sensible itself coming to itself.[32]

With this terminological move from the "body" to the common "flesh" Merleau-Ponty dislodges creativity or self-transcendence as a particular attribute of the human body and returns transcendence to the carnal world of which this body is an internal expression.[33] If we now consider the world to which Merleau-Ponty's work refers to be *this* world—that is, the Earth—this move from the "body" to the communal "flesh" suggests that for a genuine perception the human body is radically interior to the *Lebenswelt*—the life-space, or biosphere—of a worldbody which is itself in transcendence, self-creative, even—with us—alive:

> One can say that we perceive the things themselves, that we are the world that thinks itself or that the world is at the heart of our flesh. In any case, once a body-world relationship is recognized, there is a ramification of my body and a ramification of the world and a correspondence between its inside and my outside, between my inside and its outside.[34]

Here Merleau-Ponty's investigations anticipated recent work in the sciences and converge with new findings in biology, psychology, and global ecology. I will here mention only one of these developments. The "Gaia hypothesis" was first proposed in the mid-1970s by scientists striving to account for the actual stability of the Earth's atmosphere in the face of a chemical composition recently discovered to be very far from equilibrium. Geochemist James Lovelock and microbial biologist Lynn Margulis have hypothesized that the Earth's atmosphere is being metabolically generated and sensitively maintained by all of the organic life on the planet's surface acting collectively, as a single global physiology. The Gaia hypothesis (named for the mother of the Gods in Greek mythology, she whose name is at the root of such words as "geology" and "geography") provides a ready explanation, as well, for the newly recognized evidence that the Earth's surface temperature has remained virtually constant over the last three and a half billion years despite an increase in the Sun's heat of at least thirty percent during the same period. The hypothesis, in short, maintains that the Earth's biosphere acts as a vast, living physiology that regulates its temperature and internal composition much as one's own body metabolically maintains its own internal temperature and balances the chemical composition of its bloodstream.[35] The sensible world that surrounds us must, it would seem, be recognized as a sensitive physiology in its own right.[36]

But let us turn back to Merleau-Ponty, whose work on the "ontology from within" was cut short more than two decades before these developments. There is another, equally important implication of Merleau-Ponty's move from the lived-body to the flesh. For by shifting transcendence, which had been thought to be an exclusively human domain, to the whole of the sensible world of which we humans are but a part, Merleau-Ponty dissolves the traditional division between the human animal and all other organisms of the Earth. The human sentience is indeed unique, but if we follow closely Merleau-Ponty's final writings we will begin to suspect that there are *other* sentient entities in the biosphere—indeed, that each species, by virtue of its own carnal structure has its own unique sentience or "chiasm" with the flesh of the world.[37]

Why then, one might ask, do we not read much more about the flesh of other animals in the pages of *The Visible and the Invisible*? I would answer first that, given the interrupted nature of Merleau-Ponty's text, this absence is not crucial. Since a new recognition of other animals follows directly from his thesis, such a recognition would eventually have emerged. It is clear, nevertheless, especially from our own reluctance to affirm this implication of Merleau-Ponty's work, that to confront and accept this implication twenty-five years ago was to move against the accumulated bias of the entire Western philosophical tradition—a literate tradition that has its origins in the exaltation of a divine ideality beyond the sensory world, a tradition that continues, in the scientific age, as an exaltation of the divine human existent over and above the "mute" and "chaotic" world of nature. It is a tradition that has been formulated almost exclusively by male scholars working within those havens of literacy—the academy and the city—that have been increasingly removed from all contact with the wild and coherent diversity of nonhuman nature. As a result, it is hardly surprising that Merleau-Ponty himself had difficulty accepting the most subversive implication of his phenomenology. It is his reticence on this point—the fact that his thought never quite leaves the city—that is the real stumbling block of his unfinished work.[38]

Let us examine this point more closely. In his course on *Consciousness and the Acquisition of Language*, Merleau-Ponty asserts that his progenitor, Husserl, was unable to abandon the Cartesian conception of a pure transcendental consciousness, the Cartesian *cogito* which Husserl returns to or "recalls each time we would believe him to be on the verge of a solution."[39] In a similar way we may now discern that Merleau-Ponty, having dropped the Cartesian postulate of a pure consciousness in favor of a bodily subjectivity, found *himself* caught within the more tacit Cartesian assumption of a massive difference between the human body, which for Descartes was open to the intervention of the soul, and all other animal bodies, which for Descartes were closed mechanisms incapable of any awareness.[40] In Merleau-Ponty's final writings we witness him on the threshold of opening his own rich conception of an embodied intersubjectivity to include the incarnate subjectivity of other animals, although never explicitly crossing this threshold—at least not in the fragments we have. Merleau-Ponty comes upon this deep Cartesian opposition between humans and other animals, begins to dismantle it, but at the time of his death had not yet stepped through this opposition into a genuinely ecological intercorporeality.

Or had he? In his final working note—the note from which I have taken the plan for this paper—Merleau-Ponty writes that his discoveries

> must be presented without any compromise with humanism, nor with naturalism [that is, the naturalism of the "natural" sciences], nor finally with theology.[41]

Humanism is the key word here. To accept no compromise with *humanism* was difficult for Merleau-Ponty, for he was, in many ways, a committed humanist, and we can witness him grappling with this compromise throughout his late notes. But then in that same instruction to himself he writes:

> Precisely what has to be done is to show that philosophy can no longer think according to this cleavage: God, man, creatures. . .[42]

This is a powerful statement. From this last note it seems clear that Merleau-Ponty *knew* that he was out to heal the deep wound between humans and the other animals. However, his recalcitrant humanism was letting this happen only very slowly in his writings.

A single example may serve to illustrate Merleau-Ponty's dilemma. In a note from May 1960 he writes that

> the flesh of the world is not self-sensing (*se sentir*) as is my flesh—it is sensible and not sentient—I call it flesh, nonetheless, in order to say . . . that it is absolutely not an object.

He then goes on to assert that it is only

> by the flesh of the world that in the last analysis we can understand the lived body.[43]

Here we are left with an immense and ultimately untenable gap between the flesh of the world which is "sensible and not sentient" and my flesh which is "self-sensing." It is Merleau-Ponty's recalcitrant humanism that strives to maintain this distinction at the same time that his emerging ecological realism is struggling to assert the primacy of the world's flesh: "it is by the flesh of the world that in the last analysis we can understand the lived body." But it is simply because he is neglecting to consider other animals at this juncture that Merleau-Ponty is still able to assert that the flesh of the world is not self-sensing, for clearly other animals are a part of the perceived flesh of the world, and yet they have their own senses; following Merleau-Ponty's thesis of the reversibility of the sensing and the sensed, they are clearly self-sensing. As soon as we pay attention to other organisms we are forced to say that the flesh of the world is both perceived *and perceiving*. It is only by recognizing the senses of other animals that we can begin to fill up the mysterious gap Merleau-Ponty leaves in this quote. Or, to put it another way, only by recognizing the full presence of other animals will we find our own place within Merleau-Ponty's ontology. (Plants, as well, will come to assert their place, but our concern is first with the other animals because they are our link, animal that we are, to the rest of the Flesh.) In this regard, it is essential that we discern that Merleau-Ponty's thought does not represent the perpetuation of an abstract anthropocentrism, but rather the slow and cautious overcoming of that arrogance. It is only by listening, in the depths of his philosophical discourse, to the gradual evocation of a densely intertwined organic reality, that we will fully understand how it is that the flesh of the world is "absolutely not an object."

Looked at in this way, a great deal of *The Visible and the Invisible* is already about other species, whether or not Merleau-Ponty was aware of their influence. No thinker can really move from his/her bodily self-awareness to the intersubjectivity of human culture, and thence to the global transcendence that is "the flesh of the world," without coming upon myriad experiences of otherness, other subjectivities that are not human, and other intersubjectivities. Indeed, the immediate perceptual world, which we commonly forget in favor of the human culture it supports, is secretly made up of these others; of the staring eyes of cats, or the raucous cries of birds who fly in patterns we have yet to decipher, and the constant though secret presence of the insects we brush from the page or who buzz around our heads, *all of whom make it impossible for us to speak of the sensible world as an object*—the multitude of these nonhuman and therefore background speakings, gestures, glances, and traces which impel us to write of the self-transcendence or the "invisibility" of the visible world, often without our being able to say just why. It is likely that Merleau-Ponty, had he continued writing or had he written this text a few years later and witnessed the growing cultural respect for the nonhuman world as both active and interactive, would have had much less difficulty describing his experience of the "invisible" nature of the visible world and the reversibility between humanity and being.

But Merleau-Ponty did not live to read Rachel Carson's revelations about pesticides and the natural world, or the more recent disclosures about animal vivisection and the infliction of animal pain and terror on a grand scale that goes on within contemporary agribusiness and the cosmetics industry—the violent pain and death that unfolds throughout the technological world in its forgetfulness of what he called "Wild Being." It is possible that these disclosures would have been as unsettling to his thought, and as crucial for his rethinking of philosophy, as were the revelations concerning Stalin's purges when these were disclosed in Europe.[44] They would have accelerated his recognition of the nonhuman others and have helped him to welcome these wild, mysterious perceptions into an ontology that was already waiting for them.

Logos

It is the body which points out, and which speaks; so much we have learnt in this chapter. . . . This disclosure of an immanent or incipient significance in the living body extends, as we shall see, to the whole sensible world, and our gaze, prompted by the experience

> of our own body, will discover in all other "objects" the miracle of expression.[45]
>
> That the things have us, and that it is not we who have the things. . . . That it is being that speaks within us and not we who speak of being.[46]

In *Phenomenology of Perception*, Merleau-Ponty carefully demonstrated that silent or prereflective perception unfolds as a reciprocal exchange between the body and the world. Further, he showed that this constant exchange, with its native openness and indeterminacy, is nevertheless highly articulate, already informed by a profound logos. These disclosures carried the implication that perception, this ongoing reciprocity, is the very ground and support of that more explicit reciprocity we call "language."[47] Merleau-Ponty's continued focus upon the gestural genesis of language, and upon active speech as the axis of all language and thought—a focus which, as James Edie has written, distinguishes Merleau-Ponty from all other philosophers "from Plato on down"[48]—further served to ground language in the deep world of immediate perception, in the visible, tangible, audible world that envelops us, and of which we are a part.

In a more recent paper Edie maintains that Merleau-Ponty had no place, in his philosophy of language, for a depth linguistic structure such as that which Noam Chomsky has discovered in the years since Merleau-Ponty's death.[49] Now, it is true that Merleau-Ponty did not discern any surface and deep structure in the fashion of Chomsky's investigations, but I believe that this is because he was in the midst of uncovering a more primordial structural depth within language—one which has yet to be understood by other linguists and philosophers. It is that dimension in language correlative to the actual depth of the perceptual world, the deep structure of the sensory landscape.

By starting to show, as he does in his final chapter on "The Intertwining—The Chiasm,"[50] how thought and speech take form upon the infrastructure of a living perception already engaged in the world, Merleau-Ponty carefully demonstrates that language has its real genesis not inside the human physiology but, with perception, in the depth—the play between the expressive, sensing body and the expressive physiognomies and geographies of a living world. If we follow Merleau-Ponty's argument and agree that language is founded not inside us but in front of us, in the depths of the expressive world which engages us through all our senses, then we would not hunt for the secret of language inside the human physiology. We would not hunt, as Chomsky has suggested, for the ultimate seed of language within the human DNA.[51] For if the sensible world itself is the deep body of language, then this language can no longer be conceived as a power that resides within the human species, at least no more than it adheres to the roar of a waterfall, or even to the wind in the leaves. If language is born of our carnal *participation* in a world that already *speaks to us* at the most immediate level of sensory experience, then language does not belong to humankind but to the sensible world of which we are but a part. That,

I believe, is how we must read Merleau-Ponty's parting note on language: "Logos . . . as what is realized in man, but nowise as his *property*."[52] If we set this insight alongside what we have already found regarding the Visible and Nature—(1) that man, or woman, is entirely included within the visible, sensible world, and (2) that the sensible world which we are within and of, and which we may suspect is this Earth, is itself sensitive and alive, constituted by multiple forms of embodied awareness besides our own—then some interesting conclusions emerge. We begin to recognize, for instance, that our language has been contributed to, and is still sustained by, many rhythms, sounds, and traces besides those of our single species.

Since its inauguration in the Athenian *polis*, European philosophy has tended to construe language as that power which humans possess and other species do not. From Aristotle to Aquinas, from Descartes to Chomsky, "language" has been claimed as the exclusive and distinguishing property of humankind; man alone has privileged access to the Logos. Yet this exclusivity rests upon a neglect of the experiencing body, a forgetting of the gestural, carnal resonance that informs even our most rarefied discourse. In this way, it has fostered an abstract notion of language as a disembodied, purely formal set of grammatical and syntactic relations.

At least one contemporary linguist has called this entire tradition into question. Harvey Sarles, in his book *Language and Human Nature*, asks, "Is language disembodied, or just our theories about language?"[53] Sarles argues forcefully that the assumption that language is a purely human property, while providing a metaphysical justification for the human domination of nonhuman nature, nevertheless makes it impossible for us to comprehend the nature of our own discourse. Sarles asserts that "to define language as uniquely human also tends to define the nature of animal communication so as to preclude the notion that it is comparable to human language."[54] However, Sarles claims,

> Each ongoing species has a truth, a logic, a science, knowledge about the world in which it lives. To take man outside of nature, to aggrandize the human mind, is to simplify other species and, I am convinced, to oversimplify ourselves, to constrict our thinking and observation about ourselves into narrow, ancient visions of human nature, constructed for other problems in other times.[55]

Independent of Merleau-Ponty, yet entirely congruent with Merleau-Ponty's investigations of twenty years earlier, Sarles outlines the basis for a more genuine linguistics grounded in a recognition of the "knowing body,"[56] or elsewhere, the "body-as-expression in interaction."[57]

Any such Merleau-Pontian approach to language—any approach, that is, that discloses language's gestural, soundful basis in bodily receptivity and response to an expressive, living world—opens us toward an understanding of the subtle relationship between language and landscape. If it is this breathing body that speaks, writes, and thinks—if it is not an immaterial ego but rather this sensible, sensitive body that dwells and moves within language—then language is at no point a structure of wholly abstract, ideal, or mathematical relations. For it is haunted by all those carnal things and styles to which our senses give us access. Language that has its real genesis in the deep world of untamed perception is language that is born as a call for and response to a gesturing, sounding, speaking landscape—a world of thunderous rumblings, of babbling brooks, of flapping, flying, screeching things, of roars and sighing winds. . . . That is why Merleau-Ponty could write, in the last complete lines of *The Visible and the Invisible*, that "language is everything, since it is the voice of no one, since it is the very voice of the things, the waves, and the forests."[58] It is even possible that this language we speak is the voice of the living Earth itself, singing through the human form. For the vitality, the coherence, and the *diversity* of the various languages we speak may well correspond to the vitality, coherence, and diversity of Earth's biosphere—not to any complexity of our species considered *apart* from that matrix.

In any case, we can now hypothesize, following this unique philosopher, a fundamental dynamic behind the ecological and psychological crisis in which human culture now finds itself. As long as humankind continues to use language strictly for our own ends, as if it belongs to our species alone, we will continue to find ourselves estranged from our actions. If as Merleau-Ponty's work indicates it is not merely *this* body but the whole visible, sensual world that is the deep flesh of language, then surely our very words will continue to tie our selves, our families, and our nations into knots until we free our voice to return to the real world that supports it—until we allow it to respond to the voice of the threatened rainforests, the whales, the rivers, the birds, and indeed to speak for the living, untamed Earth which is its home. The real Logos, after Merleau-Ponty, is Eco-logos.

Conclusion

> Can this rending characteristic of reflection come to an end? There would be needed a silence that envelops the speech anew. . . . This silence will *not be the contrary* of language.[59]

What then does Merleau-Ponty bring to the new field of ecology? He brings it a clarified epistemology, and the language of perceptual experience. His work suggests a rigorous way to approach and to speak of ecological systems without positing our immediate selves outside of those systems. Unlike the language of information processing and systems theory, Merleau-Ponty's corporeal phenomenology provides a way to describe and to disclose the living fields of interaction from our experienced place *within* them.

The convergence of Merleau-Ponty's aims with those of a genuine philosophical ecology cannot be too greatly stressed. I have shown the equivalence between the dimensions of the "world" he discloses and the actual Earth. His *Lebenswelt* is identical to the biosphere, or to the myriad bioregions, of a truly rigorous ecology. He anticipated, I believe, that his perceptual analyses would lead to a clarified description of other embodied forms, other presences which move at rhythms altogether foreign to our own. In one note he writes:

> It would be necessary in principle to disclose the "organic history" under the historicity of truth. . . . In reality all the particular analyses concerning Nature, life, the human body, and language will make us progressively enter into the *Lebenswelt* and the "wild" being, and as I go I should not hold myself back from entering into their positive description, nor even into the analysis of the diverse temporalities.[60]

Finally, Merleau-Ponty points directly to an Eco-logos by repeatedly referring to the autonomous "*Lebenswelt* Logos," to that "perceptual logic" which reigns underneath all our categories and "sustains them from behind":

> The *sensible world* is this perceptual logic . . . and this logic is neither produced by our psychophysical constitution, nor produced by our categorial equipment but lifted from a *world* whose inner framework our categories, our constitution, our "subjectivity" render explicit.[61]

It is this mute perceptual logic, recovered in language, that gives birth to ecology. Until today's fledgling ecological science addresses itself to the experience of perception it will remain uncertain of its motives, and unable to find its voice.

Acknowledgments

A thankful appreciation is extended to Claude Lefort, David Allison, and Anthony Weston for many helpful comments, as well as to the members of the Merleau-Ponty Circle, before whom an earlier version of this paper was presented in 1983.

Notes

1. Maurice Merleau-Ponty, *The Visible and the Invisible*, ed. Claude Lefort, trans. Alphonso Lingis (Evanston, Ill.: Northwestern University Press, 1968).
2. Ibid., p. 130.
3. Ibid., p. 219.
4. Merleau-Ponty, *Phenomenology of Perception*, trans. Colin Smith (London: Routledge and Kegan Paul, 1962).
5. Ibid., p. 256.

6. Ibid., pp. 230–233.

7. Merleau-Ponty, "Eye and Mind," trans. Carleton Dallery, in *The Primacy of Perception*, ed. James Edie (Evanston, Ill.: Northwestern University Press, 1964).

8. Merleau-Ponty, *The Visible and the Invisible*, p. 231.

9. Ibid., p. 130.

10. This common visual experience of the horizon, so rarely attended to, is, I believe, the primary source of Merleau-Ponty's unique metaphor of the two leaves in "The Intertwining—The Chiasm" and in his working notes: "Insertion of the world between the two leaves of my body; insertion of my body between the two leaves of each thing and of the world" *(The Visible and the Invisible*, p. 264). Merleau-Ponty is here affirming Husserl's assertion that, phenomenologically, the perceptual field or landscape has numerous "internal horizons" as well as the "external horizon" that envelops it.

11. Ibid., p. 214.

12. Ibid., p. 229

13. Ibid., p. 13.

14. Edmund Husserl, "Foundational Investigations of the Phenomenological Origin of the Spatiality of Nature," in *Philosophical Essays in Memory of Edmund Husserl*, ed. M. Farber (Boston: Harvard University Press, 1940).

15. Martin Heidegger, "Building, Dwelling, Thinking," in *Basic Writings*, ed. D. F. Krell (New York: Harper and Row, 1977).

16. Merleau-Ponty, *The Visible and the Invisible*, p. lvi.

17. Ibid., p. 267.

18. Ibid., p. 233.

19. Ibid., p. 213.

20. Ibid., p. 233.

21. Ibid., p. 259.

22. Ibid., p. 42.

23. Ibid., p. 265.

24. Ibid., p. 237.

25. Ibid., pp. 134–135.

26. Ibid., p. 267.

27. See, for instance, Gary Madison, *The Phenomenology of Merleau-Ponty* (Athens: Ohio University Press, 1981).

28. Merleau-Ponty, *The Visible and the Invisible*, p. 216.

29. Quoted in Morris Berman, *The Reenchantment of the World* (New York: Bantam, 1984), p. 147.

30. Merleau-Ponty, *The Visible and the Invisible*, p. 257.

31. Ibid., p. 259 (translation amended).

32. Ibid., p. 114.

33. The issue is in fact far more subtle and complex than this admittedly surface analysis suggests. What is immediately evident, however, is that in *The Visible and the Invisible* Merleau-Ponty supplements his earlier perspective—that of a body experiencing the world—with that of the world experiencing itself through the body. Here he places emphasis upon the mysterious truth that one's hand can touch things only by virtue of the fact that the hand, itself, is a touchable thing, and is thus thoroughly a part of the tactile landscape that it explores. Likewise the eye that sees things is itself visible, and so has its own place within the visible field that it sees. Thus to *see* is at one and the same time to feel oneself *seen*; to touch the world is also to experience oneself touched by the world. Clearly a pure mind could neither see nor touch things, could not experience anything at all. *We* can experience things, can touch, hear and taste things, only because, as bodies, we are ourselves a part of the sensible field and have our

own textures, sounds, and tastes. Merleau-Ponty coins the term *reversibility* to express this double or reciprocal aspect inherent in all perception: surely I am experiencing the world; yet when I attend closely to the carnal nature of this phenomenon, I recognize that I can just as well say that I am being experienced *by* the world. The recognition of this second, inverted perspective, when added to the first, leads to the realization of reversibility: "I am part of a world that is experiencing itself," or even, "I am the world experiencing itself through this body."

34. Ibid., p. 136.

35. James Lovelock, *Gaia: A New Look at Life on Earth* (Oxford: Oxford University Press, 1979).

36. For a much more in-depth study of the relation between the Gaia hypothesis and Merleau-Ponty's philosophy, see David Abram, "The Perceptual Implications of Gaia," *The Ecologist* 15, no. 3 (1985); and Abram, "The Mechanical and the Organic: On the Influence of Metaphor in Science," in *Scientists on Gaia*, ed. Stephen Schneider and Penelope Boston (Cambridge, Mass.: The MIT Press, 1991).

37. *Chiasm* is the term Merleau-Ponty selects to describe the blending, the reversible exchange between my flesh and the flesh of the world that occurs in the play of perception. This interweaving, this ongoing exchange *between divergent aspects of a single Flesh*, is to be found at every level of experience; it exists already in the body's own organization as the synaesthetic intertwining between one sense and another, and even within each sensory system, between the left and the right side of that sense—as in the "optic chiasm."

38. That Merleau-Ponty's thought *was* searching for roots beyond the confines of the city is attested by his increasing fascination with painting, and with the painter's relation to the natural landscape. See "Cezanne's Doubt," in Merleau-Ponty, *Sense and Non-Sense*, trans. Hubert and Patricia Dreyfus (Evanston, Ill.: Northwestern University Press, 1964); and "Eye and Mind," in *The Primacy of Perception*.

39. Merleau-Ponty, *Consciousness and the Acquisition of Language*, trans. Hugh Silverman and with a foreword by James Edie (Evanston, Ill.: Northwestern University Press, 1973), p. 43.

40. Rene Descartes, *Discourse on Method*, trans. Laurence Lafleur (New York: Macmillan, 1986), pp. 36–38.

41. Merleau-Ponty, *The Visible and the Invisible*, p. 274

42. Ibid., p. 274.

43. Ibid., p. 250.

44. See Maurice Merleau-Ponty, *Humanism and Terror*, trans. John O'Neill (Boston: Beacon Press, 1969), and *Adventures of the Dialectic*, trans. Joseph Bien (Evanston, Ill.: Northwestern University Press, 1973).

45. From the chapter on "The Body as Expression and Speech," in Merleau-Ponty, *Phenomenology of Perception*, p. 197.

46. Merleau-Ponty, *The Visible and the Invisible*, p. 194.

47. Ibid.

48. See James Edie's "Foreword" to Merleau-Ponty, *Consciousness and the Acquisition of Language*, pp. vii–viii.

49. James Edie, "Merleau-Ponty: The Triumph of Dialectic over Structuralism," a paper presented at a conference of the Merleau-Ponty circle at SUNY Binghamton in 1982.

50. Merleau-Ponty, *The Visible and the Invisible*, pp. 130–155.

51. Cited in Edie, "Merleau-Ponty: The Triumph of Dialectic over Structuralism." Note that Chomsky is the major contemporary proponent of the view that language belongs to the human species alone.

52. Merleau-Ponty, *The Visible and the Invisible*, p 274.

53. Harvey Sarles, *Language and Human Nature* (Minneapolis: University of Minnesota Press, 1985), p. 228.

54. Ibid., p 86.

55. Ibid., p. 20.

56. Ibid., p. 249.

57. Ibid., p. 20. More recently, the bodily infrastructure of language has been forcefully demonstrated by the work of George Lakoff and Mark Johnson. See especially their joint volume, *Metaphors We Live By* (Chicago: University of Chicago Press, 1980), and Lakoff's excellent book, *Women, Fire and Other Dangerous Things: What Categories Reveal about the Mind* (Chicago: University of Chicago Press, 1987). The research of these authors is definitive, I believe, in establishing the centrality of nonverbal perceptual and kinesthetic experience in the genesis and development of human language.

58. Merleau-Ponty, *The Visible and the Invisible*, p. 155.

59. Ibid., p. 179.

60. Ibid., p. 167.

61. Ibid., pp. 247–248.

Chapter Five

Hannah Arendt and the Politics of Place: From Earth Alienation to *Oikos*

David Macauley

Introduction

> If we want to be at home on this earth, even at the price of being at home in this century, we must try to take part in the interminable dialogue with its essence.
>
> —HANNAH ARENDT, "Understanding and Politics" (1953)

Hannah Arendt was a highly unconventional, erudite, original, and independent political thinker, whose work spanned a range of subjects from the growth of banality and thoughtlessness in modern life to their expression in totalitarianism; from an examination of philosophical concepts such as authority, freedom, and judgment to their place in political phenomena like violence, civil disobedience, and revolution; and from biographical portrait of villain (Adolf Eichmann), heroine (Rahel Varnhagen), or friend (Karl Jaspers and Walter Benjamin) to historical assessment of Zionism, cultural alienation, and the civil rights movement. She is recognized widely as a thinker of subtle if controversial distinctions between, for example, work and labor; wealth and property; *vita activa* and *vita contemplativa*; the public, private, and social

First published in *Capitalism, Nature, Socialism* 3, no. 4 (December 1992): 19–45; reprinted here with additions and minor corrections.

spheres; and force, violence, and power. Her work draws on an equally diverse and eclectic background including phenomenology (particularly Heidegger), fragmentary historiography, the German intellectual tradition, storytelling and journalistic techniques, and Greek and Roman philosophy.

Controversy was a hallmark of her life and her *opus*. Her preference for the American over the French Revolution, her defense of private property, and her critique of Marxism served to distance her from much of the traditional Left, while her support of workers' councils, admiration for participatory politics, and impassioned assaults on imperialism, bureaucracy, and mass culture alienated her from much of the orthodox Right. She has been viewed variously as conservative, radical, elitist, anarchist, and even antipolitical, though she refused on principle to found or join a school of thought or movement. "Social nonconformism," she once remarked, "is the *sine qua non* of intellectual achievement,"[1] and in this regard she both experienced and courted the role of pariah. Her writings, however, are marked by an intense, if at times obscure, brilliance and by an unwavering commitment to the recovery of freedom in public life and the reclaimed capacity for political action, topics explored in *The Human Condition*, perhaps her most fundamental study.

Arendt's early uprootedness and her continued experience with the problem of modern homelessness found a second life in her writings, undoubtedly influencing her thinking on earth and world alienation. Briefly, she was born of Jewish descent in Hanover, Germany, in 1906; studied during the 1920s with three towering intellectuals, Jaspers, Heidegger, and Husserl; and then fled to France in 1933, where she was active in the anti-Nazi opposition and conducted an extensive inquiry into the origins of modern anti-Semitism, work later incorporated in her first major book, *The Origins of Totalitarianism* (1951). In 1941, she emigrated to the United States, where she taught at the New School for Social Research (home to many exiled intellectuals), continued to probe and write about modern political culture from the perspective of classical political theory, and lived out the rest of her life. She died suddenly in 1975 while working on *The Life of the Mind*, a tripartite study of thinking, willing, and judging. In that work as in others, she speaks of the "condition of homelessness as being natural to the thinking activity,"[2] a philosophical phenomenon, or perhaps fact, which no doubt she was weighing against the political realities of *apatrides* (stateless persons), "blood and soil" ideology, imperialism and the drive for *Lebensraum*, as well as her own desire to overcome or articulate her feeling of *Unheimlichkeit* and to be at home in the world.

In this essay, I offer an historical and ecological perspective on Arendt's understanding of earth alienation, nature, and related matters and develop this discussion in a critical manner directed toward contemporary relevance. Though differing with selected aspects of Arendt's views, I find her political analysis and sincere attempts to bring us closer to feeling at home in an increasingly fractured world, and on an ever-fragile earth, to be engaging, learned, and of lasting value.

Earth Alienation

Why has man rooted himself thus firmly in the earth, but that he may rise in the same proportion into the heavens above?
—THOREAU, *Walden*

In the prologue to *The Human Condition*, Arendt writes of the launching in 1957 of the first satellite, an event, she asserts boldly, that is "second in importance to no other, not even to the splitting of the atom."[3] With the projection of this man-made, earth-born, and once earthbound object into the depths of outer space, she locates both a symbolic and an historic step toward realizing the hubristic dream of "liberating" us from nature, biological necessity, and earthly "imprisonment." This desire to escape the earth (and our success in so doing) signifies to Arendt a fundamental rebellion against the human condition, of which the earth is the "very quintessence," and marks our departure into the universe and a universal standpoint taken deliberately outside the confines and conditions in which we have lived from our genesis. This monumental action, too, can be viewed as a prelude to and encapsulation of Arendt's own thinking about the realm of nature, for it is here that she establishes a stark distinction—or, more exactly, opposition—between *earth* and *world* and calls attention to an alienation which, she claims, we experience from both spheres. Arendt also shows an early concern with the subject of *dwelling*—on-the-earth and in-the-world—an activity she speaks of elsewhere as *homelessness* and *rootlessness*, and she signals a preference for turning toward or returning to an older conception of the natural and the political, namely, a Greek one. Thus, she announces her intention to "trace back modern world alienation, its twofold flight from the earth into the universe and from the world into the self, to its origins."[4]

In the initial pages of *The Human Condition*, Arendt reveals a penchant for resorting to phenomenological, historical, and, later, etymological accounts of politics and "what we are doing" within and to the world and earth, and for employing spatial metaphors and descriptions in the process. In fact, the satellite which carries us from our home and earthly place into a cosmic space and new Archimedean point is merely the first such vehicle Arendt invokes to launch us into consideration of a politics of the spatial and placial. She examines public and private space, spaces of appearance (the *polis*) and places of disappearance (the death camps), the inner space and life of the mind, and outer space and its conquest by modern science and technology. In assessing such thoughts on nature and the earth and their relevance for contemporary ecological and political thought, it is necessary to situate her views historically by positioning them against the Greeks (to whom she looks), Marx (whom she criticizes), Heidegger (from whom she borrows), and the Frankfurt School and its heirs (whom she neglects). In this way, one can perhaps better measure her contributions and failings, her blindnesses and insights.

The phenomenon of earth alienation, as Arendt conceives of it, is an interesting but curious and problematic notion. It is typified strangely by an historical expansion of known geographic and physical space which, ironically, brings about a closing-in process that shrinks and abolishes distance. Earth alienation stands in contrast, though not complete opposition, to world alienation. Both originate, in her view, in the sixteenth and seventeeth centuries. According to Arendt, there were three great events which inaugurated the modern age and led to the withdrawal from and loss of a cultural rootedness in place and estrangement from the earth. First, the most spectacular event was the discovery of America and the subsequent exploration, charting, and mapping of the entire earth which brought the unintended result of closing distances rather than enlarging them. It enabled humans to take "full possession of [their] mortal dwelling place" and to gather into a globe the once infinite horizons so that "each man is as much an inhabitant of the earth as he is of his own country."[5] Second, through the expropriation of church property, the Reformation initiated individual expropriation of land and wealth which, in turn, uprooted people from their homes. Third, the invention of the telescope, the least noticed but most important event, enabled humans to see the earth not as separated from the universe but as part of it and to take a universal standpoint in the process. From this bellwether moment, Arendt traces our ability to direct cosmic processes into the earth, the reversal of the historic privileging of contemplation over action, a resultant distrust of the senses, and a marked tendency on the part of science to dominate nature.[6] The telescope, in short, "finally forced nature, or rather the universe, to yield its secrets."[7]

The roots of earth and world alienation seem to be related for Arendt, though two of them—the charting of the earth and the invention of the telescope—are more closely linked with her conception of earth alienation than the third. To these events, we can add the rise of Cartesian doubt, for with it our earthbound experience is called into question with the discovery that the Earth revolves around the sun, a phenomenon which is contrary to immediate sense experience. Cartesian doubt is marked by its universalizability, its ability to encompass everything (*De omnibus dubitandum*), and to leave the isolated mind alone in infinite, ungrounded space. Modern mathematics and particularly Cartesian geometry are also indicted because they reduce all that is not human to numerical formulas and truths. They free us from finitude, terrestrial life, and geocentric notions of space, replacing them with a science "purified" of these elements. In effect, they take the *geo* (the earth) out of geometry.

This movement from natural to universal science and the creation of a new Archimedean point in the human mind (a metaphor Descartes employs in the Second Meditation), where it can be carried and moved about, is at the heart of her conception of earth alienation, a distinguishing feature of the modern world.[8] It is this historic process which has enabled us to handle and control nature from outside the earth: to reach speeds near the speed of light with the aid of technology, to produce elements not found in the earth, to create life in a test tube and to destroy it with nuclear weapons. In Arendt's view, this process is responsible for estranging us so radically from our given home. In fact, she appears to take a step even further in the direction of pessimism when she claims that the earth is, in a sense, dispensable and obsolete: "We have found a way," she says, "to act on the earth and within terrestrial nature as though we *dispose* of it from outside, from the Archimedean point."[9] In her essay "The Conquest of Space and the Stature of Man," Arendt elaborates on these themes and shows the futility of humans ever conquering space and reaching an Archimedean point, which would constantly be relocated upon its discovery. She suggests that we recognize limits to our search for knowledge and that a new, more geocentric world-view might emerge once limitations are acknowledged and accepted. Arendt is not especially optimistic about such an occurrence, but feels that we must recover the earth as our home and begin to realize that mortality is a fundamental condition of scientific research. It is not only modern science which she finds culpable, though, for it was philosophers, she asserts, who were the first to abolish the dichotomy between earth and *sky* (by which she might also mean space since the earth includes the sky[10]) and to situate us in an unbounded cosmos. And so the task of reconceiving our relation to the universe also rests on the shoulders of philosophers.

Regarding the events leading to the arrival and development of earth alienation, Arendt locates several of undoubtedly major importance, but the question might be raised as to whether there are others she neglects such as the microscope, compass, clock, or computer, to name a few of the most significant. The magnetic compass, for example, allowed for the exploration and mapping of the earth as well as the discovery of America. It replaced the natural forces of the winds which previously had directed sailors and had been the subject of a whole mythology (not only for the Greeks who used the word "wind" (*anemos*) as a synonym for direction, but also for the ancient Chinese who distinguished twenty-four seasonal winds). The compass provided for an absolute reference and a new orientation in space in a manner comparable to the uniformity and universality imposed by the mechanical clock and standardized hour. Similarly, the microscope, which was an invention of the same age as the telescope, allowed humans to probe new, once invisible universes on the earth and to open up a vast and seemingly infinite space within the known world which mirrors the grand cosmos. Galileo, in fact, tried to use his telescope as a microscope, remarking in 1614, "I have seen flies which look as big as lambs."[11]

As Victor Hugo wrote in *Les Misérables*, "Where the telescope ends, the microscope begins. Which has the grander view?"[12] Thus, to explore fully our earth alienation, we may have to look to additional historic events and inventions which are imprinted indelibly on the modern age.

As to the meaning of "alienation" in the phenomena of earth and world alienation, Arendt does not employ it in a consciously Marxian or existentialist manner, though there are some similarities.[13] We get an early clue as to what she means in *The Origins of Totalitarianism*, where she writes that the "alien" is "a frightening symbol of the fact of difference as such, of individuality as such, and indicates those realms in which man cannot change and cannot act and in which, therefore, he has a distinct tendency to destroy." This remark is made in the context of a discussion of ethnic homogeneity, but it clearly prefigures her thoughts on earth and world alienation, especially since the spheres which humans cannot change at will are marked by "*natural* and always present differences" and indicate "the limitations of the *human artifice*,"[14] words which also could be read to mean "earth" and "world." A second aspect of Arendt's conception of alienation seems to involve the idea of being at home on the earth and in the world, a need and right to which Arendt clings with almost "religious commitment."[15] Alienation, in this sense, means a loss of roots and a common, shared sense of place, a realm of meaningful pursuits secured by tradition against the forces of change. Indeed, the themes of homelessness and rootlessness are at the center of Arendt's political concerns (in treatments of totalitarianism or imperialism, for example) and can be compared with Heidegger's more existential treatment, which comes to the fore with frequency in his later thought.[16]

Generally speaking, for the Greeks, the world of nature and the earth is conceived in bodily terms and metaphors; it possesses a kind of corporeality. This view was initiated by the Presocratics who depicted the earth as organism, animal or, in the case of Anaximander, as a solid cylindrical body. It was taken up by Plato in the *Timeaus* where he spoke of the bodies of planets, the world's body, and the body of the universe; was altered by Plotinus in the *Enneads*; and is alluded to or developed in subsequent philosophic history. In his late mythic phenomenology, Husserl refers to the earth as "the original ark," the "basis body" for all other bodies, while Heidegger remarks that "body and mouth are part of the earth's flow and growth in which we mortals flourish."[17] Marx, whom Arendt criticizes but also borrows from, argues that "Nature is man's inorganic body," and he applies physiological language to the natural realm, speaking about our "metabolism" with nature but also our estrangement from it in failing to exploit it fully.[18] While embodiment serves to give the earth a form analogous and understandable to humans, it also provides philosophers and other persons with a pretext for disciplining and punishing this corpos, which is usually gendered in female terms.

In contrast, Arendt tends not to construe or speak of the earth as a body or even as a place, ground, or particular location in which one lives (as in the earth beneath one's feet). She rejects Heidegger's conception of Earth as a "mythologizing confusion" which cannot serve as any kind of social foundation, though when she does speak of the earth and natural world in phenomenological language, it is strongly reminiscent of her former teacher's work and laden with terms that suggest the activities of dwelling and disclosure.[19]

Rather, the earth for her appears to be primarily, though not exclusively a *planet* (a view Heidegger explicitly rejected), as when she comments on circling, discovering, measuring, and locating a point outside it, or writes about the earth as a "globe" which can be brought into the living room, or remarks that "the most radical change in the human condition we can imagine would be an emigration of men from the earth to *some other planet*."[20] She is, of course, historicizing and politicizing the problem of earth alienation, but in another sense she is simultaneously helping to cast the earth into universal space by validating the view of it as one homogeneous whole and failing to provide an alternative conception which accounts for geographic difference and the uniqueness of living in particular places. Such depictions also may have the effect of "alienating man from his immediate earthly surroundings"[21] and of representing the world as picture—Heidegger's *Weltbild*—an event which distinguishes the modern age.[22] It might be noticed in this regard that the word *planet* is related in the Greek to *planetes*, "wanderer," and *planan*, "to lead astray," and so implies something which is either without a definite home or place or which has lost its bearings and ground.

Precursers to and Influences upon Arendt

> As their telescopes and microscopes, their tapes and radios become more sensitive, individuals become blinder, more hard of hearing, less responsive, and society more opaque . . . its misdeeds . . . larger and more superhuman than ever before.
>
> —MAX HORKHEIMER, *Dawn and Decline*

Arendt's account of earth alienation relies on the work of at least three other important thinkers—Alexandre Koyré, Alfred North Whitehead, and Martin Heidegger, two of whom she acknowledges to some extent (Koyré and Whitehead) and one whom she barely mentions (Heidegger). Koyré, whose landmark work, *From the Closed World to the Infinite Universe*, was published a year before *The Human Condition* (though presented as a lecture in 1953) looks at the process whereby humans have lost their place in the world and, more fundamentally, the world we know itself. He traces the scientific and philosophic revolution from Nicholas of Cusa, Copernicus, Galileo, and Descartes through Berkeley, Newton, and Leibniz which brought about the "destruction of the Cosmos and the infinitization of the universe," by which he means the very rapid transition from the closed, finite, hierarchical conception of the world held by the ancients (Figure 5.1) to one marked by indefiniteness, infinity, and ontological parity for the moderns. With this new world-view, he finds we have lost "all considerations based upon value-concepts, such as perfection, harmony, meaning, and aim, and finally the utter devalorization of being, the divorce of the world of value and the world of fact."[23] The direct influence upon and similarities with Arendt's conception should be obvious.

Second, Arendt looks to Whitehead and particularly his work *Science and the Modern World*, which was published in 1925. While borrowing his insights on the telescope and Cartesian thought, she disregards his thoughts on what he terms the "romantic reaction," a protest by poets on behalf of nature, perhaps because she feels that we should not be misled by very general notions like "the disenchantment of the world" or "the alienation of man," which in her opinion

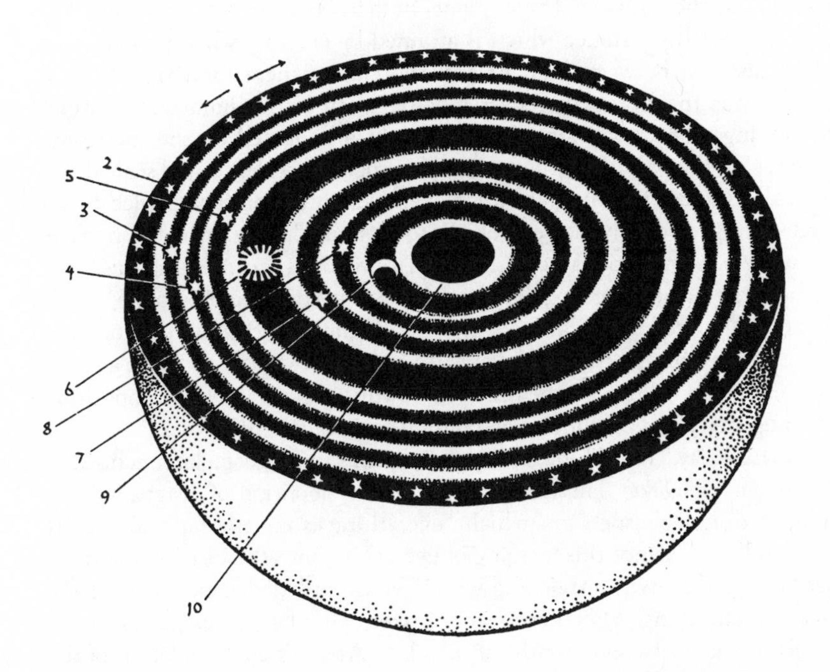

Figure 5.1. Greek conception of the spherical cosmos: (1) propelling sphere—invisible, immovable; (2) star sphere (carries other spheres with it) period of revolution: 1 day; (3) Saturn sphere—period of revolution: 29 years; (4) Jupiter sphere—period of revolution: 12 years; (5) Mars sphere—period of revolution: 2 years; (6) Sun sphere—period of revolution: 1 year; (7) Venus sphere—period of revolution: 6 years; (8) Mercury sphere—period of revolution: 3 months; (9) Moon sphere—period of revolution: 1 month; (10) Earth sphere—immovable.

often involve a romanticized view of the past (something to which Whitehead arguably does not succumb). Whitehead shows that "we gain from the poets the doctrine that a philosophy of nature must concern itself at least with . . . six notions: change, value, eternal objects, endurance, organism, interfusion."[24] Though he speaks of Wordsworth, Shelley, and Coleridge, the above concepts might also be invoked, *mutatis mutandis*, to understand his own process metaphysics and philosophy of organism, which exhibits a surprising degree of ecological sensitivity in its reworking of subject-object relations, its critical and creative return to a Greek view of mind in/as nature, its stress on purposive "evolutionary expansiveness," its ethics of immanence, and its attentiveness to the democratic consociation, communication, and coordination of organic and inorganic entities.[25]

Third and most important, Arendt's thinking on earth alienation is influenced strongly by Heidegger. This fact is apparent in her thoughts on technology and the shrinkage of the earth, which she finds the result of the invention of the airplane. The airplane, in fact, becomes for her a symbol of the loss of earthly distance which is attained by creating what amounts to a *vertical* distance between humans and the earth, where formerly *horizontal* distance was the norm, thereby removing us from our immediate natural surroundings. This abolition of distance was anticipated by the increasing tendency on the part of the human mind to survey, condense, and scale down physical distances in cartography, for example, and was accompanied by an increase in speed which allowed humans to "conquer" space. In addition to the airplane, Arendt mentions steamships and railroads as accentuating this process, though it is a bit surprising that she does not consider the automobile, which has not only altered radically our notions of time, space, and distance but transformed ruthlessly the surface of earth into a network of concrete and led to the erasure of wilderness, the rise of a sprawling suburbia, and the imposition of homogeneity and conformity on cultural and political life.[26]

In his essay "The Thing," Heidegger speaks in language later echoed in *The Human Condition*. He writes of the "restless abolition of distances," of a "uniform distancelessness" in which "everything is equally far and equally near," and asks, "Is not this merging of everything into the distanceless more unearthly than everything bursting apart?" Or again: "All distances in time and space are shrinking. Man now reaches overnight, by plane, places which formerly took weeks and months of travel."[27] Arendt's contemplation of the earth (and world) also emerges directly from the shadows of Heidegger's reflections, as does much of her thinking on homelessness, art, humanism, and the "thing-character of the world" as she terms it. Arendt often gives these subjects a new phenomenological twist or adds a needed political dimension and insight where it is woefully lacking in Heidegger's work, but it is surprising how rarely she acknowledges her debt to him and how little this borrowing is explored in the secondary literature.

In his later work, Heidegger addresses the natural world more directly and empathetically than in *Being and Time*, where the nature of Nature which is revealed to us remains a concept whose structure is determined solely by human consciousness since it is still "ready to hand" and accessible only to *Dasein*, who alone can reveal a world. In this writing after the *Kehre*, however, he shifts his thinking to suggest that animals, plants, and even things can manifest or found a world. It is here as well that he speaks of the "self-dependent," and "effortless and untiring" earth, that which provides a shelter, anchor, and orientation for Being and beings. The earth "shatters every attempt to penetrate it," he says.[28] It is self-secluding and self-concealing and is set up in a striving opposition to his conception of the world, which is said to be self-revealing. For Heidegger, it is on-the-earth, the "building bearer" that we can understand ourselves and the world we create. It is here that we can find a harmony with nature, what Rilke calls the *Urgrund* (the pristine ground), because "all things of earth, and the earth itself as a whole, flow together into a reciprocal accord."[29]

The opposition of earth and world can be seen most clearly in "The Origin of the Work of Art," where Heidegger provides a vivid description of the shoes of a peasant woman, detailing their equipmental nature, and showing how they gather a whole panoramic world into being. We are placed, in effect, inside these ragged shoes—literally put in the shoes of another, or the Other, then walked about in the world which *they* create. The earth resounds and pulsates in the "stiffly rugged heaviness" of the shoes because we can sense the moisture of the soil in them, the path upon which *they* walked, and the trudging steps which carried them along. As Heidegger says, "This equipment belongs to the *earth*, and it is protected in the *world* of the peasant woman."[30] This in-gathering of a world by entities other than humans is repeated again at several points in his writing when, in the same essay, we find a Greek temple and, in "The Thing," a jug uniting the fourfold of earth, sky, divinities, and mortals.

Arendt adopts and adapts this opposition in *The Human Condition* and in essays collected in *Between Past and Future*. Her language again resonates with that of Heidegger, as when she speaks of the "effortless earth" or "the earth, inexhaustible and indefatigable" or of "world withdrawal" and decay.[31] The emphasis on thingness and disclosure is undoubtedly drawn from his phenomenological language, too. In this regard, it should be pointed out that there is, perhaps, a certain natural affinity between phenomenological thinking and ecology, which helps to explain the strains of ecological concern in Arendt's thought. The starting point for phenomenology is that we find ourselves located in a world that reveals itself. The word "phenomenon," in fact, means "the showing-itself-in-itself," that which brings itself to light. The task, then, of phenomenology is to facilitate this encounter and to allow phenomena to be grasped originally, directly and intuitively. In this respect it is very similar to ecological thinking, which also returns us "to the things themselves," as

Husserl put it—that is, to the animals, plants, mountains, rivers, and humans who are encountered and studied within their respective environments.[32]

Contemporary Earth Alienation

> Only in the context of the space race in the first place, and the militarization and commodification of the whole earth, does it make sense to relocate that [whole earth] image as a special sign of an anti-nuclear, anti-militaristic, earth-focused politics. The relocation does not cancel its other resonances; it contests their outcome. . . . Relocated on [a] T-shirt, the satellite's eye view of the planet earth provokes an ironic version of the question, who speaks for the earth (for the fetus, the mother, the jaguar, the object world of nature, all those who must be represented)?
>
> —DONNA HARAWAY, "The Promises of Monsters"

If one can look back to the precursers of and influences on Arendt's conception of earth alienation, one can also look forward from her view to discover the contemporary relevance and transformations of her thoughts. Indeed, her warnings about earth alienation can be extended to reveal a world more deeply removed from nature and the earth than she imagined. If this process was initiated with the telescope and symbolized by the satellite, it has come to be represented continually by the photographic image of the Earth as seen from space.[33] It is this ubiquitous image of a bounded, blue sphere captured from a point outside the earth that enables us to show finally beyond all doubt that Earth is round, definable, and ultimately manipulable from a god's-eye view. In this whole-earth image, an ironic creation of a fractured and fragmented culture, we have located perhaps not simply another picture of the place we inhabit, but have attempted to *re-place*, via a universalized technological substitute, our distinct and multifarious senses of place and purpose which have been lost as we have transformed the environment. If this new representation helps us to see Earth potentially as a living organism (as the Gaia hypothesis, Presocratic thought, and Native American beliefs suggest) or as a single cell (as biologist Lewis Thomas speculates), it also "provides us with a small, comprehensible, manageable icon—an easily manipulable token Earth that we can use to replace the unfathomably immense and overwhelmingly complex reality of the world which surrounds us."[34]

Today, we see the earth no longer brought so much into people's living rooms as a globe (a multidimensional object often with textured contours), but rather, emblazoned on computer screens and televisions as an image without depth or true dimensionality. We espy a roadside metallic Atlas locating an Archimedean point and literally raising a huge scaled model of Earth up in his arms in order to sell a brand of tires bearing his name. We find a hologram image of Gaia as an advertisement for McDonald's, one of the worst corporate destroyers of the environment and culture, on the back cover of a special issue of *National Geographic* devoted to saving the rain forests and "our fragile" planet.[35] Even oppositional consciousness in the United States and Europe frequently perpetuates and uses uncritically this image of the earth in their literature and campaigns when they compete like corporations for the environmental dollar or vote, as evidenced by the commercialization of the twentieth celebration of Earth Day in 1990.[36] We are thus in an era when the image of the earth is often but a floating commodity which is appropriated like any other. In this sense it is a kind of fetish object and one might even speak of a certain "earth pornography" that now exists, as the gendered planet, the "mother of life" (as Plato called it) is not only violated literally by strip mining, deforestation, and radioactive waste, but subjected to the capricious circulation of a voyeristic media.[37] As with most commercial pornography, there is a violent "ripping" or tearing from context (natural surroundings) that has occurred and a repositioning or reterritorializing of the viewed object in a space or place in which it is not normally found, such as against a white background or on a billboard ringed by items for sale.[38]

Writing in 1948, prior to Arendt's *The Human Condition*, a British astrophysicist predicted that "once a photograph of the Earth, taken from the outside is available—once the sheer isolation of the Earth becomes known—a new idea as powerful as any in history will be let loose." Twenty years later, televised pictures were beamed back by astronauts circling the moon and the conditions for this prophecy were fulfilled. The ironic truth of the speculation, however, is that this "powerful" idea has not led to such inspiring consequences as astronomers like William K. Hartmann have suggested when they claim that "a new frontier is opening. It is interplanetary . . . [and] . . . may produce its own renaissance."[39] Since that time, we have been resolute in our determination to escape and replace our natural home with the depths of space where we have established space stations, military bases, and communication systems. Instead of, or at best concomitant with the opening of a new frontier, there has been vast shrinkage and colonization of space. The promised global village, too, is little more than a tangled economic nexus which has introduced more homogenization than harmony, more tension than tolerance, into our world. As I write, I have come across an article on the Hubble telescope in today's *New York Times* which says that this $1.5-billion-dollar disembodied eye is designed to detect objects in space that are twenty-five times fainter than the dimmest observable objects from earth. Presumably, if we keep looking with the aid of greater technology, we will find some kind of home away from home, or at least distract ourselves a while longer from problems on earth. As Michael Zimmerman writes, "Satellite photos of Earth may be instances of that 'high altitude thinking' (Merleau-Ponty) which conceives of itself as pure spirit rising above the natural world. In such photos, we see Earth reflected in the rearview mirror of the spaceship taking us away from our home in order to conquer the universe."[40]

Origins and Expressions

After seeing electricity, I lost interest in nature. Not up to date enough.

—VLADIMIR MAYAKOVSKY, *Autobiography*

Hoorah! No more contact with the vile earth!

—EMILIO MARINETTI

Arendt's conception of earth alienation finds its first expression, not in *The Human Condition* but in *The Origins of Totalitarianism*, where she writes: "Ever since man learned to master it [human nature] to such an extent that the destruction of all organic life on earth with man-made instruments has become conceivable and technically possible, he has been alienated from nature. Ever since a deeper knowledge of natural processes instilled serious doubts about the existence of natural laws at all, nature itself has assumed a sinister aspect."[41] She proceeds to claim that humans in the twentieth century are "emancipated" from nature just as those in the eighteenth century were emancipated from history. By this statement she means that objectivity has vanished and that "absolute and transcendent measurements" such as history, nature, or religion have lost the authority they once exerted. This loss of authority, a subject she explores in "What is Authority?," is held to be identical with the loss of permanence and durability in the world. In *Origins*, too, she speaks of the "shrinking of geographic distances" and the apparent fact that there is "no longer any 'uncivilized' spot on earth, because whether we like it or not we have really started to live in One World," an overstatement no doubt but not far from the truth.[42]

It might be conjectured, as Arendt merely begins to do, that nature as ground, source, or referent of meaning, order, and direction has been objectively displaced over time as the physical environment and earth have been radically altered, a process begun not recently but during the ancient Greek period. This more or less objective grounding for society (its ethics and politics) was reinvented in a removed form as God by the Neoplatonists and relocated or re-placed in a transcendent realm, away from the daily activities of those who dwell on earth, a sphere which, for thinkers like Saint Augustine, becomes that which brings forth pain and suffering. Contributing to this devaluation and displacement was a very abstractive move from the this-worldly to the other-worldly (always present or latent in Platonism) which has been responsible, in part, for encouraging domination of the earth and environment. In fact, a major source of the current ecological crisis can be traced directly to views which the Bible, Platonic, and Neoplatonic thought have fostered, particularly the idea of a radically transcendent (but not immanent) God (One, Good, or Spirit) who is separated from the natural world which plants, rocks, humans, and other animals know and dwell within. It is this transcendent being or realm which often permits, tolerates, or even sanctifies human dominion over the earth and its inhabitants, a tendency undoubtedly given impetus as the Greeks gradually abandoned animistic impulses and came to replace them with more theistic ones. As Lynn White has argued in a well-known essay, "The present increasing disruption of the global environment is the product of a dynamic technology and science which . . . cannot be undersood historically apart from distinctive attitudes toward nature which are deeply grounded in Christian dogma."[43] A dichotomy thus asserts itself between that which is perceived and protected as sacred and that which is understood and utilized as profane so that nature as it is construed (or constructed) falls almost unfailingly into the latter sphere. Our perceivable, corporeal world is thereby still a falling away from, a pale imitation of something more perfect and unattainable.

Thus, it cannot be fully said with Arendt that the Greeks completely dreaded the devaluation of nature, for a certain kind of earth alienation had already begun to set in. Arendt's claim, too, that the "notion of man as lord of earth is characteristic of the modern age" and her assertion that it contradicts the Bible, where Adam was put into the Garden of Eden to "serve and preserve it," is not strictly true for several reasons.[44] First, human dominion over and domination of the earth is much older and more deeply rooted than this statement indicates, and second, there are many passages in the Bible which attempt to establish man as master of the earth.[45]

One might argue further that the Greek view of nature to which Arendt returns was based on a very specific cultural and mythological experience of the natural world, and that with the subsequent historical encroachment, colonization, and domestication of the environment, figures like Descartes or Marx quite "naturally" experienced a diminished nature that allowed them more easily to imagine controlling it or separating humans from it. Philosophical viewpoints cannot be divorced from historical contexts, and while Greek philosophers ruminated on the origins and nature of our world, the Greek land was being desiccated and altered dramatically by human actions. Of great importance was the levelling of forests which occurred primarily between 600 and 200 B.C., during the great Classical and Hellenistic periods, to provide wood for fuel, furniture, weapons, temple roofs, and, especially, to fill the increasing demand for ships. In the *Critias*, Plato describes the result of the deforestation of Attica and points out that "by comparison with the original territory, what is left is . . . the skeleton of a body wasted by disease; the rich, soft soil has been carried off and only the bare framework of the district left."[46] Adding to this environmental destruction were the effects of intensive mining and quarrying to obtain minerals and ores. Finally, the most pervasive source of degradation was the grazing of domesticated animals who roamed on four-fifths of the uncultivated land, destroying many indigenous plants and laying the land open to erosion.[47]

Thus, while there has been a loss or at least recession of a ground to social and political thought and life, there are nevertheless certain dangers to "re-grounding" an ethics or politics in nature. The history of philosophy is rife with examples of attempts which end in oligarchy, slavery, religious hierarchy, and other forms of domination. Platonic pronouncements such as, "For whereas all that is against nature is painful, what takes place in the natural way is pleasant," can easily lend themselves to misinterpretation and misuse as a debate ensues over what is "natural."[48] Arendt is aware of this fact and does not seek false homes, roots, or grounds, but attempts to think without the security of firm foundations. The question remains whether some kind of a ground or place can be located.

Between Earth and World

> In setting up a world, the work sets forth the earth. . . . World and earth are essentially different from one another and yet are never separated.
>
> —HEIDEGGER, "The Origin of the Work of Art"

Against Arendt's conception of the earth and earth alienation stands the world and with it the phenomena of worldlessness and world alienation, which cannot be examined here, except as they relate directly to the earth and earth alienation. In this regard, the practice of agriculture is illustrative because it provides a perfect example of the *in-betweeness* which Arendt chararcteristically explores, for the tilling of soil is an activity which can be situated between the spheres of earth and world, nature and culture, and labor and work as she conceives them. The transformation of wilderness into cultivated land is paradigmatic of the ostensible transition from the biological cycle to the human artifice. In Arendt's view, agriculture readies the earth for "the building of the world," but the land is not fully a "use-object" since it must be labored upon endlessly, thus necessarily remaining outside the world of work and not being subject to true violation or violence (a point she does not make particularly clear).[49] In other words, agriculture does not effect a complete reification in her opinion because a tangible thing does not come into being. In "The Crisis in Culture," Arendt approaches the subject a little differently than in *The Human Condition* by drawing attention to a contrast between the attitude of the Greeks and Romans toward nature, culture, and agriculture, and altering her position (or disposition) somewhat in the process, perhaps inadvertently. The Romans, she argues, first saw nature as closely linked with culture, a word whose origin (from *colere*) is Latin and means to tend, preserve, dwell, and care for. Second, they viewed art as a kind of agriculture, a way of cultivating the offerings of the earth. Agriculture was therefore a peaceful and natural activity for them. The Greeks, on the other hand, were more influenced by the art of fabrication and saw agriculture as one of the ways of taming and dominating nature, of tearing from the earth what the gods had hidden. In her explication of these historical views, Arendt seems to prize the Roman perspective over the Greek (somewhat untypically), noting that the Romans preserved the Greek heritage (i.e., cultivated it) and that the Greeks did not even know what culture or cultural continuity were because they were too busy taming nature rather than tending it. She concludes the essay with the assertion that the Romans were the "first people that took culture seriously the way we do."[50]

Despite its many merits, Arendt's account seems not to have recognized or foreseen the true destructiveness of the very historical practice of agriculture for both the earth (in terms of soil depletion, erosion, forceful re-shaping of natural contours and boundaries) and the human world. We may still be "under the spell of the Roman heritage" as she puts it, but we have gone far beyond even the Greek view of agriculture as a "daring, violent enterprise" because we no longer exhibit any respect for the ability of the earth to respond to our attempts to dominate and domesticate it, or recognize the element of "violence" we inflict upon it.[51] Modern agriculture, with the hand of capital, has transformed the land into property and a reified commodity, a fact which is absent from Arendt's depiction. Agriculture, too, as Mircea Eliade has rightly noted, "provoked upheavals and spiritual breakdowns" of a magnitude we can hardly imagine and initiated what physiologist Jared Diamond has termed "a catastrophe from which we have never recovered."[52] In fact, there is reason to believe that it contributed to the development of private property and work as a distinct category of life separate from other spheres and activities. As Arendt herself points out, the word for "tilling" later came to mean "laboring" in Hebrew (*leawod*), an association implying servitude on the part of humans.[53] Agriculture, it might be argued, also enhanced or accentuated social stratification, violence toward women and animals, and the destruction of wilderness areas.[54] In contrast to gathering and hunting which it systematically replaced, it is marked by regularization, routinization, and repetition of daily activities, the rise and spread of numerous diseases such as tuberculosis, and the narrowing of food choices. Unlike the dream of escaping *Gaia* upon which Arendt comments, agriculture is a flight *into* the earth rather than away from it, and can be said to be a different but related species of earth alienation (one which should be noticed by those contemporary naturalists, ecologists, and "avant-gardeners" who see in a return to the land a panacea for social and political problems). With the transition of *agri-culture* to *agri-business*, we have lost an element of the past (in the tie to culture) and moved dangerously into a future where the tilling of the soil is now spoken of almost exclusively in economic terms. In the process, we have experienced an even greater sense of "world alienation." If, then, agriculture is a kind of space *in-between* Arendt's conception of earth and world, it may also occupy that same position between the two spheres of alienation she delineates.[55]

The Relation of Earth and World Alienation

Give me a place to stand and I will move the world.
—ARCHIMEDES

[Man] found the Archimedian point, but he used it against himself; it seems that he was permitted to find it only under this condition.
—FRANZ KAFKA

This brings us to a final question: the way that world and earth alienation relate to one another more broadly, a subject as complicated and controversial as it is ambiguous in Arendt's thought. Arendt remarks that at least in part out of the despair of world-alienation arose "the tremendous structure of the human artifice we inhabit today, in whose framework we have even discovered the means of destroying it together with all non-man-made things on earth."[56] In this characterization, the phenomenon of worldlessness helped to ignite and continues to fuel our distancing from the earth and so seems to precede earth alienation. Such a depiction, however, is too clean and neat, even as a reading of Arendt's writing, and the matter is undoubtedly more complex (though it may not be as consternating as a "chicken and egg" debate about origins). In contrast to Arendt's discussion of earth alienation which has an approximate historical beginning, the Frankfurt School located the conquest of nature in human reason itself. In *Eclipse of Reason*, Horkheimer asserted that reason's "disease" is that it was born from the desire to dominate nature, and he went so far as to conjecture that "the collective madness" which reigns today "was already present in germ in primitive objectification, in the first man's calculating contemplation of the world as prey."[57] This formulation, in turn, has been subject to trenchant criticism and near-reversal by social ecologists who find the emergence of the idea of dominating nature in the domination of humans by other humans.[58] Without entering into an extremely complex historical and political controversy, little can be said on this debate except that Arendt's thesis may be questioned on the grounds that it is structured too "telescopically" (so to speak) on a few events and inventions and does not consider other forms and sources of alienation from or control over nature.

Nevertheless, she holds that world alienation, which has determined the course of development for modern society, is of "minor significance" compared with earth alienation, which is the distinguishing feature of modern science. George Kateb in turn has qualified this remark with the claim that world alienation is presently a "more actual and widespread" experience and that we are only at the early phases of earth alienation, a pessimistic but possibly accurate portrayal.[59]

Two last critical comments might be offered with respect to her account. First, it is at least questionable whether, as she claims, inner-worldly alienation (or "inner worldly asceticism," in Weber's terms) has absolutely nothing to do with earth alienation, for as I have attempted to point out, the roots of our separation from nature have much to do with the attitudes and institutions which Christianity established and has fostered. Second, at the very heart of Arendt's thought about the earth and world there lies a fundamental ambivalence or ambiguity, which often borders on becoming a contradiction, in the way she looks at the natural and artificial. On the one hand, she finds that the more stable, permanent human world is withdrawing and decaying, giving us the impression that either the realm of nature is expanding and threatening to devour the world of things, or that the human-made artifice is becoming more closely integrated with the natural environment. However, elsewhere she speaks of the fact that little is left of nature; everything natural has become artificial. It is not apparent how these views stand with respect to each other.

Arendt's Conception of Nature

> When we speak of a picture of nature . . . we do not actually mean any longer a picture of nature, but rather a picture of our relation to nature.
>
> —Werner Heisenberg (1958)

Arendt approaches nature and the natural world through her admittedly "unusual" distinction between labor and work. In developing this distinction, she speaks of nature as belonging to a household (*oikia*) and emphasizes its circularity, repetition, and endless, changeless, deathless quality. Nature is thereby linked with the realms of necessity, futility, and animality as well as labor while it stands in contradistinction to the activity of work and the condition of worldliness. Arendt finds labor to be part and parcel of biological processes and life itself (the life of the body and the earth) and argues that in this sense, the "animal" in the concept of *animal laborans* is fully justified. She invokes Marx's remarks on labor and nature to emphasize her point, noting with him "man's metabolism with nature," then finds with Locke a basis for private property in the laboring body whose activities, she implies, should not be checked because to do so would "destroy nature." In so doing, though, she is seemingly unaware that private property (an "enclosure from the common") has always been a major threat to wilderness (and society). In contrast to labor, work is unnatural, artificial, and characteristically human. Our existence is likewise described as unnatural and lying outside the "ever-recurring life cycle." Work provides for an objective, stable world of things which are unlike all natural surroundings, and the human condition to which it corresponds is worldliness. *Homo faber* "works upon" and transforms nature rather than mixing with it. His job, Arendt says, "is to do violence to nature in order to build a permanent home for himself."[60]

Therefore, in every way, we find Arendt linking labor with nature and work with the world, radically separating the two realms from each other, and in the process making associations, distinctions, and divisions which become very difficult and often untenable for her to maintain. These difficulties are evident in her discussions of rhythm and the "thing-character" of the world[61] but from "the viewpoint of nature," as she sometimes puts it, they are most relevant and pronounced in her thinking on necessity, animals, and agriculture. Arendt follows Locke and Marx in characterizing nature as the "realm of necessity" which must be overcome (subjugated and mastered for Marx) in order to reach the "realm of freedom," a conceptual distinction which goes back to Aristotle's *Politics*. In Marx's words, man must "wrestle with nature," "bringing it under [his] control" in order to subdue the "eternal necessity" which it imposes.[62] Arendt likewise attributes to nature a necessitarian quality and seems to condone *Homo faber's* "violence" against the natural world as a form of "protection," because nature "invades" the human artifice, "threatening" the world's durability.[63] "In the world of nature," she writes in *On Violence*, "there is no spontaneity, properly speaking," excluding it as she excludes play from the world and *Homo ludens* from the human condition.[64] Arendt's concept of nature is therefore as "blind" as Marx's. Though she speaks about "nature's fertility," there is little room for it to exist as a fertile idea with social or political import. In this sense, it is a *de-natured* rather than liberatory concept, as Marcuse attempts to make it in his work.[65]

Contrary to Arendt's claims, nature has been a source of freedom, value, and even objectivity—for example, in the writings of Peter Kropotkin, Hans Jonas, Murray Bookchin, and Alfred North Whitehead—and more important, it might serve as at least one guide, among others, for reconstructing social life.[66] Freedom is arguably not a concept restricted solely to the "artificial" world of humans; it is already present in the realm of nature, in which humans find themselves first embedded, though increasingly removed. The political task, then, is not only to enlarge social and political freedom but to reintegrate ourselves with the freedom which exists naturally and to create an ecological sensibility which permeates the human world that has been sharply divorced from the realm of nature. The rift between necessity and freedom is of the same kind as the stultifying dualism which has been established between, for example, "objective" and "subjective" and which has been challenged only rarely with depth and creativity by thinkers in the critical utopian tradition, such as Charles Fourier or Ernst Bloch.

Arendt, it seems, excludes freedom from nature and the possibility of discovering freedom in our experience of, or relationship with, the natural world because she construes it only as a narrow political concept—it is the *raison d'etre* of her particular conception of politics—and one which must manifest itself in a public space and in the agonistic field of action. It is humans alone who have received the "two-fold gift of freedom and action." Nature, on the other hand, is something which for Arendt, like Heraclitus, remains hidden, and the "life process is not bound up with freedom but follows its own inherent necessity." "It can be called free only in the sense that we speak of a freely flowing stream," she says in "What Is Freedom?"[67] Freedom is won by transcendence of and alienation from nature, the processes of life, and the "curse" of necessity.[68] That nature is conceived in these terms—necessity, futility, and darkness—almost by definition excludes the possibility of locating freedom within it. In fairness to Arendt, one need note that she was understandably suspicious of the invocation and spiritualization of nature in political discourse, given the atavistic conceptions put to use by the National Socialists and others, including perhaps her mentor, Heidegger.[69] She holds, for example, that totalitarian thinking found intellectual support in the Darwinian theory of evolution and that tyranny was depicted as an extension of "natural" processes.[70]

If it is true that the Greeks did not know "freedom" as we know it (because it was an exclusively political notion and they were a people submerged in politics), it is also true that neither do organic societies (who in most ways are pre-political in Arendt's sense) since there is no sharp conceptual split between "freedom" and "domination" (or necessity) for them. Freedom in a broader, though no less important sense is a fundamental, even if unconscious condition of their social life. Moreover, there is no pronounced division between nature and culture (or wilderness and society) since the two spheres are not distinguished as such. To the extent that they are, there is a healthy cross-fertilization between them. The freedom they *find* in nature informs and guides the freedom they *found* in society.[71]

From Greek to Green: Techne, Politeia, and Communitas

> Orbiting the earth aboard Friendship 7 in February 1962, astronaut John Glenn noticed something odd. His view of the planet was virtually unique in human experience. . . . Yet as he watched the continents and oceans moving beneath him, Glenn began to feel that he had seen it all before. Months of simulated space shots in sophisticated training machines and centrifuges had affected his ability to respond. In the words of chronicler Tom Wolfe, "The world demanded awe, because this was a voyage through the stars. But he couldn't feel it."
>
> —LANGDON WINNER, *The Whale and the Reactor*

If Arendt's conception of nature is problematic, her thought is more directly relevant to green political theory because of its critique of modern science and technology and its vision of a politics compatible with the current ecological challenge to capitalism, productionism, and industrialism. First, Arendt offers an insightful analysis of modern technology and science. She argues that in a society given to the primacy of laboring, "the world of machines has become a substitute for the real world." However, such a "pseudo world," as she calls it, "cannot fulfill the most important task of the human artifice, which is to offer mortals a dwelling place more permanent and more stable than themselves."[72] In this regard, Arendt distinguishes between tools and machines; the former are guided by the hand and exist within the stable world of work, but are replaced by the latter, which are associated with the repetitiveness of labor and which reinforce or redirect bodily processes. This substitution alters our understanding of means and ends and tends to undermine a sense of public permanence. Nevertheless, in her view, the problems related to technology are not tied to the question, are we are slaves or masters of our machines? They are rather twofold: first, whether machines still serve the world or tend to destroy it, and, second, the concern that future automation will intensify and magnify life processes and rhythms, wearing down the durability of things.

With the historical ascendency of the notion of process, Arendt discovers a loss of worldly objectivity and a tendency to privilege action over all other activites, including contemplation, fabrication, and laboring. This new predilection is especially dangerous because it is typified by humans acting *into* nature and bringing with them an element of unpredictability, particularly in the area of technology, where nature and history have been wedded and interfused. Arendt finds in Whitehead's formulation, "Nature is a process," the most persuasive argument for this intervention[73] and claims that it is one of the axioms of all branches of modern science.[74] Arendt had the foresight to see and to caution us as well about what might now be termed a broadly postmodern view of nature and technology.[75] Such a perspective is a logical outgrowth of nature as process and is characterized early on, as she suggests, by the ability to start new natural processes and to "unchain" natural forces. For Arendt, this "stage" of modern technology begins with the discovery of nuclear energy and goes far beyond former periods when (1) nature's power was used to substitute for or to augment human power (e.g., windmills), (2) technology merely imitated natural processes (e.g., the steam engine), and (3) we changed and "denaturalized nature for our own worldly ends" so as to sever earth from world (e.g., electricity).[76]

Arendt had intimations of a time when life will be made "artificially," and "the last tie through which even man belongs among the children of nature" will have been cut.[77] The attempt to make life in the test tube and to create superior human beings with advanced technology, she avers, is of the same nature as our desire literally to escape the earth, so that it can be seen as part of the phenomenon of earth alienation. She thus anticipates the rise of biotechnology and perhaps even Foucault's biopolitics, along with their darker implications. For just as fire technology allowed humans to solder, forge, melt, and heat *objects* for particular ends, genetic engineering now enables us to stitch, splice, edit, program, and delete *living beings*. Instead of depicting nature in mechanistic metaphors and locating its intelligence in a divine mind (the Renaissance view) or speaking of it in terms of human progress and evolutionary process (the modern view), near-autonomous science and technology have begun to reduce the natural world to mere pieces of information and to characterize it in the language of cybernetic feedback loops, self-organizing programs, and the like.[78] One wonders invariably if there might be a more reductive and earth-alienated phase waiting in the offing for, as Arendt remarks, "there is no reason to doubt our present ability to destroy all organic life on earth."[79] The use of scientific and technical knowledge, then, remains for her an important *political* question for democratic society as a whole, not one simply for professional politicians and scientists. In fact, to the extent that *techne* and metaphors of fabrication have infiltrated the political realm, substituting "making" for "acting," they have contributed to the rise of technocracy and acted so as further to debase public life.[80]

Second, Arendt's thought is in line with the left-green perspective in terms of its critique of capitalism, unchecked consumerism, industrialism, and productionism. With the rise of the social sphere (which blurs the distinction between the public and private realms) and the triumph of *animal laborans*, Arendt locates the emergence of an unencumbered consumer society. Such a society knows no natural or sustainable limits to growth. It is in this sphere that Arendt finds conformity, normalization, flight from the public world, and loneliness. The social sphere is marked by an "unnatural growth of the natural" that devours the objectivity and excellence of public life, protects private accumulation (not property), and leads to a rule of no-one (but not no rule). Consumer society and its fictitious common household, the nation-state, are for Arendt bound intimately with an economy of waste "in which things must be almost as quickly devoured and discarded as they have appeared."[81]

Arendt further criticizes the "anthropocentric utilitarianism" of man qua *Homo faber* who, as "measure of all things," treats nature and the "things themselves" as mere means and as valueless material for *his* own consumption or production-related ends.[82] In this vein, the contemporary green critique points up the necessity of changes not only in patterns of consumption, but also in the modes of production. However, Arendt resists the turn to Marx, whom she calls the "Darwin of history." Her differences with Marx stem from his understanding of labor, which he conflates with work in defining productivity, and his conception of power which, in her view, disregards its distinctly political dimension. At a conference just prior to her death, she commented on the "cruelty" of capitalism and remarked that she did not share Marx's "great enthusiasm" for it, which she finds present in the opening pages of the *Communist Manifesto*.[83] Arendt acknowledges that Marx correctly foresaw the withering away of the public realm, but unlike Marx, she seeks to recover political action and to protect the public sphere as civic community. She notes finally that a material condition for the industrial revolution was the elimination of forests, recognizing like Carolyn Merchant the links between industrialism and the "death of nature."[84] It is still perplexing that she did not consider the insights of critical theory and the Frankfurt School on these and other matters.[85]

On a last note, it can be said that Arendt belongs in a broad sense to the communitarian and, particularly, the civic republican tradition, which is as old as Aristotle but as recent as the ideas of Cornelius Castoriadis and Murray Bookchin or, in a different light, Alasdair MacIntyre, Charles Taylor, Michael Sandel, and Michael Walzer.[86] Unlike some communitarians, however, Arendt is critical of forms of political community rooted in custom and racial, religious, or ethnic identity. Rather, she conceives of the political community as based upon active citizenship, collective deliberation, and civic engagement. Arendt's neo-Aristotelianism (as with Hans Jonas, Ernst Bloch, and Bookchin) also implies a commitment to social *and* natural equilibrium, aesthetic, and ethical balance, and face-to-face public relations, meaningful foci for an emerging ecological society. In this philosophical tradition, community remains coextensive with smaller and prior forms of association, but can reach out potentially to touch larger spheres, including the natural world. In contrast to liberalism, communitarian thought often stresses the *polis* and the particular rather than the *cosmo-polis* and the universal; virtue ethics as opposed to Kantian or utilitarian morality; conceptions of the *good* and good life over notions of the *right*; freedom as against a more narrow view of justice; and the embedded self instead of atomistic individualism. Like many left-greens and communitarians, Arendt identifies herself with or praises the revolutionary tradition, direct political action and direct democracy (rather than representation), decentralization, forms of organization such as the council system (rather than political parties), and *potestas in populo*. Given these orientations, her writings can serve to inform and inspire green political theory and action.

Conclusion: Out of Space and Into Place

> To know the spirit of place is to realize that you are a part of a part and that the whole is made of parts, each of which is whole. You start with the part you are whole in.
>
> —GARY SNYDER, *The Practice of the Wild*

While Arendt shows a keen interest in and insight into the spatial, she does not explore fully a politics or phenomenology of the placial, which is regrettable since place, in an important sense, precedes space in that it makes or clears room for its possibility.[87] In this regard, earth alienation involves less a conquest of space, as Arendt suggests, than it does a loss of place.[88] Her emphasis on the spatial may perhaps be explained by the fact that spatial metaphors and images are often used in traditional descriptions of how the mind and thought stalk and stumble toward clarity, a subject which always preoccupies Arendt *the philosopher* (a characterization she rejects), who feels the necessity to withdraw from the world in order to think about it.[89] She writes, for example, "Before men began to act, a definite space had to be secured and a structure built where all subsequent actions could take place, the space being the public realm of the *polis*."[90] Later she holds that the *polis* is not even a physical location and describes it as a sort of ambulatory or peripatetic entity that materializes in a "space of appearance." "Wherever you go, you will be a *polis*" were the memorable words of the Greeks. Though this statement is accurate, it does not reflect completely the extent to which the *polis* provided a locus of orientation. In a sense, it functioned as a sphere of political placing which was analogous to the *chora*, the Platonic receptacle and placer or superlocator.[91] Heidegger indicates, too, that the Greeks did not even have a word for "space" because they experienced the spatial on the basis of place (*topos*) and not as extension.[92] In his understanding, *chora* "signifies neither place nor space but that which is occupied by what stands there." He speculates that *chora* is "that which abstracts itself from every particular, that which withdraws, and in such a way precisely admits and 'makes place' for something else." Elsewhere he writes, "spaces receive their being from locations and not from 'space.' "[93]

Finally, in addition to the critical comments on Arendt's thinking about nature and the earth offered earlier, several others might be advanced. First, while Arendt delineates public and private space and defines each in relation to the other, she ignores the importance of *natural* space (or place) and its political or prepolitical dimensions. Natural spaces and places might be defined loosely as those areas generally outside or between the *oikos* and the *polis* which are neither strictly public nor private, but which often ground, embed, or even enclose the agonistic *and* cooperative political spaces, a phenomenon which can be grasped by observing the way that different peoples have settled by and been shaped by the surrounding environment. These areas include the woods, rivers, seas, caves, and sky but also parks, lakes, and beaches where human-human and human-nonhuman contact is frequent. In these spaces and places, there are flourishing communities which merit the respect of humans who enter and often disrupt or destroy them. Thus, while it is right to say that a forest or field can be a public space when "action in concert" occurs there,[94] it is important to recognize that they are already complex and diverse locations which offer us more than "raw materials," "resources," or a *res publica*, even if they are by nature nonpolitical (though the disposition we have toward such places is often very political).

Second, it must be borne in mind that Arendt writes from a strongly urban, intellectual perspective that sometimes emphasizes cosmopolitan values and which is frequently very distant from a more intimate understanding of the earth and natural environment.[95] Coupled with this orientation is an extremely strong attachment to the Greek tradition which occasionally blinds her to important anthropological insights into property, wealth, or art, for example, which might call into question her lapses into cultural conservatism. Finally, while Arendt is able to raise important questions, she is unable to provide many practical solutions concerning earth and world alienation and, given contemporary problems, those at which she hints may not appear to us to be sufficiently radical. In this regard, the nascent schools of thought and the corresponding social movements in bioregionalism, socialist ecology, ecological feminism, and social ecology have something to offer.[96] In a day when much of humanity is more concerned with exploring "outer space" than recovering our own sense of earthly place, this kind of thinking can help us to reconceive our relationship with nature and the world and enable us to establish a new, more participatory and cooperative perspective on the natural home we inhabit. In the end, it might be said that Arendt did not so much initiate a dialogue with the earth *itself* as she suggests in "Understanding and Politics" (and as the Orphic Rilke did in the *Duino Elegies*), but rather kindled a valuable inquiry into the origins and meaning of our changing relation to and transformation of our given home. The task falls to the rest of us to continue and to deepen this discussion.[97]

Acknowledgments

I would like to thank Seyla Benhabib and John Ely for their comments on an earlier version of this essay.

Notes

1. Quoted in Elisabeth Young-Bruehl, *Hannah Arendt: For Love of the World* (New Haven, Conn.: Yale University Press, 1982), p. xv.

2. Hannah Arendt, *The Life of the Mind* (New York: Harcourt Brace Jovanoch, 1978), p. 199.

3. Arendt, *The Human Condition* (Chicago: University of Chicago Press, 1958), p. 1.

4. Ibid., pp. 2, 6.

5. Ibid., p. 250.

6. For an interesting discussion of the role of the telescope in scientific practice and its impact on a new epistemology, see Paul Feyerabend, *Against Method: Outline of an Anarchistic Theory of Knowledge* (London: Verso, 1975), especially chaps. 9–11. In line with Arendt, Feyerabend argues: "The first telescopic observations of the sky are indistinct, indeterminate, contradictory and in conflict with what everyone can see with

his unaided eyes. And the only theory that could have helped to separate telescopic illusions from veridical phenomena was refuted by simple tests" (p. 121).

7. Arendt, *The Human Condition*, p. 290.

8. In his search for rock-solid certainty and an Archimedean *punctum firmum*, Descartes loses—of all things—his footing. He finds wandering (*aberrare*) to be a form of erring, that is, a deviation from truth. Yet it is precisely his wandering ruminations and wondering ramblings (his methodical questioning quest) which allow him to constitute the *cogito* and to affirm his own awareness, which in turn settles his search. His meditations rely upon a *methodos*, or literally a path or way, which guides, coordinates, and eventually places him securely in the world, something which could not be accomplished by pure thought alone, which is no-where and belongs to no-body.

9. Ibid., p. 262, emphasis added.

10. Arendt, "The Conquest of Space and the Stature of Man," in *Between Past and Future: Eight Exercises in Political Thought* (New York: Viking Press, 1968). In a more literal and phenomenological sense, we live not *on* the earth as we are wont to think, but *within* it, since the earth includes the heavens (or sky and atmosphere) as well as the soil and the sea, a point brought out by the Gaia hypothesis and suggested in Merleau-Ponty's later writings. See David Abram, "Merleau-Ponty and the Voice of the Earth," Chapter Four, this volume. Compare John Muir's journal entry of 16 July 1890: "Most people are *on* the world, not in it—have no conscious sympathy or relationship to anything about them—undiffused, separate, and rigidly alone like marbles of polished stone, touching but separate."

11. Quoted in Daniel J. Boorstin, *The Discoverers* (New York: Vintage Books, 1985), p. 327.

12. Victor Hugo, *Les Misérables*, in *The Novels of Victor Hugo*, trans. J. Beckwith (Philadelphia: George Barrie, 1892).

13. Marx distinguished four kinds of alienation in the labor process: (1) alienation from *nature* and the sensuous external world expressed in the relation with the product of one's labor; (2) estrangement from *self* and one's activity; (3) alienation from the human *species-being*; and (4) estrangement from other humans. Arendt's thinking cuts across some of these distinctions, though she probably would not acknowledge Marx's third sense since she is skeptical toward essentializing notions of human nature. There are also certain vague similarities to Heidegger's use of *unheimlich* (uncanny, or literally not-at-home) and *fremd* (alien or strange) in *Being and Time*.

14. Arendt, *The Origins of Totalitarianism* (New York: Meridian, 1960), p. 301, emphasis added.

15. George Kateb, *Hannah Arendt: Politics, Conscience, Evil* (Totowa, N.J.: Rowman and Allanheld, 1983), p. 158.

16. For example, at an address delivered in his hometown of Messkirch in 1955, Heidegger begins by speaking of homeland and asks, "Does not the flourishing of any genuine work depend upon its rootedness in the soul of a homeland?" In his "Letter on Humanism," he suggests that "Homelessness . . . consists in the abandonment of Being by beings," and in "Building Dwelling Thinking," he raises the question, "What if man's homelessness consisted in this, that man still does not even think of the real plight of dwelling as the plight?"

17. Edmund Husserl, "Foundational Investigations of the Phenomenological Origin of the Spaciality of Nature," in *Husserl: Shorter Works*, ed. Peter McCormick and Frederick Elliston (Notre Dame, Ind.: University of Notre Dame Press, 1981); Martin Heidegger, *On the Way to Language*, trans. Peter D. Hertz (New York: Harper and Row, 1971), p. 98.

18. Karl Marx, *The Economic and Philosophic Manuscripts of 1844*, in *The Marx-Engels Reader, 2d ed.*, ed. Robert Tucker (New York: W. W. Norton and Company, 1978), p. 75.

19. Arendt, "Existenz Philosophy," *Partisan Review* 81 (Winter 1946): p. 51.

20. Arendt, *The Human Condition*, p. 10, emphasis added.

21. Ibid., p. 251.

22. See Martin Heidegger, "The Age of the World Picture," in *The Question Concerning Technology*, ed. William Lovitt (New York: Harper and Row, 1977).

23. Alexandre Koyré, *From the Closed World to the Infinite Universe* (Baltimore: John Hopkins Press, 1957), p. 2.

24. Alfred North Whitehead, *Science and Modern World* (New York: The Free Press, 1925), p. 88. Whitehead remarks that "the nature-poetry of the romantic revival was a protest on behalf of the organic view of nature, and also a protest against the exclusion of value from the essence of matter of fact" (p. 94).

25. See especially Whitehead's *Process and Reality* (London: Free Press, 1978).

26. It is interesting to note that the average, motorized city dweller's speed is about twice that of a pedestrian's, but that when the social time necessary to produce the means of transport is added to time spent in transit, the average global traveling speed of modern man is less than that of Paleolithic people. For critical thoughts on the automobile, see Kenneth P. Cantor, "Warning: The Automobile is Dangerous to Earth, Air, Fire, Water, Mind and Body," in *The Environmental Handbook*, ed. Garrett De Bell (New York, Ballantine Books, 1970); "Aberration: The Automobile," *The Fifth Estate* 21, no. 2 (1987); and Paul Goodman, "Banning Cars from Manhattan," in *Utopian Essays and Practical Proposals* (New York: Vintage Books, 1962). For a defense of walking as an ecological practice, see David Macauley, "A Few Foot Notes on Walking," in *The Trumpeter* 10, no. 1 (Winter, 1993): 14–16.

27. Martin Heidegger, "The Thing," in *Poetry, Language, Thought*, trans. Albert Hofstadter (New York: Harper and Row, 1971), pp. 166, 165.

28. Heidegger, "The Origin of the Work of Art," in *Poetry, Language, Thought*, p. 47.

29. Ibid., p. 47.

30. Ibid., p. 34.

31. Arendt, *Between Past and Future*, pp. 42, 213.

32. In the later works of Husserl, Heidegger and Merleau-Ponty, one can discern a gradual transition from themes related to "world" to those concerned with "earth" and the construction of various ontological and epistemological bridges between the two realms, such as phenomenal field, landscape, home-world, basis body, place and dwelling. The connections between phenomenological and ecological thought are just beginning to be explored. See Neil Evernden, *The Natural Alien* (Toronto: University of Toronto Press, 1985); Michael Zimmerman, "Toward a Heideggerean Ethos for Radical Environmentalism," *Environmental Ethics* 5, no. 2 (1982); and Monika Langer, "Merleau-Ponty and Deep Ecology," in *Ontology and Alterity in Merleau-Ponty*, ed. Galen Johnson and Michael Smith (Evanston, Ill.: Northwestern University Press, 1990). On Husserl, see Maurice Merleau-Ponty, "Husserl's Concept of Nature," trans. Drew Leder in *Texts and Dialogues*, ed. Hugh J. Silverman and James Barry Jr. (Atlantic Highlands, N.J.: Humanities Press, 1992).

33. Heidegger remarked, "I was worried when I saw pictures coming from the moon to the earth. . . . The uprooting of man has already taken place. . . . This is no longer the earth on which man lives." Heidegger, "Only a God Can Save Us: *Der Speigel's* Interview with Martin Heidegger," *Philosophy Today* 30 (Winter 1976): p. 277.

34. Yaakov Garb, "Perspective or Escape? Ecofeminist Musings on Contemporary

Earth Imagery," in *Reweaving the World: The Emergence of Ecofeminism*, ed. Irene Diamond and Gloria Orenstein (San Francisco: Sierra Club Books, 1990), p. 270. Garb argues that a "distancing, disengaged, abstract, and literalizing epistemology is quintessentially embodied in the whole Earth image where the visual mode of understanding is applied to the entire planet" (p. 267).

35. McDonald's has recently stooped to the hypocrisy of printing a glossy pamphlet entitled "We-cology," which tries to portray to children that it is an ecologically responsible corporation.

36. For a recent discussion of the commercialization of nature, see L. M. Benton, "Selling the Natural or Selling Out?" *Environmental Ethics* 17, no. 1 (1995): 3–22.

37. My use of the term "earth pornography" is intentionally speculative and provocative, designed to initiate thought about the frequent use of earth imagery by ecoactivists and the media. It is vaguely analogous to Rosalind Coward's use of the term "food pornography" in *Female Desires* (New York: Grove Press, 1985) and stands in contrast to explicitly sexualized *body* imagery. On the one hand, the image of earth as living organism and nurturing mother functioned for many years as a restriction on certain human interventions into nature as, for example, mining (Carolyn Merchant, *The Death of Nature* [New York: Harper and Row, 1979]). On the other hand, "Sex-Typing a gender-free entity also reinscribes an anthropomorphism that *alienates* Earth by trying to render it in our image" (Patrick Murphy, "Sex-Typing the Planet," *Environmental Ethics* 10, no. 2 (1988): 155–169).

38. See also David Macauley, "On Women, Animals and Nature: An Interview with Eco-feminist Susan Griffin," *APA Newsletter on Feminism and Philosophy*, 1990 (3): 116–127; and Macauley, "Echoes to Ecofeminism," *APA Newsletter on Feminism and Philosophy*, 1991 (1): 82–85.

39. William K. Hartmann, "Space Exploration and Environmental Issues," *Environmental Ethics* 6 (Fall 1984): 228.

40. Michael Zimmerman, *Contesting Earth's Future: Radical Ecology and Posmodernity* (Berkeley: University of California Press, 1993), p. 75.

41. Arendt, *The Origins of Totalitarianism*, p. 298.

42. Ibid., pp. 235, 297.

43. Lynn White, "The Historical Roots of Our Ecologic Crisis," *Science* 155: 1203–1207.

44. Arendt, *The Human Condition*, op. cit., p. 139n.

45. See Jeanne Kay, "Concepts of Nature in the Hebrew Bible," *Environmental Ethics* 10, no. 4 (1988): 309–327.

46. Plato, *Critias*, in *Collected Dialogues*, ed. Edith Hamilton and Huntington Cairns (Princeton, N.J.: Bollingen, 1961), p. 1216 [111b]. This denuding of the earth in Greece, as elsewhere, has accelerated greatly in recent years, and now less than one-tenth of the country is covered with trees.

47. For a broader picture of historical attitudes toward nature, see J. Donald Hughes, *Ecology in Ancient Civilizations* (Albuquerque: University of New Mexico Press, 1975).

48. Plato, *Timaeus*, trans. F. M. Cornford (New York: Macmillan Publishing, 1959), p. 102 [81e].

49. Arendt, *The Human Condition*, p. 138.

50. Arendt, *Between Past and Future*, p. 226.

51. Ibid, p. 213.

52. Quoted in John Zerzan *Elements of Refusal* (Seattle, Wash.: Left Bank Books, 1988), pp. 63, 76.

53. Arendt, *The Human Condition*, p. 107.

54. For a wide-ranging critique of this institution, see Zerzan's *Elements of Refusal*, pp. 63–76; and Paul Shepard, *Nature and Madness* (San Francisco: Sierra Club Books, 1982).

55. For some promising attempts to restore a balance between land and culture and to move toward a more ecological practice of agriculture, permaculture, aquaculture, and the like, see Richard Merill, ed., *Radical Agriculture* (New York: New York University Press, 1976); Wendell Berry, *The Unsettling of America: Culture and Agriculture* (New York: Avon Books, 1977); Wes Jackson, *New Roots for Agriculture* (San Francisco: Friends of the Earth, 1980); and Peter Kropotkin, *Fields, Factories and Workshops* (New York: Harper and Row, 1974).

56. Arendt, *Between Past and Future*, pp. 53–54.

57. Max Horkheimer, *Eclipse of Reason* (New York: Continuum, 1974), p. 176.

58. See Murray Bookchin, *The Ecology of Freedom* (Palo Alto, Calif.: Chesire Books, 1982).

59. Arendt, *The Human Condition*, p. 264; Kateb, *Hannah Arendt*, p. 173.

60. Arendt, *The Human Condition*, pp. 85, 112, 7, 304.

61. Arendt excludes rhythm from the realm of work (*Human Condition*, p. 145), ignoring the role it often plays in the activities of writing, creating music, sculpting, or woodworking, for example. The sharp distinction between labor and work which is drawn in terms of life expectancy is also suspect. Her example of the table and bread (*Human Condition*, p. 94) is no longer so revealing, for today we have bread and bread products whose life outlasts that of our throwaway furniture.

62. Karl Marx, *Capital*, vol. 3, in Tucker, *The Marx-Engels Reader*, p. 441.

63. Arendt, *The Human Condition*, p. 100.

64. Arendt, *Between Past and Future*, p. 158n. For a consideration of the central place of play in the human condition, see Johan Huizinga, *Homo Ludens* (Boston: Beacon Press, 1950); for its importance for political thought and action, see Francis Hearn, "Toward a Critical Theory of Play," *Telos* 30 (Winter 1976–1977): 145–160. One of Arendt's few sympathetic comments on the subject, in an otherwise scornful treatment, occurs in her essay, "Thoughts on Politics and Revolution," where she writes of the generation of radical students in the 1960s and their spontaneity: "It turned out that acting is fun. This generation discovered what the 18th century had called 'public happiness,' which means that when man takes part in public life he opens for himself a dimension of human experience that otherwise remains closed to him and that in some way constitutes a part of complete happiness." Arendt, *Crises of the Republic* (New York: Harcourt Brace Jovanovich, 1972), p. 203.

65. Marcuse's fruitful but problematic treatment of this subject is most explicit in his chapter "Nature and Revolution," in *Counterrevolution and Revolt* (Boston: Beacon Press, 1972), though he considers it elsewhere at times.

66. See David Macauley, "Evolution and Revolution: The Ecological Anarchism of Kropotkin and Bookchin," in *Anarchism, Nature, and Society: Critical Perspectives on Social Ecology*, ed. Andrew Light (New York: Guilford Press, in press).

67. Arendt, *Between Past and Future*, pp. 171, 150.

68. Arendt, *The Human Condition*, pp. 120–121.

69. On the controversial relation between certain strains of ecological thought and fascism, see Anna Bramwell, *Ecology in the 20th Century: A History* (New Haven, Conn.: Yale University Press, 1989); Robert Pois, *National Socialism and the Religion of Nature* (London: Croom Helm Publishers, 1986). For a related discussion of Heidegger, see Michael Zimmerman, "Martin Heidegger: Antinaturalistic Critic of Technological Modernity" (Chapter Three, this volume); David Macauley, "Greening Philosophy and Democratizing Ecology" (Introduction, this volume).

70. Arendt, "Ideology and Terror, a Novel Form of Government," *Review of Politics* 15 (1953): p. 303.

71. For an interesting discussion of primitive societies and the politically relevant insights we might glean from them, see Stanley Diamond *In Search of the Primitive: A Critique of Civilization* (New Brunswick, N.J.: Transaction Books, 1974). Diamond has carved out a small field in what he terms dialectical anthropology, a unique blend of critical social theory and anthropology.

72. Arendt, *The Human Condition*, p. 152.

73. Arendt, *Between Past and Future*, p. 62.

74. Arendt, *The Human Condition*, p. 296.

75. In contrast to earlier historical *representations* of nature, a postmodern view appears to conceive of nature in nonanalogical, informational terms.

76. Arendt, *The Human Condition*, p. 148.

77. Ibid., p. 2.

78. At a meeting of the National Institutes of Health Genetic Engineering Committee in 1985, for example, scientists were describing the horse as a temporary encasement for the genetic information housed in the animal because at one point in history no horses existed and, in their view, at some future point horses will not exist. They saw the horse as a set of programs or instructions that operates through negative feedback to maintain homeostasis.

79. Arendt, *The Human Condition*, p. 3.

80. For an examination of the conjuction of *techne* with *politeia*, see Langdon Winner, *Autonomous Technology* (Cambridge, Mass.: The MIT Press, 1977); and *The Whale and the Reactor* (Chicago: University of Chicago Press, 1986). One should note the surprising absence of a discussion of technology in Arendt's *Origins of Totalitarianism* and its relation to or role in the Holocaust. For the only study focused on Arendt's perspective on science and technology of which I know, see Peter Tijmes, "The Archimedean Point and Eccentricity: Hanah Arendt's Philosophy of Science and Technology," *Inquiry* 35, nos. 3/4 (December 1992). Tijmes offers an interesting but, I believe, untenable treatment of Arendt on world alienation and related matters, claiming with Helmuth Plessner that "alienation is not a phenomenon of modernity, but a constitutional feature of human beings" (p. 401). Even if such a view has merit, it downplays the qualitative changes in and historical heightening of this phenomenon brought about or accentuated by the role of economic arrangements, modern technology and scientific developments.

81. Arendt, *The Human Condition*, p. 134.

82. Ibid., p. 155.

83. Arendt, "On Hannah Arendt," in *Hannah Arendt: The Recovery of the Public World*, ed. Melvyn Hill (New York: St. Martin's Press, 1979), pp. 334–335.

84. Arendt, *The Human Condition*, p. 148. Arendt takes a step further in *The Origins of Totalitarianism* to claim provocatively that "Nature has been mastered" (p. 302). For another view of Arendt's critique of productivism in relation to green political theory (which appeared too late to consider here), see Kerry Whiteside, "Hannah Arendt and Ecological Politics," *Environmental Ethics* 16, no. 4 (1994): 339–358.

85. The Frankfurt School and its heirs' image of nature and the natural world is a mixed and ambivalent one, which begins with a penetrating but problematic contribution by Adorno and Horkheimer, moves to the utopian but technologically optomistic views of Marcuse and, trails into the contemporary anthropocentrism of Habermas's communicative ethics. For a consideration of Marcuse's and Habermas's views, see C. Fred Alford, *Science and the Revenge of Nature: Marcuse and Habermas* (Gainesville: University Presses of Florida, 1985). For a perspective on Arendt and critical theory, see

Gerard P. Heather and Mathew Stolz, "Hannah Arendt and the Problem of Critical Theory," *The Journal of Politics* 41 (1979): 2–22.

86. A distinction can be made between "communitarians" and "civic republicans" in terms of the types of community, kinds of identity, and conceptions of the political to which they turn. Two of the most cited works of the new communitarians are Michael Sandel, *Liberalism and the Limits of Justice* (Cambridge, England: Cambridge University Press, 1982); and Alasdair MacIntyre, *After Virtue* (Notre Dame, Ind.: University of Notre Dame Press, 1981).

87. For example, J. J. Gibson claims that "places are ecological layouts." "Where abstract space consists of points, ecological space consists of places—locations or positions." *The Ecological Approach to Visual Perception* (Hillsdale, N.J.: Lawrence Erlbaum Associates, 1986), pp. 200, 65.

88. There is a growing literature which makes connections between place and ecophilosophical ideas. See Gary Snyder, *The Practice of the Wild* (San Francisco: North Point Press, 1990); Paul Shepard, *Nature and Madness*; and Edward Casey, *Getting Back Into Place* (Bloomington: Indiana University Press, 1993). Interestingly, Charles Molesworth has found a basis for comparing Arendt's *The Human Condition* with the ecopoetic and political views of Gary Snyder (author of *Earth Household*, *Turtle Island*, *Riprap*, and other works) in terms of their distinctions between earth and world, use of placial vocabularies, and location of an Archimedean point. See Molesworth, *Gary Snyder's Vision: Poetry and The Real Work* (Columbia: University of Missouri Press, 1983), pp. 123–128.

89. Arendt quotes Aristotle as saying that "wherever on earth somebody devotes himself to thinking, he will attain the truth everywhere as though it were present." She comments that "philosphers love this 'nowhere' as though it were a country (*philochorein*) and they desire to let all other activities go for the sake of *scholazein* (doing nothing, as we would say) because of the sweetness inherent in thinking or philosophizing itself." (Arendt, *The Life of the Mind*, p. 200.)

90. Arendt, *The Human Condition*, p. 194.

91. Plato's receptacle is that "in which" physical things are constituted from forms, unlike Aristotle's substrate as that "out of which" (*ex hou*) they come.

92. For Aristotle, to be is to be somewhere, to be emplaced or seeking to return to one's natural place (*topos oikeios*), as it could be said of water qua rain or the other elements. This view was so powerful that it held sway for two thousands years until it was *re-placed* by Newton's law of universal gravitation, which does not strictly explain an experience or locate a force (that can be seen, heard, or touched), but simply states its effects.

93. Martin Heidegger, *An Introduction to Metaphysics*, trans. Ralph Mannheim (Garden City, N.Y.: Harper and Row, 1966), p. 66; and "Building Dwelling Thinking," *Poetry, Language, Thought*, p. 154.

94. Seyla Benhabib, "Hannah Arendt und die erlösende Kraft des Erzählens," in *Zivilisationsbruch: Denken nach Auschwitz*, ed. Dan Diner, (Frankfurt: Fischer Verlag, 1988), pp. 150–175.

95. For an interesting and provocative exchange on the subject of Jews and their historical relationship to nature, a topic which bears at least peripherally on Arendt's discussion of earth alienation, see Steven Schwarzschild, "The Unnatural Jew," *Environmental Ethics* 6 (Winter 1984): 347–362; and the reply by Jeanne Kay, "Comments on the Unnatural Jew," *Environmental Ethics* 7 (Summer 1985): 189–191.

96. One possible response, among others, to the problem of dwelling and earth alienation is bioregionalism. According to Kirkpatrick Sale, the bioregion is an area "defined by its life forms, its topography and its biota, rather than by human dictates; a

region governed by nature, not legislature." Bioregions are distinguished by particular kinds of soils, flora, climate, water, landforms, and fauna and can be subdivided further into ecoregions, georegions, or morphoregions which fit into each other like Chinese boxes. Such regions may be as vast as the Ozark Plateau or Sonoran Desert, which covers about 100,000 square miles, or as specific as the Connecticut River Basin. See Kirkpatrick Sale, *Dwellers in the Land* (San Francisco: Sierra Club Books, 1985); and for a discussion of bioregionalism and its narrative dimensions, see Jim Cheney, "Postmodern Environmental Ethics: Ethics as Bioregional Narrative," *Environmental Ethics* 11 (Summer 1989): 117–134.

97. For further thoughts on Arendt in relation to nature, animality, and ecological theory, see David Macauley, "From Greek Ideas to Green Ideals: Hannah Arendt and Eco-Politics," unpublished manuscript. Interestingly and on a final optimistic note, the Clinton Administration has just announced that it plans to use some of its Cold War spy satellites as ecological eyes. These heavenly "technobodies" will mind nature and such earthly phenomena as glaciers, deserts, tropical rain forests, sea ice, and clouds in order to gather information about long-term global climactic changes and ecological problems. See William Broad, "U.S. Will Deploy Its Spy Satellites on Nature Mission," *New York Times*, 27 November 1995, pp. A1, B5.

Chapter Six

Ernst Bloch, Natural Rights, and the Greens

John Ely

Ernst Bloch: Between Neo-Aristotelian Political Thinking and Marxism

To establish the claim that Ernst Bloch is the only Marxist political thinker, I invoke three names in twentieth-century German political thought: Carl Schmitt, Leo Strauss, and Hannah Arendt. The first is the single greatest German political thinker of this century, though a conservative antirepublican. The other two develop their political theory, Aristotelian in tenor, once they leave Germany for the United States. Students of Strauss and Arendt have widespread influence in political theory while Marxism, like "bourgeois sociology," has avoided systems based on a concept of the good or the ideal and focused on a *functional analysis* of the *state* instead, an analysis expressly nonnormative in character. This is not unlike the criticism that Strauss and Arendt make of Marxism, along with political "science" in general.

Bearing Strauss and Arendt in mind, political theory can and ought to be distinguished from political science and sociology, including Marxist political science and sociology. Beyond emphasis on historical depth and hermeneutic integrity, political theory in Arendt's or Strauss's sense contrasts with political science because it not only describes political life, but asks the question what form of polity best provides a rich, flourishing human life. In one form or another, it invokes a teleological framework. Political theory in this sense does not treat premodern forms of thinking as antiquated, passed by, or obsolete. It supports practical measures that draw on such premodern, precapialist forms of thought and that preserve or retrieve institutional forms dealt with by such thinking that are old or lost.

In contrast to political science, which tends to conceive itself as a "science" differentiated from ethics, practical theory maintains a deep and abiding relationship to philosophy. Indeed, political theory and philosophy emerged together as considerations of political life brought forth the need for philosophy.[1] The question, what is a "polity," "state," or the "political," cannot be answered sensibly without simultaneously asking what is the good polity and the virtuous form of public life. The *locus classicus* of political theory in this sense is Aristotle.[2]

Despite strong differences, both Arendt and Strauss thought politics in an "Aristotelian" manner, to use a disputed term by second-generation critical theorist, Jürgen Habermas. Together, they and their students constitute one of the most significant gravitational poles in the revival of what is called "practical philosophy" or "neo-Aristotelian" political thinking, though one can include others in this tendency (Alisdair MacIntyre, Charles Taylor, William Gallston).[3] The core of this tendency in practical philosophy consists in arguments about the unity of morals and institutions, ethics and "ethos."[4] Arendt, Strauss, and their followers, like the contemporary followers of Aristotle, emphasize a way of thinking about politics in which ethics, custom, and law are based in everyday customs, habits, practices, and intentionalities, that is, in a *way of life*. They look critically at conventionalism, utilitarianism, theories of social relativism, and positivism in jurisprudence and the human sciences. For them, no sharp distinction need be made between the morally right and the legally required. In one form or another, they are advocates of a doctrine of natural law, right, or justice with classical roots.[5] Common features of this neo-Aristotelian tendency include claims of intellectual and ethical solidity as against relativism. Ethics or *Sitten*, they argue, are not only embedded in the forms of life but also in our understanding of the world, indeed connected as well in the nature of things: morality interrelated with cosmic piety. Such ontological requirements, raised by contemporary neo-Aristotelians in terms of the problem of ecology, are increasingly relevant today.[6]

Finally, despite a largely classical emphasis on embedded institutions, personality, reason, and limit, the neo-Aristotelian perspective in political theory is "romantic" in the sense that, in greater or lesser degrees, it is critical of modernity. It is critical of morals based on life, necessity, and survival rather than the ethics of a good life, and it seeks to base freedom and happiness on practice rather than utility, possession, and consumption. The return to forms of life and associated forms of thinking is evident in the use of words like "retrieval," "recovery," "recollection," "resurrection," "rebirth," "revival," and finding that which has been "lost" or "buried."[7]

In this sense, neo-Aristotelianism fits the rubric of what Georg Lukács, in a famous Marxist critique of expressionism, called a "romantic anti-capitalist" perspective,[8] namely, a perspective criticizing capitalist modernity from the standpoint of (idealized) precapitalist forms of life.[9] Lukács also articulated this position in an unpublished review (written in Moscow in 1935) of a book by an old friend of his, *Heritage of Our Times* by Ernst Bloch.[10] Lukács, one of the premier Marxist thinkers of the twentieth century, criticized Bloch for attempting to combine his romantic anticapitalism with a left-wing, Marxist politics. While Lukács questioned Bloch's Marxism, however, Bloch himself never questioned his own adherence to Marxism; indeed, he defended the policies of the Stalin-led Soviet Communist Party and the international Communist movement throughout the thirties. In 1952, soon after assuming Hans-Georg Gadamer's chair in philosophy at the East German University of Leipzig, he published a long essay defending the idea of an "Aristotelian Left"[11] and he continued working on a long treatise on the idea of a Marxist natural law.

Left and Right Aristotelianism

An Aristotelian perspective constitutes a unified pattern of thought, but its compatibility with Marxism is highly questionable. Without exception, all neo-Aristotelians have criticized, even condemned Marxian conceptions of history as megalomaniacal and boundless, reducing politics to society or history and a *telos* of the good to a conception of history based on expansion of the forces of production. Introducing Ernst Bloch as a contributor to the revival of Aristotelian political theory would appear at first to provoke an impossible burden of contradictions. Bloch was a Marxist. Arendt, Strauss, and those influenced by them, as well as other strands of neo-Aristotelianism, have been resolutely critical of Marxism, which for them is a key facet of the decline of political thinking in the modern world. For Arendt, Marx's work marked the "definite end" of political thinking in the modern world.[12] Strauss, less focused on Marx, regarded him as the proponent of an eschatological conception of history whose vision no longer contained "political society" and which, in terms of this doctrine of history, is one step from Heidegger's nihilism.[13]

Yet such antagonism must not hinder an illuminating connection. Bloch, with similar roots in philosophy and from the same generation of exiled German-Jewish intellectuals as Strauss and Arendt, provides an interesting comparison and contrast to them. Arendt and Strauss were both critics of Marxism and focused upon politics later in life. Bloch was an active Marxist (though acknowledged universally to be an iconoclastic one) and commentator on politics most of his life. Nonetheless, he, too, wrote a major treatise on political theory late in life. Despite the fact that he is not usually dealt with as a political thinker, one can, without flinching, argue that he is *the only major figure in Marxism whose thought developed during the first half of the twentieth century* to have concerned himself directly with political theory. *Natural Right and Human Dignity*, an integral part of his self-constituted system of philosophy,[14] presents natural law thinking as a history of "normative" conceptions of popular justice. Bloch defends this concept in communist form; in contrast to *all* other Marxists, he does not "analyse" the "function" of natural law.[15] He delves into the past history of this idea as part of an attempt to realize its promise in our world. He was, to be sure, first and foremost a philosopher who built in political philosophy as part of his system. His opus is not predominately concerned with political thinking per se. Nonetheless, like Arendt and Strauss, he produced his major work in political thinking in the wake of World War II, and it can be read as a confrontation with problems that events of the first half of this century raised for the possibility of a just, or free and equal, political order.

Bloch's status is apparently anomalous. He is a Marxist advocate of the doctrine of natural law or natural right. He is a Marxist proponent of a "left-Aristotelian" concept of nature, a sophisticated defense of Aristotelian ontology. Thus, Bloch's anomalousness represents at the same time a fundamental ambivalence. He is caught between an Aristotelian philosophy and a Marxian theory of history. On the one hand, he remained heretical in terms of his Aristotelianism within the Marxian tradition. On the other, this Aristotelianism (in many senses overly general, abstract, and institutionally weak) was advanced by him as one compatible with the Marxian conception that the process of history itself provides redemption. In this sense, Bloch himself *embodies* a contradiction, an incompatibility between Marxist historicism and Aristotelian critique. Yet, it is precisely his ambivalence and anomalous status which, amid the tremendous intellectual and political flux regarding contemporary understanding of Marxism and the so-called crisis of the Left,[16] means that his work and perspective help to disrupt worn grooves and calcified categories. By situating Bloch at the crossroads between a collapsing Marxist historicism and the emergence of a left neo-Aristotelianism, he becomes a striking and important predecessor to the Left's new project and form as *green politics*.

Bloch's Left Aristotelianism

Bloch is connected to the project of the new social movements and green politics by neo-Aristotelianism. His left Aristotelianism (or at least, the parameters of a left-Aristotelian project which he offers) emerges from the fringes of history in *two key empty spaces* generated by the ossification of Western Marxism. First, it emerges in the space left by an absent theory of "nature," the "objective" or "substantive rationality" which could provide a practical alternative to the negative dialectic of "subjective" or "instrumental" rationality characterizing the central aporia of critical theory. Second, it offers a substantial theory in the second empty space, namely the absence in critical theory (and in Marxism generally) of a theory of politics. This *deficiency*, to use an Aristotelian term, of a political theory in the "normative sense"—a theory of the institutions providing the good life—is the most serious and tragic characteristic of Marxism and its history of crisis in the twentieth century. Indeed, the crisis of historicism *is* the crisis *not* of the disappearing "revolutionary subject" so typically the focus of "Western" Marxism, but of a *deficiency* in *political theory* in Marxism, including a "Western" Marxism which shared the overall narrative of proletarian revolution.[17] Although critical theory provides an intersection where the crisis of progressive historicism intersects with the deficiency of Marxism in political theory, it offers a substantive theory of neither politics nor reason and nature.

In some respects, Western Marxism is defined by its dissociation from political engagement.[18] From the point of view of "political theory" in the Aristotelian sense of an Arendt or a Strauss, the Western Marxists of Bloch's generation (Lukács, Karl Korsch, Antonio Gramsci, Max Horkheimer, T. W. Adorno, Walter Benjamin, Franz Neumann, Otto Kirchheimer) would be regarded as sidelights in the overall Marxist reduction of politics to history. Their approaches, like the vast majority of Marxisms, have a functional conception of politics without a teleological backing. Marxism, including Western Marxism, never contemplated political theory in breadth or depth. It never sought to explore political theory's central problem: namely, the form and nature of *political life* or, more specifically, the form and nature of a *free and equal political mode of life*.

Of all the deficiencies in Marxist theory relevant in today's world, its failure to address political theory stands out most sharply. Given the claim (to paraphrase John Plamenatz) that German Marxism lost its promise when it became "Russian communism," the failure of Western Marxism becomes particularly acrid when one locates Germany as its geographic locus (from Luxemburg through Lukács and Korsch to critical theory). Germany, a country perpetually short of political theoreticans during the period following Hegel (the Napoleonic period), failed particularly to generate political theory during the first half of the twentieth century. This is precisely the period in which the entire world saw Germany as the strong, massive core of international revolutionary socialism, with the largest communist party in the world and in the Communist International. That the jurist Carl Schmitt is the only major German figure in political theory during this period is strong evidence of the sharp inability of Marxism to generate intelligent discussion of political categories. The looming presence of Schmitt on the right (and not only there!) is evidence of the absence of substantial political theory on the German Left.

While Bloch is not the single missing, unread source of a Marxist political theory, he alone began charting paths where critical theory identified an intersection. His treatment of the history of natural law and his left-Aristotelian philosophy offer a point from which the project of a reconstructed Left must depart as the last fig leaf drops away from the failed project of Marxist emancipation in the twentieth century. Bloch wrote a major historical work in political theory not only because of an archaic, somewhat Thomist outlook, but also because he realized the weakness of twentieth-century Marxism's reliance on the radical ideals of right and democracy associated with the French Revolution. He sought to address this deficiency after the foundation of the German Democratic Republic. While Bloch's philosophical roots are in most respects alien to Marxism (though he seems to have developed an early and passionate love of Hegel, which he passed on to Lukács, among others), for this reason—despite his marginality—his is a particularly good address from which to consider this deficiency in Marxist political theory.

Critical theory's romantic anticapitalism resided above all in a history of the progressive colonization of a substantive rationality of ends (in which "reason" was a "principle inherent in rationality")[19] by an instrumental reason, which progressively replaced ends with means until the capacity to contemplate *telos* evaporated completely. But for critical theory, despite Horkheimer's sympathy for neo-Thomism,[20] the possibility of a reactualized and emancipatory theory of *substantive reason* was caught in the pincers of the bourgeoisie's rationalizing positivism and National Socialism's blood and soil, leaving mere wisps of a mimetic promise in the ethereal reaches of modernist aesthetic practice. Here, Bloch's persistent ontological systemizing offers an alternative path from an intersection, the so-called crossroads of anticapitalism where a concern for a "natural subject" and a pantheistic, alternative vision of modernity arose in several ambiguous versions. Bloch offers a dual lesson, not only for Marxism but also for the revived neo-Aristotelianism of our time.

In contrast to most contemporary practical Aristotelians, Bloch was scholastic in outlook, idiosyncratically metaphysical from his early years in Heidelberg throughout the century. In this sense, his conception of natural right stands upright upon the ideal of a resurrected nature, and ecological rationality as well.[21] Nature itself is the unfinished building site of the unfinished human project, a "co-productive" element with its own "subjectivity." Hence, for him, the problem of natural right cannot be simply extracted from the problem of *ontology* in general.

This point is crucial to the dilemmas of contemporary neo-Aristotelian communitarianism. This tendency agrees that modernity and its disembedded conventionalism leaves us directionless,[22] a view it shares passionately with critical theory. While it argues convincingly that virtue is and must be embedded in a way of life, it is left with an ironically postmodern dilemma: Whose way of life? Refusing to ground virtue and polity in nature, in the world, it becomes de facto a kind of strong ethnos nationalism rather than a project of emancipation. We must choose between these traditions.[23] However, the opposition of Aristotelianism to conventionalism, to systems theory, indeed to all forms of goalless functionalism, leads to one of its primary aporias. To offer an Aristotelian or natural-right approach to understanding the dynamic of modern society (as, in different ways, Taylor or Dworkin do)[24] is to beg the question what modern society's specific qualities and virtues are, and whether they are desirable (whether they are, indeed, virtues). The strong communitarian Aristotelianism of MacIntyre agrees with Taylor and virtually all other Aristotelian political thinkers in distinguishing between a salvageable ethics and politics, and an outdated or mistaken ontological basis (in MacIntyre's phrase, "metaphysical biology"). One must take Aristotle's ethics without his science of nature and being in a more general sense, a doctrine incompatible with the "virtues" of modern society.

In this context, Bloch's position as a notable voice in the left-wing philosophy of nature is illuminating because he raises the issue of "nature" in Aristotle while calling the modern mechanical paradigm into question. Due to its refusal to take metaphysics or ontology seriously, neo-Aristotelianism without a "metaphysical biology" is largely incapable of generating universal standards for ethics and polity, becoming essentially a pluralistic relativism of differing kinds of communities and ways of life, each with its own traditions. It loses any legitimate universal claims (claims, one might add, central to the development of humanism as a specifically civic phenomenon) as well as the notion of philosophical reason as a universal quality of the "political animal."

Bloch's great contribution to an Aristotelian project is his insistence on "metaphysical biology." As he once put it, dialectical materialism must carry with it a theory of "dialectical matter." In this sense, his philosophy runs as counter to the mainstream of neo-Aristotelian political theory as it does to the historicist principles of Western Marxism. Indeed, Marxism has always been weak or even theoretically incoherent with respect to nature, which tends to veil the question whether it can or does ground itself ontologically.[25] Bloch, however, was a fellow traveler, even a member of that tradition of German natural philosophy (Jacob Boehme, Paracelsus, "Doctor Faust" and Goethe, Friederich Schelling, Hegel, Hans Dreitsch, Nicolai Hartmann, Jonas) which was the most important single tendency in natural philosophy in the modern epoch, having a consistent critique of mechanistic and nonteleological concepts of the order of natural existence. Bloch himself characterized this tradition, in a typically anachronistic manner, as "left Aristotelianism," a tradition of materialist teleology or matter-primary metaphysics which stems from the debates between the science-oriented Islamic Aristotelians in the twelfth century and the authoritarian Christian Aristotelianism of Europe (the feudal ideology of Thomas Aquinas). Bloch saw this former, "leftist," pantheist materialism developing from Islamic Aristotelianism into early Renaissance radicalism, especially in the work of Giordano Bruno but also in Paracelus and Boehme. Left Aristotelianism was a "process philosophy," a theory of material development with a "utopian"-imaginary facet of what Bloch called a theory of "being-in-possibility," but this "transcending" facet of a teleological system was always firmly rooted in or concretely mediated by real political circumstances. Bloch's reading of the history of natural law emphasized an inhering historical connection between radical democratic and populist movements, on the one hand, and pantheist rather than transcendent doctrines of spirit on the other (such as Gnostic heresy, especially in the medieval Mediterranean, Catharism, freemasonry, Anabaptism, and Winstanley).[26]

Ernst Bloch, proponent and critic of Marxian historicism, needs to be read in the force field constituted by the trianglar impulses between history, nature, and politics. In this tension, the points are Western Marxist historicism, neo-Aristotelian political theory, and a philosophy of nature that defends *substantive rationality*.[27] In this respect, Bloch is a window on the crisis of Germany, the crisis of the German Left, and the crisis of socialist theory in the first half of the twentieth century—a place from which one can view the crisis of Marxian *historicism* (including its Western Marxist variants). In various forms this sort of historicism provided the theoretical orientation for Marxism. These largely overlapping crises can be read as a crisis of absence—that of *political* theory, on the one hand, and from Bloch's metaphysical perspective, that of a thoughtful investigation of substantive reason. In this respect, he forms a crucial link between the problem of a deficiency of political theory in Marxism and a philosophy of ecological rationality.

Five Premises in Bloch's Treatment of Natural Right

The centerpiece of Bloch's metaphysical and Aristotelian position is a dialectic between *natural* law, which always has a "primal," "original" referent, and *social* utopia, which looks forward. This dialectic is elaborated most extensively in Bloch's treatise on natural law, though it was written in conjunction with a larger utopian-historical treatise, *The Principle of Hope*, and a treatise on natural philosophy, called *The Problem of Materialism*. Together, they constitute the components of Bloch's political, social-historical, and natural philosophy. Though only suggestive of a portion of the wide historical scope of Bloch's treatise on nature law, let me distill five of his central thoughts, sketching out some of the major principles combining these aspects of his philosophy.

(1) One can begin by quoting the first sentence of one of the aphorisms with which he opens his books, an aphorism entitled, "That so-called sense of justice." "One who has lost," Bloch writes, "does not want to admit it," that is, feels the "justice" of the one in charge, who has decided his case, to be injustice. In this old circumstance, we find a common experience. "It is precisely in those who are not crooks," Bloch continues, "but who are merely economically weaker and disadvantaged by others in matters of power that we find an ancient,

almost proverbial mistrust of the courts."[28] Bloch's point can be stated more generally. If the history of humanity is one of class exploitation and domination by institutions of organized coercion, then the exploited and the oppressed (and their defenders) will not refer to *positive* but to *natural* law in their struggle for a just social order and human dignity. For socialist political thought, the history of class societies and states is also one of oppositional images, hopes, fragments of speeches, occasionally even doctrines of natural rights.

(2) The second principle runs against the grain of many approaches, recent or past, left or right, Aristotelian or otherwise, to invoke natural-right thinking and classical political theory against functionalist tendencies of modernity and positivist political science and jurisprudence. For Bloch, invocation of natural law does not mean assertion of a premodern, aristocratic ethos or even a premodern institutional matrix against a modern one. As he reads it, natural law is a countertradition of margins and fringes, of intellectuals thinking and struggling for the rights of those who do not have their rights. In contrast to Strauss's enthronment of Plato as the "discoverer" of natural law, Bloch emphasizes the role of the Sophists as the first opponents of both tradition and "enactment."[29] Like Ernest Barker or Sheldon Wolin, Bloch sees the rise of ideas of natural justice lying in the civic vitality that produced rhetoric as a profession. The Sophists, not Plato, introduced *physis* (nature) into Athenian political discourse.

(3) Natural law is solid and ontological. It is a primal, fundamental, and constant yearning of the exploited and subjugated for dignity. It is the freedom and training to "walk upright" and a claim for the just order. It thus appears as a practicable doctrine or idea in creative, cooperative relation to a forward-looking moment in human yearning and the human imagination, that which needs to be attained and must be reached for rather than retrieved: social utopia. His frequently misunderstood study of natural law, he writes, rests on a messianic concept similar to that of Benjamin—a dialectic between a primordial ideal state or divine power (natural justice, Eden) and the idea of recovery as a "leap" or "burst" into the present. The "just price" is an old idea, he tells us, of associated producers, yet a "new" and modern one, indeed, one not-yet-achieved, utopian. Here, Bloch's attitude is comparable to that of Arendt, who emphasized thinking "in the gap between the past and the future," who smiled warmly at actors in the age of the democratic revolution who "ransacked the past for its republicanism" in their search for models for new, free institutions.[30] This very real historical dynamic repeats itself again and again in times of reform and revolution, though it is too often regarded condecendingly as "backwardness," lack of clarity, or millenialism.

(4) An historical critique of the ideological role of ideas of nature and natural law must always be kept firmly in mind. Bloch is one of the most stimulating practicioners of such a critique. For example, consider predictions of a world-historical "turning point" or "paradigm shift" from a mechanical

world-view, such as can be found in the influential popular writings of Fritjof Capra. Bloch would argue that such predictions are premature so long as the modern "world system" (to use Immanuel Wallerstein's term), that is, the constantly expanding law of value underlying the mechanical interpretation of nature, continues unchecked. That "nature" is in this sense a social construct, however, does not free us from the project of "constructing" it in a fashion adequate to a world of free and equal human institutions. Nor does it free us from carefully studying the "natural" limits and properties of such institutions. History is not just a "waste bin" of past forms of class domination. In its margins and fringes, it is a storehouse of ideas, images, and memories about such limits and properties, waiting to be ransacked.

(5) In contrast to most current protagonists of neo-Aristotelianism or natural law, Bloch retains concern for nature as a whole, demanding integration of an empirically and theoretically reasonable interpretation of the cosmos in which human experience is embedded. Recall Bloch's anachronistic reading of Averroist "left Aristotelianism." Averroism interpreted or modified Aristotle's doctrine of final causality to emphasize the not-privative development nisus or the inherent developmental inertia of matter. When developed consistently, Bloch argues, such a doctrine begins to undo the authoritarian, ultimately monarchist, epistemology of rule inhering in a doctrine of eternally fixed, final causes. At the same time, an Aristotelian doctrine underscores the goal-directed nature of natural development, the presence of ends in things, and the graded, gradual emergence of mind from nature. On the one hand, such an approach avoids an unreflected rejection of Aristotelian "metaphysical biology" (found in the work, for example, of Arendt or MacIntyre) a general problem of political theory leading to unsophisticated views of ecological issues. On the other hand, in the realm of environmental ethics, it provides a means to avoid the reductionism and tendential antihumanism of a merely "ecocentric" or "biocentric" approach, a point to which I return later.

The center of Bloch's argument about natural law—or better, his central problem—is to question the extent to which natural law can be successfully *historicized*. Stoic natural law, for Bloch, marked an historical tendency to the extent that it justified slavery. In this sense he was a true modern: "classical" natural law meant early modern bourgeois doctrines such as those of Hugo Grotius, Johannes Althusis, Thomas Hobbes, John Locke, Christian Thomasius, and Rousseau. Bloch took account of Christian or Thomist natural law only in a few respects. For the most part, one can agree with Bloch that it served to legitimate the state, as in the doctrine of "relative" natural law (that which existed after the fall of man) in Aquinas.[31] Nonetheless, the transcendent tension between future redemption of body and mind and the misery of this world (including the "kingdom of Ceasar") built in a crucial historical tendency. In this case, Bloch's theory of natural law remains *scholastic* in a key respect; namely, he deals with the "natural law" category in a *religious* manner as well. He elaborates and exaggerates Marx's comparison of law and religion. For Marx, law was at no time merely juridical. It was in the first place a social and political phenomenon with a specific character of state-sanctioned imposition of norms. If, as a totality, such issues were rooted in the last instance in production relations, it is also the case that in the ideological sphere, both law and religion utilized *utopian* images to cover over the missing justice of productive life. Such images contain a distinct "surplus," Bloch argues. With *law*, as with religion, one must be concerned (using Marx's twist on Hegel's phrase), "to pluck the rose from the cross of history."[32]

Bloch focuses particularly on those traditions which emphasized the problem of natural law in terms of *Auftrag* or delegation. This slippery boundary of reducing or minimalizing civic activity to voting for representatives occured, he argues, in the division in bourgeois thought after the French Revolution between the "citoyen" and the "bourgeois." The problem of domination by an elite reappears in the later development, he adds, between "comrade" and "functionary."[33] This underscores the degree to which "function" in the positivist sense of a Comte or Parsons is rooted in legal-bureaucratic systems and is coincident with the *eclipse* of the communicative-democratic conception of "comrade."

Bloch addresses the question of how one *historicizes* natural law in various layers. First, drawing from the tradition of the sociology of knowledge, Bloch observes the manner in which cosmological and ontological categories in the sciences have been framed historically by the conditions out of which they emerge, then reproduce the required relations of hegemony in ideological forms. This historic confrontation between different groups and factions is carried out as well in terms of contesting the interpretation of that nature in which human beings are embedded. By the same token, the *mediation* between differing conceptions of the "state" or "order" of "nature" upon which the social contract is justified, also varies historically.

These historical ligatures must be taken into account in order to generate a theory of natural law with historic subtlety. One must take care, however, in formulating this problem or one will make much too quick a leap into simply using the "path of history," in Hegelian fashion, as an Aristotelian mode of justification.[34] To be sure, from an "Aristotelian" perspective, modern history can be understood rather well in terms of the systemic, functionalist logic of valorization (profit for the sake of profit) intrinsic or built into modernity's central institution, the world market. Its *ethos* is the substance of atomistic individuals compelled by the mechanically experienced and conceived laws of natural motion, assuming a natural and individual hedonism in the "subject" just as do the "laws" of "supply" and "demand" in the economic system. However, this raises the problem of the extent to which an "Aristotelian" manages to find a *virtue* in such an accumulation logic.[35] The Hobbesian, mechanically conceived individual and world do indeed approximate the character of the world market as institution, but from Bloch's perspective, an ethos residing on the fringes—the subaltern view of "right" opposing the order of exploitation—is more important. In this sense, Bloch's historicism is in part a history of the ideas of the oppressed; its "normativity" is a generalized desire, cultivated and politicized by collective practices and movements of subaltern groupings, to end exploitation and domination.

"Modern neo-Aristotelianism," Schnädelbach argues, "replaces what Aristotle conceived in a static manner as a lived-ethos, with 'reason in history.' The central task of a philosophical ethics is then to interpret reason in its practical reality."[36] However, Schnädelbach notes as well that Hegel actually constituted the *end*, not merely the modern form of Aristotelian natural law.[37] The neo-Aristotelians differ from Hegel in that they "forego the guarantees offered by a philosophy of history, yet are thus forced to place even greater trust in accumulated practical experience (i.e., tradition) than would a Hegelian."[38] In the case of Hegel his leftist followers, one sees the negative effect of capital accumulation (and the emergence of the state) on what counts as history. In contrast to Marxists, no neo-Aristotelians unguardedly endorse the logic of modernity as the expression of reason or right in a Hegelian sense of "history." If Hegel and Marx too quickly replace natural with historical law, a left Aristotelian keeps track more sensitively of older anticapitalisms. This makes the *mode* of historicization much more complicated. This *historicization* becomes, rather than a systematic account of alienated power and events, a *search* for tendencies and movements which raised oppositional voices and an elaboration of their potentiality—a look at fringes and nearly lost ideas.

Bloch mediates these problems. On the one hand, his focus on historical materialism allows him to account for capitalist development more than other Aristotelians, despite the fact that he does not have an explicit theory of accumulation in his work.[39] From a left-Aristotelian point of view, however, this mediation can be constructed only as positive utopias *against* the grain of history seen through the eyes of the bourgeoisie, with a thorough skepticism toward Whiggish history and its aggressive confidence that the past is gone, good riddance. This tendency connects into the traditions of subalternity for its conceptions of emancipation and thus draws a different, animated, layered, and fragmented relation to history, one which must be placed together with the more "objective" development largely coincident with the expansion of the forces of production.

Left Aristotelianism as Green Political Theory

In Germany from the mid-seventies through the eighties (i.e., during the period of the emergence of the Greens from the ecology and alternative movements), one observed a kind of vacuum dialectic between left-Schmittianism on the one hand and the rarefied atmosphere of neo-Kantian legal positivism (Habermas and his followers) on the other. This generated a "negative" dialectic in German political history. Left-Schmittian tendencies sought to create a crisis of the "state of exception" in the context of German nuclear and security policies and, in some cases, tendential support of left terrorism.[40] Justification of the monopoly of force or violence on the part of the strong state led to a counterjustification of movement militance.

Given the absence of a solid *republican* alternative to Weber's influential concept of the legal state, the evident dialectic between state violence and movement violence unfolded, a development whose ultimate result was the expansion of the security state and escalation of political violence on the part of both parties competing for legitimacy—veiled by the *shared understanding* of a Hobbesian anthropology of struggle, whether between classes or between state and "scene." On the one hand, Weber's positivist and unitary concept of law is abstract in the sense of a neo-Kantian set of formal rules or a "basic" norm, reemphasized by Habermas in his legal theory. The reduced, formal legal principle, explicitly basing sovereignty in a noncorporative, unembodied politic, creates an ethereal state of legality prone to "deficits" of legitimacy.[41] The ultimate sanctioning of coercion, cause for alarm as Habermas's most recent texts appeared, has been present in Habermas's critique of various new social movments from the 1968 generation onward. A coercive, positive nation-state, combined with formal Kantian legality, provides a constantly reappearing problem of legitimacy.

This vacuum dialectic could be hindered, if not in part negated. First, a creative, substantial dialectic between Greens and "new social movements" with Social Democracy requires the elaboration of a radical democratic version of the civic republic, a rearticulated vision of direct democracy found in the *politics* of the new "social" movements and influencing parts of the Greens. Second, Social Democratic reform needs to be open to such ecological conceptions rather than merely rejecting them as "conservative."

The Greens represent a reemergence of that kind of political formation which the Independent Social Democratic Party represented at the end of World War I, one with a mixture of reformist and radical democratic views and a broad left program tinged with romantic anticapitalism. The Greens' innovative role in reforming the German Left is in large part responsible for the fact that the German Left remains the strongest in global terms.

In this regard, the Greens represent a version of the goal-oriented dialectic between ideal form and real reform which Bloch envisioned for the USPD. Such politics require a strong utopian moment, as he always emphasized. This utopianism is repeated again and again in Green Party statements, speeches, and programs, and it is therefore not surprising that Bloch has influenced the Green Party. From the beginning, through Rudi Dutschke's early support for the Greens[42] and the New Left background of many of its activists, Blochisms have appeared and reappeared in the Party's discourse. Early green activists reproached the historylessness of the party regarding the issue of ecology, and used Bloch's concept of nonsynchronicity to analyse modern "value-conservative" elements in the Green Party orbit.[43] More typically, Blochian noninstrumental dialectic of reason is used as an epigram,[44] or a figure serves to guide a convention, speech, or article, such as "concrete utopia"[45] or "walking upright,"[46] the former as shorthand for the entire spectrum of changes or "reconstruction" which the Greens advance and the latter expressing the value of "civic courage."

The old expressionist dream of a "turning point" or new beginning, an *Aufbruch*, reappears in the Green Party with explicit reference to Bloch. The *Aufbruch* or New Beginning faction used Bloch's notion of natural right as "freedom and order" in their conception of the "democratization of everyday-life," combined with his "Socrates and propaganda" notion of political commitment.[47] The Green Party itself is an experiment that is "not-yet": "Without cultural and political breakthroughs, and new orientations, the Greens would never have arisen. . . . But we can't evade the necessity of 'getting aboard ship' as Ernst Bloch put it. Only insofar as we understand new beginning in this [Blochian] sense, can the Greens' model of a polity achieve something of its wishes."[48]

Bloch is invoked in regard to utopia and revolution and occasionally also to nature, but the Greens have been characterized by general disinterest in theorizing the word ecology.[49] This is accompanied by an absence of interest in a theory of democracy in the sense of *right* and the institutionalization of political counterforms.[50]

The first failure follows from the general absence of republican traditions in Germany, and the second from the problems of a conservative philosophy of nature in German history, though both have been countered by Habermas's interventions against such approaches as well as the way in which the introduction of the idea of "civil society" into German debates actually managed to cloak the issue of counterinstitutions in liberal jurisprudence.[51] Habermas's combined critique of both embodied democracy and the role of natural philosophy in ethical questions is one of the strongest attacks on libertarian ecology. In a "new world order" of untrammelled liberalism, even Marxist thinkers reach for a bourgeois doctrine of rights (based on Kant) as the basis for a theory of "natural relations."[52] In this sense, a left Aristotelianism addresses two issues for green politics.

(1) It raises embodied democracy in the context of a republican theory of city-states, city-state and communal federations, town meetings, syndics, councils, revolutionary committees, and assemblies. Such a history allows for a critique of the nation-state from the perspective of the civic republican, underscoring as Arendt did the inhering nationalist ground generated by the nation-state for the violent ethnic and nationalist politics of the Right.

(2) Green politics requires an account of ethics for ecological questions, one ultimately requiring a vitalist, Aristotelian concept of the gradual emergence of mind from nature within a system of ends. A left-Aristotelian concept of *entellechia* or *process* serves as the only obvious radical democratic basis for that "substantive" or "objective" reason which the Frankfurt School correctly required to overcome the dynamic of instrumental reason (the "dialectic of Englightenment").

Thus, left Aristotelianism addresses both issues of *polity* and *nature*.

Left-Aristotelian Politics as Counterinstitutions

Not surprisingly, civic republican views were not influential among German leftists or Greens until after 1989,[53] even though the moral positions of conservative Thomists affected the Party's rhetorical critique of technology.[54] Consideration of the problem and theory of democracy has not been a focus. Michels's sociology of the Social Democratic Party was out-of-print in Germany in the seventies and eighties, and his analysis of organizations had been replaced with a systems-theory focus on transparency and the question of steering mechanisms.[55] The plebiscitarian and culturally rooted view of American democracy, stressing the ward system, forms of associative richness, and sources of that "solidarity" which has supported liberal conceptions,[56] are not part of the Habermasian reception of American constitutional and procedural law. Not surprisingly, civic republican thinking in order to help or give shape to the new social movements is weaker among the German Left, unlike the situation in the United States today (where fledgling green formations were named "committees of correspondence"). The development of an Aristotelian position on democratic citizenship and natural *entelecheia* is required to give substance to new-social-movement actors. In Germany in particular, Greens tended to reach with Schmittian gloves for a Weberian definition of legal power, upheld by the positivist establishment.

The libertarian tradition of natural law, including left Aristotelianism, is superior given its broader, substantial conception of right. This attitude sees the problem not only in terms of legal institutions (e.g., the courts, lawyers, executive bureaucracy and parliament), but also in terms of "legal" constitutional *associations* in general, that is, the multiplicity of civic fellowships and groups.[57] Gierke further underscores the role of the organized plurality of institutional forms *between* the individual and anonymous forms of society (its market, the state, and the ever-present threat of compulsion). Gierke provides the institutional concept to establish solidarity as organized comradeship, as well the means for rationally developing substantive grounds for opposing fellowship and consociation to the formally rational "fictive personalities" of the corporation. As Max Weber did later, Gierke draws a structural contrast between "voluntary" and "compulsory" forms of organization, between *Genossenschaft* and *Anstalt*. A participatory-federalist concept of politics institutionally avoids the anomie-filled world-market *forces* and the bureaucratic apparatus organizing the "system." As a feature of market laws and bureaucratic states, force becomes the key term for *politics* as an oppositional game takes on the character of a confrontation between friend and enemy (a characteristic of both two-class versions of capitalist development and modern [post-Hobbesian, post-Bodinian] notions of the state). Note that in the German Left, the theory of historical materialism and the vacuum dialectic between movements and the state are evidence of the ubiquity of a modern concept of positive law, sovereignty, and "force" based on a Newtonian model. Freedom is based largely on practice, so much so indeed that the term freedom is hardly used by those who have it. Freely participating citizens are precisely that kind of order or rule which *is* the polity in Aristotelian terms, and is therefore characteristic of what is considered as political. "For Aristotle, politics is all that which has to do with power and rule which is at the same time the ethical institutionalization of human existence, and as such has freedom as its content. This makes it, in contrast to other forms of rule, politics."[58]

This contrasts to functionalism. Functionalisms tend towards the "existential" decision rather than judgment; they see only goalless analysis with *no positive teleology* in history and hold no particular trust or "interest" in the goal of communism or an elaborated institutional vision of the just society and its corresponding virtues.[59] The *pessimistic* functionalism of Kirchheimer or Neumann builds an antagonistic Schmittian conception of politics into the functionalist logic determined by a Marxian depiction of accumulation without any utopian surplus whatsoever. Yet, this model constitutes form for Marxist legal theory. Nonnormative analysis of the accumulation process, tied with Weberian or Schmittian definition of the state and the political (legal monopoly of force, friend/enemy distinction) are highly prevalent, not only in West German but in contemporary Marxism in general.[60]

In contrast to modern functionalist theories of the state, the Greens' economic programs are characterized by the moral rhetoric of both utopian and civic republican (Arendt, Wolin) or "alternative" economic (Polanyi) discourses. The Greens' economic programs, both of which were cast as "concrete utopian" proposals, had a left-Aristotelian tone. The Greens' second program for a proposed "reconstruction of industrial society" is framed in terms of polity rather than political economy. In contrast to the functionalist language in which social scientists such as Joachim Hirsch and Andrei Markovits, Claus Offe, or Herbert Kitschelt write about such positions,[61] the Greens themselves use an explicitly noneconomic, nonfunctionalist language. Their "economic program" has a *qualitative* feel reminicent of Polanyi. The range of changes in the nature of society required by ecological "reconstruction" means that this project becomes a largely civic concept of economy and ecology in which industrial decentralization and reduced division of labor play key roles. However, local and federal bodies with stringent requirements of positional goods and social costs serve to reward ecologically responsible processes of production fairly. The model suggests a mix between local and syndical sovereignties as proposed in Gierke's confederalism. "Basis democratic economics" would seem to be based on such largely reformist models,[62] just as "labor as free, self-determined activity" is the goal of labor policy.[63] The "reconstruction" idea is a comprehensive one in which the social state is administered and ordered in civic units, integrating a wider variety of changes in society, economy and science. In this sense, it is conceived as a "concrete, realizable utopia," a "utopia which is oriented on the long-term reconstruction of the economy, changes of a fundamental, radical nature. But this recognition is realizable only when the necessary political will exists." [64]

In terms of ecology and green politics, a left-Aristotelian position addresses two controversial issues in political theory. First, it provides a critical perspective on the claim that our age has seen the "end of the subject," whether this claim stems from Foucault's critique of Kant, from poststructuralist approaches to literary theory, or even (more particularly) Francisco Varela's claim that systems theory as a basis for cognition provides a better alternative.[65] Such claims might encourage further administrationalization of society, as observed decades ago by critical theory. However, the left-Aristotelian perspective notes the *heterogenous* roots of the concept of subject, that is, subject or subordinate as in the sense "king's subject," the original meaning of *sub-iectio*, "thrown down," "submitting," "subordinate." Left Aristotelianism insists on making the *citizen* and not the "*subject*" its central category for acting, participating individuals, as the basis of its version of "autonomy."

Second, left Aristotelianism provides the central category to mediate between an ideal natural law based on original justice and participatory counterinstitutions, and a teleological theory of nature, namely the ideal of symbiosis. Symbiosis, in the confederal theory of association elaborated by Althusius, is the legal category for voluntary associations or *Genossenschaften—ius symbiotica*. Symbiosis serves as image of confederal civic and cooperative political organization as well as a key category in biology. Both libertarian thinkers and alternative biologists advocated symbiosis as a means of understanding evolution, including Peter Kropotkin in *Mutual Aid*, Lynn Margulis in *Symbiosis in Cell Evolution*, and Murray Bookchin in *The Ecology of Freedom*.

Given this counterinstitutional perspective in political theory, the problematic rejection of an embodied concept of popular sovereignty advocated by Habermas comes into relief. Participation occurs of necessity in "bodies" or "assemblies," though the matter dispersion or organization of such bodies is a complex matter for confederalist theory. However, democracy in this sense can hardly be maintained if claims to embodied sovereignty are replaced by one merely "diffused" across the population of a nation-state.[66] Second, such a participatory concept of association opposes the idea of an "identity" within a given nation-state. A civic federalist, dispersed concept of sovereignty rejects the modern form of nation-statist *ascription* of the "subject's" "identity" along with the focus on "primary socialization" in a particular ethnos as the basis for an individual's "identity" or orientation. (In this manner as well, left Aristotelianism looks skeptically on any concept of a "cultural ecology," such as is advocated by members of the German far Right.) The basis for belonging to a group in the associationist theory of Althusius, Gierke, Arendt, and Bookchin lies in the *practical, nonascriptive* categories of public life and meaningful labor.

The Left requires from both sides a more substantive, institutionally elaborated conception of political freedom. Both those affirming Schmitt's theory of politics, transformed by Benjamin into a negative philosophy of history, and those defending a formal universalism fail to offer a reconstructive understanding of the elements and institutions of the good life—one which finally transcends this fatal dialectic, perhaps the most decisive result of Marxism's deficiency in political thinking. Here one can only underscore the view of those, like Offe, who see "red-green" as the project for maintaining the historic project of the Left, given the rise of the new social movements.[67] The critique of Habermas is thus not meant in fundamental hostility to the Social Demcratic tradition, but rather to a Social Democratic Kantianism which defends modernity so aggressively that it can only see antimodernism in those new social movements (above all ecology) leading down the dark path to "particularism," or that sees the harsh hands of Cardinal Ratzinger in "fundamentalisms."[68] Only a social democratic perspective that opens itself up to green politics, recognizing its radical democratic and republican potential, can overcome its own problems. Likewise, it requires a green politics which sees the transformation of the "structural violence" of state and accumulation logic not as some seductive and naive belief in democracy as "the identity of ruler and ruled," or the mythos of the "state of exception." Rather, it needs to emphasize the formation of democratic counterinstitutions, civic virtue, balance and limit, and the "reconstruction" not merely of industrial "society," but the foundation of a new, confederalist and universal association, expanding beyond the grounds of the national state while reinvigorating public life from below.

Left-Aristotelian Nature and Ecology

Left-Aristotelian approaches circumvent the political deficiency expressed in neo-Kantian formalism and unreflective left conceptions of democracy. Left Aristotelianism provides the corresponding capacity to engage theoretically in the problem of ecology. The ecology movement and the Greens in Germany stand today between old choices in political theory painted in new colors. The choice repeats itself in intellectual life (just as in mass culture, all books are immediately made into films). Habermas's procedural approach bypasses all the central points of emerging debate on "nature" by conceiving the entire realm of nature as involving "systemic," technical problems requiring administrative and professional solutions.[69] This approach cannot grasp the scope of the problems in such technological developments as nuclear-power stations and is surprised by both the degree of protest and declining state legitimacy that accompanies such protest. For Habermas, ecological protest—which ultimately most threatens the modern logic of capital accumulation and historically is the most central of the new social movements—is "ambivalent" in its meaning and politics, and antagonistic to socialism.[70] He is unable to distinguish between aesthetic colonization of the life-world and the administrative problems of overcomplexity, seeking to go "behind the modern division of labor."[71]

The inaccessibility of the German state to civic protest[72] causes the German "trauma" of law and violence. The intellectual tendency to draw sharp divisions between parliamentary and extraparliamentary (inherent in Kantian formalism) bares too quickly a Weberian conception of law. At the same time, such an attitude is unable to muster the theoretical requirements for economic, social, political, and scientific solutions to ecological problems. Problems of genetic technology, the political economy of deforestation,[73] the parameters of the politics of "risk," the ozone layer, thermodynamic gradients, and "ecological economics,"[74] the rights of sea lions—all remain opaque to a Habermasian theory of communicative action and procedural legitimacy.[75] In the meantime, real problems of environmental ethics and legislation, which cannot be addressed sensibly without a firm left-Aristotelian basis, go unaddressed.

Here, the Blochian theory of natural subjectivity and the teleologically directed emergence of the world—the developing intersubjective dialectic between the human house and its natural building site—plays a crucial role. As Bookchin, Spaemann, Löw, Hösle, Kummer, and others have emphasized, an ecological ethic without a clear developmental conception of nature and mind always tends to be drawn into a reductionist, biocentric attitude, insofar as it cannot distinguish between degree.[76] Left Aristotelianism maintains the doctrine of a graded emergence of mind from nature, but as part of an open system which is, in Bloch's expression, "not yet." From an ecological perspective, only an *Aristotelian* concept of development, grafted onto evolution conceived symbiotically,[77] provides the basis for possible judgment on ecological issues, even within other moral frameworks such as utilitarianism and procedural (neo-Kantian) theories of animal rights. Bookchin emphasizes Aristotle rather than Hegel here because he detaches an ecological conception of natural *entelechia* from the modern dialectic of history, that driven by the commodity nexus.[78] The ecological understanding of biotic and ecocommunities, their development and maturation, and patterns of stability and negentropic articulation, provides the required framework for making sense of moral systems of animal rights, whether Kantian or utilitarian.[79] Further, general goods (those referring directly not to the logic of the individual's pain or autonomy, but rather to collective groupings, whether an ecosystem or the general welfare of one community) need a development perspective not merely "individually" and ontogenetically, but also collectively and phylogenetically *as well*—in terms of the various forms of ecosystem, community, or standards of maturation. The rationale for preserving a species, a level of atmospheric CO_2, or ecosystemic and species diversity, or the right to an abortion—all require collective basis for judgment. Such standards are required as well to constitute institutional forms strong enough to recognize the globally destructive character of capital accumulation without being seduced by the various forms of conservative and right-wing calls for a stronger state as the solution.[80]

Here, liberal environmentalism is a maze of potential obfuscation. The work of Christopher Stone is illuminating in this respect. In his well-known essay "Should Trees Have Standing?" he emphasized the irrationality in certain extant forms of juridical personality, such as the "intentions" of ships. He argued, however, for the importance of articulating other forms of natural and social forms which could be granted status as real or fictional legal persons.[81] Only a theory of morally and legally grounded collective, including ecologically collective personalities will make it possible for incorporated communities and ecocommunities to oppose the externalization of "social costs" by those other kinds of "fictional" legal persons, multinational corporations.

In subsequent work, Stone opposes the "moral monism" of Kantian and utilitarian positions, which have only human interest or volitional capacity as the basis for moral judgment, to a "moral pluralism" which can give ethical claims to other beings within the context of a "verticalism" of various planes or levels of moral judgment. This moral pluralism, however, is actually a "monist" (i.e. holistic) Aristotelian attitude recognizing the gradual and graded emergence of mind out of nature, but clothed in liberal rhetoric. "Moral pluralism" conceives "moral activities as partitioned into several distinct frameworks each governed by distinct principles and logical texture."[82] This sense of graded or layered development ("verticalism")[83] is important, but Stone leaves in the air the form of ethical reasoning which will provide for the "unification of layers." How is one layer or one corporative group's claims in court to be judged against another? Further, how are they related vis-à-vis the integrating institution of the world-market (about which Stone remains silent)? This institution is crucial to the logic of many kinds of corporations, and provides the systematic capacity to "externalize costs" vis-à-vis other groups and communities. These unanswered questions are no doubt related to Stone's odd use of pluralism; historically, it is not the "pluralist" but the "monist" position which has sought to generate criteria for the unification of layers. Ernst

Haeckel, who coined the word ecology, was also the founder of the "monist federation."

The distinction between a left-Aristotelian conception of nature and a Hegelian-Marxian conception of historical stages of the world-spirit can be grasped only insofar as one has a nonsynchronous, hermetic, and mediated conception of history. One cannot simply reject past systems of knowledge. One must perceive layers of historical reality and potentiality in every present and recognize the tradition of the oppressed, the undercurrent of subaltern views riddling every age's dominant ideologies, the gnostic heresies of very epoch, and the pantheist sentiments appearing as attributes of democratic and plebian movements. Further, left Aristotelianism advocates differentiation and plurality of competencies and sovereignties, but *not* on the basis of a Kantian divide between morality and law or pure and practical reason. Rather, in each sphere of reason (action in various productive, political, intellectual, and aesthetic pursuits), ethical principles must be part of the institutional framework and practices, and concern must be given to how these relate to one another. Differentation in this sense must be sharply opposed to differentiation in the functional-positivist sense of Luhmann in which the ends of differentiation or the ends of the differentiated spheres remain uncontemplated because a framework of ends is explicitly rejected. Differentiation is also different from the Habermasian conception. In his case, it is the very division into morals and laws, "life-world" and "system," which has served as the crucial precondition for setting "free" (*freisetzen*) the goalless logic of capital accumulation and "state-making." An Aristotelian conception of differentiation conceives of the distinction of realms in a manner in which questions of ethics, ends, virtue, and the purposes of institutions and forms of consociation are a concern of all realms of knowledge, its production, and the institutions in which life and political life take place. The liberal idea of rights is too weak to grasp ecosystematic issues beyond the level of mere individual rights, such as thermodynamics or the concept of an ecocommunity.[84]

The best foundation for a *developmental* philosophy of nature remains both historicized and mediated with modern history as an accumulation of power and value without being either reductionist or "teleological" (*telos* in the sense of Hegel's doctrine of destiny, his backward-looking view which *recognizes* historical necessity after the fact but cannot determine it). By historicizing the forms of cosmological speculation and mediating them with subaltern conceptions of natural right, Bloch provides preliminary historical material with which to articulate an emancipatory and ecologically considered philosophy of nature. This investigation begins by recognizing the problem of a "heritage in nature." As Bloch argues, we must take the forms designating those outstanding elements of the great tendencies in the essence of nature and in individual concepts of nature which emerged historically (the primitive, animistic, magical, qualitative) "*beyond ideology*."[85] One has to look beyond the "mechanistic relativism of today, as the decline of physics from the point of view of the object produces a sensitivity for the possibility of earlier absolutes as relative absolutes in nature so that the dialectics of nature would contain not only the mechanistic natural science of yesterday, but also earlier elements, particularly the qualitatively graded elements." [86] In *The Problem of Materialism*, Bloch elaborates this in explicitly ecological terms. One must never forget the "coproductive" aspect of the relation to nature, evident in the old notion of the physicist. "How is it that so much of the physical objectivity or all of it corresponds to the mathematical natural science of capitalist society? Is not the pulse of nature being taken here in a way similar to that in which an ideological pulse would be taken so that the pulse of nature can hardly be counted without being disturbed?"[87] Modern capitalism, like modern science, regards nature instrumentally, making "calculatory and abstract thinking" the "final conclusion of the knowledge of nature." Teleology is overthrown, and the goalless commodity logic dominates the realm of nature as well, a "formalization of the mere functional connections without content and viewpoint," without a "final conclusion."[88]

Bloch's continued attack on bourgeois mechanism is important because while Habermas weakens the Frankfurt School's original critique of instrumental reason vis-à-vis outer nature, Bloch continues the critique of mechanical science that is increasingly central to contemporary ecological questions. But while Bloch recognizes the importance of the mechanical world-view or paradigm, he does not view its development outside the positivism (in the form of analytical philosophy) which still dominates the human and natural sciences[89] or the parameters of the capitalist world market. Such science is dominated by the economistic premises of terms like "laws," "functions," "fitness," "survivorship," and "adaptation."

"Ecology," argues Bloch, becomes "so urgent in our time" because it identifies the point where a qualititive approach to humanity's intercourse with nature requires articulation. In so far as the "symbiosis" between humanity and nature is upset by industrial destruction, a reexcavation of nonmechanist doctrines of nature is important: "the earlier connections to nature, the primordial animistic, the oriental-magical, even the class-qualitative, revealed a theory about objects that is not to be completely rejected . . . , and these connections are not entirely submerged in mere ideology." [90] These are important because the historical understanding of the forms of ideology not only produces a "surplus," argues Bloch, but also because one is still required to develop standards and principles of nature for understanding "scientific realities," and human "value-choices," and the *risks* which interpenetrate them. One needs standards and criteria of ecological judgment and practices, including a modulated conception of nature with its diverse but overlapping *entelechias* and goals, ultimately a universe with ends, a gradient in the emergence of mind from nature, however incomplete or unachieved, embedded within itself.[91] Bloch rightly underscored a materialist interpretation of Aristotle as one compatible with contemporary scientific research stemming from Averröes. His concept of a nonprivative, potential matter, especially in his biological writing, is central to his biological notion of "end," whose "central explanatary concepts are 'nature,' '*phusis*,' and 'potential' (*dunamis*)."[92] Life, for Aristotle, is inherently an affair of realizing potentials for which "necessity" can only be hypothetical.

The left-Aristotelian doctrine of nature and ends converges where the determinate dominance of the "final cause" is rejected, where the potentiality, or the "being-in-potentiality" of material cause becomes more than mere privation. Mind emerges from nature, in gradient, overlapping communities formed by symbiosis and coevolution, with an inhering moment of diversity. Recognition of this irreducible moment of diversity in the form of a developing gradient of complexity is only one of the defects of an individual-oriented natural ethics based on utilitarian or neo-Kantian animal rights. Neither approach is able to give a neutral, instrumental basis to "right" in the form of methodological individualism, and neither can grasp the relation of individual to group, "systems," and "communities" without an embodied notion of sovereignty. In the realm of political metaphysics and the *entellechias* of differing groups and orders of human consociation, as Gierke noted, this privileging of the "material" in this sense is privileging the doctrine of popular sovereignty over that of princely rule. By placing the cause of generation in matter itself, the rule of matter by cause, from without (the model of instrumental rule or domination) is overthrown. Those things by nature are those whose cause or reason of generation lies within themselves. Just as the chair or the slave exists "for the sake" of something else, the tree or the citizen has the cause of their motion within themselves, "for the sake of themselves."[93] Numerous examples of such arguments evidence themselves in ecological debates. The question over standing in environmental law attempts to give place in court to entities based on their good for themselves rather than as grounds for tort to another human while as in the great tradition of natural law, under universal criteria of justice and right, the left-Aristotelian principle recognizes both autonomy and diversity.

Conclusion

Left Aristotelianism and civic republicanism are the crucial form in which the new social movements can play an influential role in providing barriers to capital while maintaining openness to new utopian projections. From the perspective of Bloch, Gierke, Arendt, and Bookchin, there is a need for a coherent conception of public life and direction, supported by a teleologically founded ontology and philosophy of nature. The role of natural law in the history of contesting the status of "nature," as well as the interpretation of natural rights by different interests and class layers. Bloch offers one of the strongest leftist positions from which to consider the problem of ecology, since he above all others built his concept of natural law into a process-oriented, cosmological world picture with clear ecological elements. Taking up the issue of nature in Aristotelian terms (and *not* leaving it to be made visible first by conservative Aristotelians) is unquestionably the best perspective from which to give an ethical foundation to ecological politics. In this respect, arrival at a crossroads of romantic anticapitalism has been reached again by the confrontation of the Left with ecology: Left Aristotelianism is the outlook which serves both socialist and ecological demands.

Acknowledgments

I would like to thank Martin Jay, Robert Meister, Sigrid Müller, John Schaar, Karl-Heinz Wiegand, and Peter Zudeick for comments and help in preparing this essay, though the views expressed here are my own.

Notes

1. See for example, Sheldon Wolin, *Politics and Vision* (Boston: Little Brown, 1960), pp. 28ff.; Hannah Arendt, *The Human Condition* (New York: Doubleday, 1959), pp. 13, 20; Leo Strauss, Introduction to *History of Political Philosophy* (Chicago: University of Chicago Press, 1963), p. 1.

2. See Aristotle, *Ethics*, 1094a13–20, 1095b15–31, 1103a15–19, 1180a15–25; *Politics*, bk. 1, chap. 1–2, bk. 2, chap. 1, 1323a14–1324a; *Physics*, bk. 1, chap. 2. Unless otherwise noted, references to Aristotle are from *The Complete Works of Aristotle*, 2 vols., ed. J. Barnes (Princeton, N.J.: Princeton University Press, 1984).

3. See Joachim Ritter, *Studien zu Aristoteles und Hegel* (Frankfurt am Main: Suhrkamp, 1967); Murray Bookchin, *The Rise of Urbanization and the Decline of Citizenship* (San Francisco: Sierra Club Books, 1987); Cornelius Castoriadis, *Crossroads in the Labyrinth* (Cambridge, Mass.: The MIT Press, 1984), especially the concluding chapter; *Gesellschaft als imaginäre Institution* (Frankfurt am Main: Suhrkamp, 1984). Neo-Aristotelianism is closely related to the contemporary emergence of hermeneutics as method in the social sciences, especially in the work of the preeminent proponent of this approach, Hans-Georg Gadamer, who proposes Aristotle's conception of *phronesis* or practical wisdom as an explicit alternative to methods based on modern (natural) science. See Gadamer, *Wahrheit und Methode: Grundzüge einer philosophischen Hermeneutic*, in *Gesammelte Werke* vol. 1 (Tübingen: JCB Mohr, 1986, 1987) pp. 317ff.; and "Probleme der praktischen Vernunft," in *Gesammelte Werke*, vol. 2, pp. 319ff.

4. Aristotle, *Ethics*, bk. 1, chap. 2; *Politics*, bk. 1, chap. 2. See Herbert Schnädelbach, "What is Neo-Aristotelianism?" *Praxis International* 7, nos. 3/4 (Winter 1987): 228. As Ritter argues, "practical philosophy" in its Aristotelian sense is "politics as 'ethics.' " Ethics, in this sense, can only be understood as *embodied* in institutions—"me-

diated concretely through the 'habitual' institutional life-world and the forms of the speech and action which constitute it." *Studien zu Aristoteles und Hegel*, p. 110.

5. To be sure, particularly with respect to the comparison of Arendt(ians) and Strauss(ians), those making the comparisons are typically unsympathetic toward both positions. Indeed, "neo-Aristotelianism" is a coinage of critics such as Jürgen Habermas and like-minded thinkers. See Habermas, *Zur Rekonstruktion des historischen Materialismus* (Frankfurt am main: Suhrkamp, 1976), pp. 295ff.; "Modernity and Post-modernity," *New German Critique* 22 (Winter 1981): 295ff.; Richard Bernstein, *Beyond Objectivism and Relativism: Habermas, Science, Hermeneutics* (Philadelphia: University of Pennsylvania Press, 1985), pp. 189ff.; Schnädelbach, "What is Neo-Aristotelianism?" p. 228; Seyla Benhabib, "In the Shadow of Aristotle and Hegel: Communitarian Ethics and Current Controversies in Practical Philosophy," *Philosophical Forum* 21, nos. 1/2 (1989–1990): 2ff. In this case, it is the combination of a rootedness of morality in *ethos* or a way of life, arguments for natural law, and a critique of modernity in those terms which constitute the key facets of "neo-Aristotelianism."

6. The largest majority of neo-Aristotelian political and ethical thinkers, including Strauss, Arendt, and their followers (for example, Bernard Crick, Stuart Hampshire, Sheldon Wolin, John Schaar, Hannah Pitkin, Alisdair MacIntyre, Joachim Ritter, Manfred Riedel, A. P. d'Entreves), have largely ignored ecology, and are skeptical of or even hostile to the idea of reconstructing Aristotelian natural philosophy along with practical philosophy. However, important contributions from both conservative and radical democratic perspectives have been made which seek to extend an Aristotelian conception of *telos* into contemporary debates about the relationship between humanity and outer nature. For conservative perspectives, see Hans Jonas, *The Phenomenon of Life* (New York: Harper and Row, 1966) and *The Imperative of Responsibility* (Chicago: University of Chicago Press, 1984); and Robert Spaemann *Philosophische Essays* (Stuttgart: Reklam, 1983) and *Die Frage Wozu?* (Munich: Piper, 1985). For a radical democratic, indeed libertarian communist perspective, see Murray Bookchin, *The Ecology of Freedom* (Palo Alto, Calif.: Cheshire, 1982). For a discussion of these themes, see John Ely, "An Ecological Ethic? Left Aristotelian Marxism vs. the Aristotelian Right," *Capitalism, Nature, Socialism* 1, no. 2 (Summer 1989): 143–156. For a thoughtful but critical reconstruction of Arendt's views of nature, see David Macauley, "Hannah Arendt and the Politics of Place: From Earth Alienation to *Oikos*," Chapter 5, this volume.

7. It is important to emphasize that this issue constitutes the most controversial aspect of neo-Aristotelian thinking. Despite their proclivity for classical Greek or premodern thought, many neo-Aristotelians studiously avoid romantic formulations. Nonetheless, the most strident, neo-Aristotelians have argued that the progress of modernity is rife with "disastrous consequences" and the dangers of rampant "nihilism" (Strauss) or "unnatural growth" and "world alienation" (Arendt), to be avoided through "recollection" (Strauss) or the possibility of regaining that political realm which has been lost through the invasion of the public realm by society" and the "failure of thought and remembrance" (Arendt). See Leo Strauss, *Natural Right and History* (Chicago: University of Chicago Press, 1953) pp. 3, 7; Hannah Arendt, *The Human Condition*, pp. 43, 224ff.; and Arendt, *On Revolution* (New York: Viking, 1966), pp. 219, 223.

For other "romantic" formulations of neo-Aristotelians, see Gadamer, who argues with respect to the crisis of modernity, including the ecological crisis: "It could be that that which is imprinted in us as a guiding image of humanity, imprinted in us from a long antique and Christian history, lives yet," to which we may, more "than we see today," make a "self-conscious return." "Zu Problemen der Ethik," in *Gesammelte Werke*, vol. 2, p. 255. See also Gadamer, *Wahrheit und Methode*, vol. 1, p. 3; MacIntyre, on the "vindication" of a "premodern view of morals and politics . . . against modernity" in "Aristotelian terms" *After Virtue* (South Bend, Ind.: University of Notre Dame Press, 1981), p. 111; and J. Budziszewski on the need to "resurrect" a "buried" nature, *The Resurrection of Nature* (Ithaca, N.Y.: Cornell University Press, 1966), pp. 17, 23.

8. Georg Lukács, "Expressionism: Its Significance and Decline," in *Essays in Realism* (London: Wishart, 1971), pp. 82ff. (originally published in *Internationale Literatur*, 1934).

9. See Michael Löwy and Robert Sayre, "Figures of Romantic Anti-Capitalism," *New German Critique* 23 (Spring/Summer 1984): 43.

10. Georg Lukács, "Ernst Blochs '*Erbschaft dieser Zeit*' " (1935), in *Ernst Bloch and Georg Lukács: Dokumente* (Budapest: Lukács Archiv, 1984).

11. Ernst Bloch, *Avicenna und die aristotelische Linke*, in *Gesamtausgabe*, vol. 7 (Frankfurt am Main: Suhrkamp, 1977), pp. 479ff. Further references to the *Gesamtausgabe* are abbreviated GA.

12. Arendt, "Tradition and the Modern World," in *Between Past and Future* (Cleveland: Meridian, 1963), p. 17; see also *The Human Condition*, p. 13. Nothing seems more organic than the opposition between Marxists and Aristotelians, whether one thinks of Jacques Maritain, A. P. d'Entreves, and other Thomists, or Arendt and Strauss, or MacIntyre and Charles Taylor, or "civic republicans" like Schaar and Wolin, or ex-Marxist libertarians like Castoriadis and Bookchin.

13. Strauss, "Philosophy as a Rigorous Science," in *Studies in Platonic Political Philosophy* (Chicago: University of Chicago Press, 1983), pp. 32f.

14. Bloch is one of the few figures in the history of philosophy to have edited his own collected works. Though *Natural Law and Human Dignity*, trans. Dennis Schmidt (German: GA 6; English: Cambridge, Mass.: The MIT Press, 1986) was only published in 1962, it, or a text on natural law, was conceived from the beginning as the political facet of his philosophical *Summa* from the period before World War I onward. *Natural Law and Human Dignity* was composed largely during the years of his exile in the United States. That it was finally published after the World War II suggests a comparison with Arendt and Strauss, who also studied German philosophy intensely but emerged only after the war as preeminent political thinkers.

15. See Franz Neumann, *Die Herrschaft des Gesetzes* (Frankfurt am Main: Suhrkamp, 1980) and "Types of Natural Law," in *The Democratic and Authoritarian State* (New York: The Free Press, 1957); Lucio Colletti, *From Rousseau to Lenin* (London: New Left Books, 1972), pp. 144, 148; Jürgen Habermas, *Theory and Practice*, trans. John Viertel (Boston: Beacon Press, 1973), pp. 82ff.

16. "Left" is used here and subsequently in a *generic* sense to denote an insurgent, collective, an universalizable politics focusing on popular sovereignty, democracy, and social justice, as well as the critiques of hierarchical and exploitative social relations which impede these demands. Leftists must take responsibility for the fact that the demand for democracy has not always been unequivocably or coherently integrated into many elements included in a given historic "Left"—one thinks of Jacobinism, of the Trotskyist positions opposing the NEP, or of Stalinist positions supporting forced collectivization in the late 1920s. But even these circumstances are not comprehensible outside the Left's historic commitment to democracy and popular sovereignty as well as social or equalized and substantive distributive justice. The concern here, however, is to detach "left" from any specific events or collective actors in history, and to see it rather as the fragmentary but collective struggle against exploitation and domination generally.

The Left certainly must pay a huge cost for those times when the critique of exploitation opposed a critique of domination. Indeed, new beginnings, maybe even new foundations, become possible once this opposition is overcome and we can recall, as

Arendt does with reference to Herodotus, that the word *isonomia* encompassed the notions of political freedom and equality together (*On Revolution*, pp. 23f.). See also Bookchin, *Urbanization and the Decline of Citizenship*, pp. 29, 65–67. Rousseau's *Social Contract* also made clear utter incompatibility of inequality and freedom.

17. To be sure, critical theory took note of this deficiency in its polemic against a "linear" Hegelian narrative about the historic growth of forms of spirit and reason. Adorno, for example, argued that while Marx refused to "locate antagonism in human nature or in primitive times," he buried the political implications of this problem by "deifying history," locating the solution to antagonism in the historical process of economic development, the "happy end immanent in history." The revolution desired by Marx, wrote Adorno, "was one of the economic conditions of society as a whole; . . . it was not a revolution in society's political form, in the rules of the game of dominion." Marx's point, he continued ironically, "was directed against the anarchists," *Negative Dialectics*, trans. E. B. Ashton (New York: The Seabury Press, 1972), p. 322. Horkheimer too identified the crisis of progressive historicism with the need for a theory to overcome the "theory of accelerated development which has dominated 'politique scientifique' since the French Revolution." For Horkheimer, "talk of inadequate conditions" was "cover for the tolerance of oppression. . . . The invocation of scheme of social states which demonstrates *post festum* the impotence of a past era was the inversion of a theory and *politically* bankrupt. . . . The theory of the growth of the means of production, of the sequence of the various modes of production, and of the task of the proletariat is neither a historical painting to be gazed upon nor a scientific formula for calculating future events. . . . Critical Theory . . . rejects the kind of knowledge one can bank on. It confronts history with that possibility which is always concretely visible within it." Max Horkheimer, "The Authoritarian State," in *The Essential Frankfurt School Reader*, ed. Andrew Arato and Eike Gebhardt (New York: Urizen, 1978), p. 106.

18. See Perry Anderson, *Considerations on Western Marxism* (London: New Left Books, 1976).

19. Max Horkheimer, *Eclipse of Reason* (New York: Oxford University Press, 1947), pp. 4ff.

20. Ibid., pp. 58ff. See also Habermas, *Theory of Communicative Action*, vol. 1, trans. T. McCarthy (Boston: Beacon Press, 1984), pp. 373ff.

21. Bloch, *Das Materialismusproblem*, in GA 7, pp. 433ff.

22. MacIntyre, *After Virtue*.

23. In this respect, the role of a given "tradition," whether in the sense of MacIntyre or in the older sense invoked by Burke, reflects the degree to which habituation and convention are defined in their doctrines by prepolitical modes of "identity" ascription such as family, relation, language, culture, ethnicity, and nation. See MacIntyre, *Which Rationality? Whose Virtue?* (South Bend, Ind.: Notre Dame University Press, 1988).

24. Taylor, "Diversity of Goods" and "Legitimation Crisis," in *Philosophy and the Human Sciences: Philosophical Papers*, vol. 2 (Cambridge, England: Cambridge University Press, 1985); Ronald Dworkin, "Law as Principle," in *The Philosophy of Law* (Oxford: Oxford University Press, 1977).

25. Beyond the biological-Darwinian and mechanical-physical determinisms of the Second International, the important tradition is again Lukács's argument that nature is also a "social category" mediated and constructed by human activity seen historically. *History and Class Consciousness*, trans. Rodney Livingstone (Cambridge, Mass.: The MIT Press, 1971), pp. 130–131, 234ff. Cf. Alfred Schmidt, *The Concept of Nature in Marx* (London: New Left Books, 1971), the most elaborate statement within these

terms. Within this context, Marx inherited bourgeois ontological presuppositions about the nature of man, namely, that under all modes of production, he is engaged in a struggle with nature (the realm of material necessity). This point is made by Hannah Arendt, *The Human Condition*, pp. 74ff., 86f. One result is Marxism's inability to conceptualize humanity's "metabolism" with nature in anything other than bourgeois, Baconian terms of necessary "mastery" of nature, or to historicize humanity's relation to and conception of outer nature in other than instrumentalist terms. This problem of the "domination of nature" (*Naturbeherrschung*) becomes a central aporia in the work of Horkheimer and Adorno, who agreed with Marx that instrumental reason was "necessary" to "master" nature while recognizing that mastery, being nature itself, became the mastered. As Bookchin has noted, the very notion that one can "master" nature is itself ideological. Murray Bookchin, "Marxism as Bourgeois Sociology," in *Toward an Ecological Society* (Montreal: Black Rose Books, 1981).

26. Bloch, "Avicenna und die aristotelische Linke" (1952), in *Das Materialismusproblem*, in GA 10. Bloch's interest in this subject is found as well in his early fascination for theosophy, Rudolf Steiner and his followers, physics, vitalism, and the theory of the "natural subject," as well as in the role Johann Jakob Bachofen's theory of "mother-right" in his political theory. See Bloch, letters to Lukács, 22 February 1911, and 16 August 1916, in *Briefe*, vol. 1 (Frankfurt am Main: Suhrkamp, 1985), pp. 35ff., 166–168; *Naturrecht und menschliche Würde*, in GA 6, pp. 115–128/97–119. On the ties between pantheism and anarchistic movements, see Carl Schmitt, *The Crisis of Parliamentary Democracy*, trans. Ellen Kennedy (Cambridge, Mass.: The MIT Press, 1985), 66–76; Schmitt, *Politische Theologie* (Berlin: Duncker & Humblot, 1934), pp. 81–83; and Ernst Troeltsch, *The Social Teaching of the Christian Churches*, trans. Olive Wyon (London: Allen and Unwin, 1934), pp. 462ff.

27. This point marks a key overlap with the Frankfurt School's "critique of instrumental reason," since a "substantive" of "objective" reason to oppose the instrumental rationality dominating the modern world is a key to the problem of the "dialectic of Enlightenment" (the required but impossible solution, so to speak). See Wellmer, *Critical Theory of Society*; and "Reason, Utopia, and the Dialectic of Enlightenment," *Praxis International*, no. 32 (July 1983).

28. Bloch, *Naturrecht*, p. 16/3.

29. Strauss, *Natural Right and History*, pp. 81ff.; Bloch, *Naturrecht*, p. 21/7.

30. Bernstein, *Beyond Objectivism and Relativism*; Arendt, *On Revolution*, pp. 200ff.

31. Bloch, *Naturrecht*, p. 37/24.

32. Oskar Negt, "10 Thesen zur marxistischen Rechstheorie," in *Probleme der marxistischen Rechstheorie*," ed. H. Rottleuthner (Frankfurt am Main: Suhrkamp, 1975), pp. 36ff. The reference is to Marx's twist on Hegel's famous comments in the preface to his *Philosophy of Right*, characterizing reason "as the rose in the cross of the present."

33. Bloch, GA 11, pp. 454, 459; *Naturrecht*, pp. 13, 81/xxx, 65.

34. In this sense, Aristotelian "teleology" contrasts radically with a "teleology" of history, as political thinkers such as Arendt and Strauss have insisted. The "goal" or "direction" of the "system" of capital accumulation, the driving force of the world spirit, is a tendentially "blind" progress. The "cunning of history" is like Smith's "invisible hand" or the behind-our-backs quality of human historical development. It is experienced, in other words, as a functional system determined by laws beyond the control of the rational human.

35. This is a more characteristic tendency of moderate and liberal forms of communitarianism, such as that of Charles Taylor, Michael Walzer, or the natural-law doctrines of Ronald Dworkin.

36. Schnädelbach, "What Is Neo-Aristotelianism?" p. 226.

37. See Habermas, *Theory and Practice*, pp. 121ff.; and Seyla Benhabib, *Reason, Norm, and Utopia* (New York: Columbia University Press, 1985).

38. Schnädelbach, "What Is Neo-Aristotelianism?" p. 226.

39. See John Ely, "Ernst Bloch and the Second Contradiction in Capitalism," *Capitalism, Nature, Socialism* 1, no. 2 (Summer 1989): 91ff.

40. This dialectic is described in greater detail in John Ely, "Libertarian Ecology and Civil Society," *Society and Nature*, no. 6 (Fall 1994): 103–115.

41. Habermas, *Faktizität und Geltung* (Frankfurt am Main: Suhrkamp, 1992), pp. 611ff.

42. See Rudi Dutschke et al., *Mein Langer Marsch* (Reinbek: Rowohlt, 1980), pp. 154–158.

43. Ernst Hoplitschek, "Die Schwierigkeiten der Linken in Umgang mit den Werkonserativen," in *BUS-Rundbrief*, "rot und grün," no. 2/3 (1980).

44. For example, on the cover of *Anders Leben*, no. 2/3 (1981), the predecessor to *grüner Basisdienst*): "A truth is not useful because it helps, but because it is true it helps.

45. See Jutta Ditfurth, "Radikal Ökologische Politik" (1984), pp. 7ff.; and Ditfurth, *Lebe Wilde und Gefährlich: Radikal Ökologische Perspektiven* (Frankfurt: KSW, 1991), pp. 50ff.

46. The conservative journalist Karl Friedrich Fromme referred to the Greens as the "banner of the ahistorically upright" during the unification process because they advocated a nation composed of several states rather than "re"-unification. "Vor dem 2. Dezember," *Frankfurter Allgemeine Zeitung*, 30 November 1990.

47. Bloch emphasized this issue by resurrecting his original *neue Weltbühne* article in new form during the sixties in his support of the student movement. See "Erinnerung (bei Gelegenheit der Kontaktfrage Student-Masse): Socrates und die Propaganda," in *Politische Messungen*, in GA 11, pp. 402–408; and Traube, *Gespräche mit Ernst Bloch*, pp. 196ff.

48. *Manifest grüner Aufbruch* (Bonn: Die Grünen, 1988), p. 10.

49. See Joachim Raschke, *Krise der Grünen?* (Marburg: Springer-Verlag, 1990); John Ely, "Conservative Ecology," in *Telos* 97 (Fall 1993): 186–191; and Barbara Holland-Cunz, *Soziales Subjekt Natur,* (Frankfurt am Main: Campus Verlag, 1994).

50. This is most notable in the manner in which the left in the Green Party focuses on *vetoing* projects rather than on retaining or strengthening the democratic nature of their party. Ely, "The Greens between Legality and Legitimacy," in *German Politics and Society* 14 (June 1988).

51. Ely, "Libertarian Ecology and Civil Society."

52. Ted Benton, *Natural Relations* (London: Verso, 1992).

53. Especially notable in the renaissance of influence by Hannah Arendt, as well as the discussion of civil society. See Ulrich Rödel, Helmut Dubriel, and Ginter Frankenberg, *Die demokratische Frage* (Frankfurt am Main: Suhrkamp, 1990); and Ely, "Libertarian Ecology and Civil Society," pp. 115ff.

54. Hans Jonas, *The Imperative of Responsibility*; Spaemann/Löw, *Die Frage Wozu?* See also Fritz Reusswig and Michael Scharping, "Wo steht die Natur wenn der Geist rechts steht," in *Die Kunst des Möglichen*, ed. Thomas Noezel (Marburg: Springer-Verlag, 1989); and Habermas, "Röckkehr zur Metaphysik?" *Merkur* 439 (1985).

55. The green ideal of making institutions and institutional decision-making "transparent" is drawn from theories of institutional crisis associated with Habermas. See Habermas, *Legitimation Crisis*, trans. Thomas McCarthy (Boston: Beacon Press, 1973); James O'Connor, *Fiscal Crisis of the State;* and Claus Offe, *Contradictions of the Welfare State*, ed. John Keane (Cambridge, Mass.: The MIT Press, 1984).

The ideal of "transparency," a term used by Preuss, Offe, and others in the circle of green *Realpolitik*, serves frequently as a "modern" conception of system-openness for a complex state. It refers, however, to a fishbowl conception of democracy, a kind of panopticon in reverse; and it still assumes passive citizenry and a managerial approach. The glorification of the "media" as a "fourth" branch in the liberal state's "division of powers" (alongside the executive, legislative, and judicial) tends rhetorically to leave unquestioned the relation of a passive citizenry to all four such powers.

56. See, respectively, Alexis de Tocqueville, *Democracy in America* (Garden City, N.Y.: Anchor, 1969), pp 189ff.; Arendt, *On Revoution*; and Sandel, *Liberalism and the Limits of Justice* (Cambridge: Cambridge University Press, 1982). In *Faktizität und Geltung*, Habermas uses the writings of Michelman and Sunstein as his republicans: two figures whose work remains within the parameters of the liberal democratic state, and who are no more interested in direct democracy than Charles Taylor.

Neo-Aristotelians avoid the problem of neo-Kantians who separate the world of institution, the "system" from that of moral decision making "life-world"). From an Arisotelian perspective, "freedom" is defined neither by the adaptation of freedom to a functionally understood notion of "optimization" nor as freedom "from" system colonization of the life-world. Freedom is defined by political activity and political life per se.

57. Otto Gierke makes the distinction between "association" and "institution" in this sense, between *Gemossenschaft* and *Anstalt*, which in large measure approximates the changing meanings of organization from *community* to *corporation*, from a precapitalist to a systems, functionalist conception of institution. Gierke, *Das deutsche Genossenschaftsrecht*, vol. 1 (Berlin: Wiedmannische Buchhandlung, 1868), pp. 638ff., 1030ff.; and Anthony Black, "Introduction," to Gierke, *Community in Historical Perspective* (Cambridge: Cambridge University Press, 1988), pp. xvff.

58. Ritter, *Studien zu Aritoteles und Hegel*, p. 132.

59. Functionalist conceptions of association as "interest groups" have limited or deficient ways of judging the collective rationalities of different kinds of association, or the virtues and appropriate goals and intentions of its individual members. As a result, it is frequently a short distance from "preferences" to "decisions."

60. See Nicos Poulantzas, *State, Power, Socialism* (London: New Left Books, 1981); Joachim Hirsch, *Der sicherheitsstaat* (Frankfurt: EVA, 1980); Ingeborg Maus, *Rechtstheorie und politische Theorie im Industriekapitalismus* (Munich, 1986); and Alfons Söllner, "Neumann, Kirchheimer and Marcuse, Students of the Conservative Revolution," *Telos* 61 (1984): 69ff.

Structural theories of "rational choice" which are used to prove the historical necessity of social democracy (e.g., Adam Pzerworski) tend to presuppose conceptions of "personal interests" and their incommensurable clash in a manner which fails to confront these functional models and their undiscussed anthropological assumptions (need, interest, self-preservation, and morals as the plurality of "norms"). In the end, they provide no more defense against left-Schmittian conceptions that the traditional social-democratic functionalism of Renner, Neumann, or Habermas. In all we see Social Democratic jurisprudence has been corrupted by incorporating too many ("ontological" and "anthropological") characteristics of bourgeois modernity.

61. See Claus Offe, "New Social Movements: Challenging the Boundaries of Institutional Politics," *Social Research* 52 (1983); Hirsch and Roth, *Das neue Gesicht des Kapitalismus* (Hamburg: VSA, 1986), pp. 211ff; and Herbert Kitschelt, "The Left-Right Semantics and the New Politics Cleavage," *Comparative Political Studies* 23 (Spring 1990).

62. See Paul Hirst, *Associative Democracy* (Amherst: University of Massachusetts Press, 1993); and Ulrich Beck, *Risikogesellschaft* (Frankfurt am Main: Suhrkamp, 1986).

63. Die Grünen, *Umbau der Industriegesellschaft*, Nuremberg, 1986 (Bonn, 1986), pp. 10ff.

64. Ibid., p. 8.

65. Francisco Varela, "Cognition as Embodied Action," presented at the conference "Beyond Dualism," Stanford University, March 1994.

66. Habermas, *Faktizität und Geltung*, p. 661. By the same token, one must beware of conceiving "embodiment" in land-based, territorial terms in which "bio"-regionalism quickly leads to nonpolitical images of cultural ecology and species survival becomes associated with survival of native cultures and entwined in a politics of localism and rightist cultural nationalism. "Embodiment" means first and foremost *public* assemblies rather than some relationship between human and nature, micro- and macrocosmos. *Territoriality* is a concept stemming from feudalism and the development of the modern nation-state out of feudalism. See John Ely, "Libertarian Confederalism and Green Politics," in *In the Twilight of the Twentieth Century*, ed. Michael Crozier and Peter Murphy (Champaign-Urbana: University of Illinois Press, in press).

Already the tendency for ecocentric defense of liberal rather than socialist approaches to economics suggest the terms of this conjuncture. *The New York Times* emphasizes that rising ocean levels and a drier Southern Hemisphere are the results of greenhouse gas increases. While downplaying the role of the industrial north in producing such gases, it emphasizes the changes in environment—largely beneficial to northern areas and detrimental to the south and to delta areas like Egypt and Bangladesh. While the wealthier, northern areas can "adapt" to environmental changes through complex systems of dikes and extending breadbasket agricultural areas, the south will experience severe flooding in low-lying deltas, "casting many of their inhabitants on the world's mercies as environmental refugees" (William Stevens, "Dire Forecasts of Disruption From Global Warming," 18 September 1995, p. 1). The issue of taxes on greenhouse gas production is not broached, while the fear of "environmental refugees"—an act of God?—suggests the need to reestablish "local" territorial borders between northern and southern "bioregions." "Region" itself has a monarchist and feudal, not a civic or republican, genealogy.

67. Offe, "Challenging the Boundaries."

68. Habermas has never publically addressed "red-green" or the problem of ecology as an issue of social or political theory, though private circles since 1987 indicate that he sees it as the long-term prospect for the Left. Ironically, precisely in the years 1978–1980 when the Green Party formed due to the coalition of ecology and left elements, Habermas went on record, framed by a revival of Jaspers's concern about the "spiritual condition of the age," to argue that ecology was an "ambiguous force," thus far "unreceptive to socialist orientations." Social Democracy's role in compelling an unwanted nuclear-power industry on an unwilling populace and the problem of "receptivity" were not mentioned, despite recognition of some "justified anxiety" about this industry. See Habermas, Introduction to *Observations on the "Spiritual Condition of the Age"* (Cambridge: The MIT Press, 1984), pp. 9, 18.

"Fundamentalism" as a characterization of the green Left or radical faction stems from Oligarch Joschka Fischer, who started using the term to denounce the Greens soon after a tourist trip to Iran to check out the revolution. Party Oligarch Fischer has been Habermas's main dialogue partner, a dialogue which began only after the Left had lost its majority in the Party Congress.

69. See Thomas Blanke, "Rechtssystem und Moral—Vorüberlegungen zu einer ökologischen Verfassungstheorie," in *Recht, Justiz, Kritik*, ed. Boettsche (Baden-Baden: Nomos, 1985), pp. 402ff.; and, critically, see Alex Demirovic, "Ecology and Democracy," *Capitalism, Nature, Socialism* 2, no. 3 (Fall 1989): 43–48.

70. Ironically, he made this argument during the same year (1980, in *Observations on the "Spiritual Condition of the Age"*) that the Greens first formed as a decisively socialist and ecological party, integrating major segments of the German Left. John Ely, "Marxism and Green Politics," *Thesis Eleven*, no. 12 (1986).

71. Habermas, *Theorie des kommunikativen Handelns*, vol. 2 (Frankfurt am Main: Suhrkamp, 1982), pp. 579–585. Yet it is precisely such a purely procedural attitude which overloads the parliamentary-administrative system, and shears citizens of their role in forming a collective political project. The new social movements and the Greens articulate new forms of productive and political life with a carefully considered universalistic perspective and an Aristotelian-practical conception of freedom as participation *and*substantive equality (i.e., in the original Greek sense of *isonomia*). These developments fall through the cracks between "system" and "life-world." There is simply no institutional space in which to conceive these new political forms without reducing them to "cultural" phenomena ("colonization of the life-world") or "interest groups" in the political "system." Hence, not surprisingly, Habermas and his followers have not particularly investigated the new social movements, but have treated them as a black box within their models of society. See Rödel et al., *Die demokratische Frage* (Frankfurt am Main: Suhrkamp, 1989); and Cohen and Andrew Arato, *Civil Society and Political Theory* (Cambridge, Mass.: The MIT Press, 1992).

72. Roland Roth et al., *Parlamentarisches Ritual und politische Alternativen* (Frankfurt am Main: Campus, 1980).

73. See Elmar Altvater, *Sachzwang Weltmarkt* (Münster: Westfälisches Dampfboot, 1990).

74. See Jan Martinex-Alier, *Ecological Economics* (London: Basil Blackwell, 1989); and Frank Beckenbach, ed., *Die ökologische Herausforderung für die ökonomische Theorie* (Marburg: Metropolis, 1991).

75. Can Habermas's jurisprudential discussion of system and life-world or competent communication be used to address issues such as those raised in international law regarding transformations in substantial constitutional law? See Heinhard Steiger et al., "The Fundamental Right to a Decent Environment," in *Trends in Environmental Policy and Law*, ed. Michael Bothe (Switzerland, 1980), pp. 5ff.

This perspective unwittingly contributes to deepening the divide between radical civil society, the Autonomen, the radical state—that described by Boella, Agnoli, and the covey of Orwellian voices in the contemporary Federal Republic. One would refer here especially to those involved in the public sphere of the "militant" faction, who appear in *Tageszeitung* columns on the RAF: an assortment of New Left luminaries (like Karl-Heinz Roth), but also intellectuals whose work is influential in such circles, like that of Agnoli (*Transformation der Demokratie* [Berlin: 1968]); Hirsch (*Der Sicherheitsstaat*); or even Peter Brückner. This is a negative mutualism in which one's voice makes the other stronger via mutual reproaches.

76. Murray Bookchin, *The Philosophy of Social Ecology* (Montral: Black Rose Books, 1990), pp. 113ff.; Spaemann and Löw, *Die Frage Wozu?*; Christian Kummer, *Evolution als Höherentwicklung des Bewusstseins. Über die intentionalen Voraussetzmngen der materiellen Selbstorganisation* (Freiberg: Verlag Karl Alber), pp. 143ff., 248ff.; Vittorio Hösle, *Philosophie der Ökologischen Krise* (Munich: Beck, 1991), pp. 72ff. Cf. Ely, "Conservative Ecology," *Telos* 97 (Fall 1993), on the conservative and far-right elements, respectively, in the last two texts.

77. Bookchin, *The Philosophy of Social Ecology*, pp. 77–89.

78. Here Bloch, with "irrationalist" and "metaphysical" roots, has reintroduced Aristotle's conception of nature into the Left, even though, due largely to a failure to see the *corrosive* and *destructive* aspects of capitalist mechanization, he has

compromised it with his own political theology of the transition from capitalism to socialism.

79. In each case, the judgment concerning the presence of subjective autonomy, or the presence of pain, is one achieved through some notion of complexity and the development of sense, quickness, and mind. One requires first a developmental gauge in which to distinguish degrees of sentience and autonomy. See Tom Regan, "Ethical Vegetarianism and Commercial Animal Farming," in *Today's Moral Problems*, ed. R. Wasserstrom (New York: Macmillan, 1985), pp. 461ff.; and Peter Singer, *Animal Liberation* (New York: Avon, 1975), pp. 12ff.

80. See Jonas, *The Imperative of Responsibility*, for an authoritarian form based on Plato, as well as the ubiquity of statist notions of sovereignty and anthropology (need and interest) in systems theory. Worth noting here is the problem such calls for a stronger state could have for the Left. At a 1988 conference on ecology sponsored by the Greens' Heinrich-Böll-Foundation, the conservative *Frankfurter Allgemeine Zeitung* reporter Konrad Adam argued that by Hobbsian standards of security, present national-states could not guarantee the safety of their subjects because of increasing ecological dangers and "risks." The speech received general applause from the core of Frankfurt's realist-tending green intelligentsia.

81. Christopher Stone, *Should Trees Have Standing?* (Los Angeles: William Kaufmann, 1972), pp. 33ff., 47f.; and especially *n*. 125. In this regard, Gierke underscores the "fictional" quality of a joint-stock or limited liability corporation lies in the way it *externalizes* labor while transforming interest as a shared commitment into interest as banking speculation. Ecocommunities are "real," working entities in contrast to "fictional" Mobil Corp., and so on.

82. Christopher Stone, *Earth and Other Ethics* (Cambridge, Mass.: Harvard University Press, 1987), p. 13.

83. Ibid., p. 178.

84. See Martinez-Alier, *Ecological Economics*; Ely, "Left-Aristotelianism vs. the Aristotelian Right," pp. 148ff.; and Bookchin, "The Concept of Eco-Communities and Eco-Technologies," in *Toward an Ecological Society*, pp. 99ff.

85. Bloch, *Das Materialismusproblem*, pp. 425, 431. See also *Erbschaft dieser Zeit*, in GA 4, pp. 205f.

86. Bloch, *Erbschaft*, p. 205/267.

87. Bloch, *Das Materialismusproblem*, p. 431f.

88. Bourgeois mechanism hides the commodity logic by destroying the notion of independent *telos* in the forms of nonhuman nature, hence weakens the resistance of this nature (now reduced to commodifiable quantities) to the law of value. In the modern economy, *telos* occurs on the principle of a causal agency totally outside the system, reproducing the vision heteronomy for which Laplace's vision of the cosmos is so famous.

89. Elliott Sober, *Philosophy of Biology* (Boulder, Colo.: Westview, 1993), pp. 1, 13, 143ff.

90. Bloch, *Das Materialismusproblem*, p. 435.

91. See Eberhard Braun, "Natur und Utopie: Bedeutung des Naturbegriffs in Ernst Blochs Philosophie," in *Ökologie und Politik*, ed. Fachschaft Biologie Tübingen (Tübingen: VAUD, 1988) pp. 202ff.

92. Gotthelf, "Aristotle's Conception of Final Causality," in *Philosophical Issues in Aristotle's Biology*, ed. Gotthelf (Cambridge, England: Cambridge University Press, 1987), p. 208f. As Gotthelf notes, this is "the closest Aristotelian equivalent to 'law' in Modern science" (p. 211).

93. Aristotle, *Physics*, bk. 2, chap. 2.

Chapter Seven

"The Outcry of Mute Things": Hans Jonas's Imperative of Responsibility

Lawrence Vogel

Along with Hannah Arendt and Emmanuel Levinas, Hans Jonas was among Martin Heidegger's foremost Jewish students. Jonas received his doctorate under Heidegger at Marburg, fled Germany for Palestine in 1933, fought in both the Second World War and the Israeli War of Independence, joined the Graduate Faculty at the New School for Social Research in 1955, and served as the Alvin Johnson Professor of Philosophy at the New School until his death in 1993 at the age of eighty-nine.

Jonas's work displays extraordinary breadth and continuity. *The Gnostic Religion*, completed in the mid-1950s, is still a classic today. It traces the roots of metaphysical dualism at the beginning of the first millenium A.D. His second book, *The Phenomenon of Life*, discovers a Gnostic pattern of thinking at the core of modern philosophy, including Heidegger's "existentialism." Jonas argues that metaphysical dualism is responsible for the difficulty that we moderns have in thinking of nature, even human nature, as more than an object for technological manipulation. As a cure for modern "nihilism" Jonas provides an account of nature that is more in the spirit of Aristotle than Descartes, while still in keeping with modern science.

Jonas's final major work, *The Imperative of Responsibility: In Search of an Ethics for the Technological Age*, connects his speculation about nature to the domain of ethics. He thinks that unless we can think of nature as being a source of value, and not a mere resource upon which we project our interests, we will be unable to believe in the importance of limits to our technological remaking of nature. Such limits are especially pressing when we are ever more able to destroy our habitat and to alter "the human image" by exerting control over behavior, the process of dying and even the genetic makeup of life. *The Imperative of Responsibility* had a significant impact on the Green movement in Germany, has sold close to 200,000 copies in German, and won both the 1987 Peace Prize of the German Booksellers' Association and the 1992 Nonino Prize in Italy. Less well known, especially to the Anglo-American audience, is Jonas's effort to develop a Jewish theology for our time. A volume of his later naturalistic and theological essays, *Mortality and Morality: A Search for the Good After Auschwitz*, is forthcoming from the Northwestern University Press.

The Stages on Jonas's Way

For the past several decades until his recent death, Hans Jonas alerted us to an "ethical vacuum" at the core of our culture: a vacuum caused by both traditional ethics and modern natural science.[1] Traditional ethics has presumed that the effects of our actions are quite limited and that (with the exception of medicine) *techne* is ethically neutral. Ethical significance belonged to relations between humans, not between us and nature. Humanity was a constant, not an object of reshaping *techne*. And while the moral good or evil of our actions lay close at hand, the long run was left to chance, fate or providence. All of this has changed with the advent of modern technology, which has altered the very nature of human action by allowing us to affect nature, both outside ourselves and within, in ways that are long-range, cumulative, irreversible, and planetary in scale. Traditional ethics leaves us ill-equipped to account for our responsibilities when the very future of humanity is at stake.

This vacuum is intensified by the dominant scientific view of nature in the modern period: reductionistic materialism. On this view, nature is a machine, harboring no values and expressing no purposiveness. The idea that there are ends in nature is rejected as an anthropomorphic conceit. Extrahuman nature is indifferent to itself as well as to human beings who are cast adrift in it. We may matter to ourselves, but there is no larger scheme of mattering to which we belong. Though human beings may be subjects who posit ends and act in light of purposes, nonhuman organisms are mere objects: matter in motion. Eventually humans, as part of nature, become objects of their own fabrications to be shaped according to the designs of biotechnology. If nature presents us with no ethical norms, then no effort to change our own nature in the name of perfection, convenience or experimentation could count as a transgression of essential limits or a violation of a natural standard of goodness.

Herein lies the deepest root of our cultural crisis: nihilism. Lacking grounds for judging nature to be good and deprived of any stable "image of

Humanity" to which we owe reverence, we are unable to answer the fundamental ethical challenge posed by our novel powers: "Why should we care about the distant future of mankind and the planet?" Unable to justify seeing the maintenance of humankind on the earth as a categorical imperative, we are unprepared for the attitude of stewardship that we must cultivate if we are not to squander the future in the interests of a profligate present.

If nihilism is at the root of our ecological crisis, then the only sufficient response would be a philosophical critique of nihilism. This is precisely the task that Jonas sets for himself, not only in his later writings but from the very beginning. I shall show how his project unfolds in three stages: existential, metaphysical, and theological. Each stage is a response to the crisis of nihilism that Jonas diagnoses in his early essay, "Gnosticism, Existentialism and Nihilism," the epilogue to his first major work, *The Gnostic Religion.*[2] First, in *The Phenomenon of Life*, Jonas offers "an existential interpretation of the biological facts," arguing that purposive existence is not a special attribute of human beings, but is present throughout living nature. Second, in *The Imperative of Responsibility*, he provides a metaphysical grounding of our ethical obligations to nature and to ourselves as special products of its evolutionary labors. Third, in "The Concept of God After Auschwitz," he presents a theology of divine creation that is consistent with both his existential and metaphysical views.

I shall argue that Jonas does not take theology to be necessary for an overcoming of nihilism. Rational metaphysics must be able to ground an imperative of responsibility without recourse to faith. Furthermore, while many environmentalists presume that nonanthropocentrism must be a defining feature of any ecological ethics, Jonas does not believe that overcoming nihilism requires a renunciation of anthropocentrism in favor of biocentrism or ecocentrism. Instead, I shall contend, he tries to undercut the very distinction between anthropocentrism and its supposed alternatives. Finally, I shall raise some critical questions about Jonas's project. In particular, I worry that a pluralistic culture cannot bear the burden of such a substantive metaphysics. If our future depends upon citizens agreeing with Jonas's speculations, then I fear we are not up to the task.

The Phenomenon of Life: *An Existential Interpretation of the Biological Facts*

The first stage on Jonas's way is reached in *The Phenomenon of Life* through an attack on his mentor, Heidegger, whose existentialism Jonas takes to be the most powerful expression of nihilism in our time. Jonas acknowledges *Being and Time* as "the most profound and important manifesto of existentialism," but criticizes Heidegger for restricting existential interpretation to human beings.[3] He begins his critique of nihilism not by rejecting Heidegger's approach altogether, but rather by extending his teacher's categories to yield "an existential interpretation of the biological facts." From Heidegger's perspective, however, the very idea of "an existential interpretation of the biological facts" must sound like an oxymoron. It would be inappropriate to interpret nonhuman organisms existentially because only humans "exist." We are the only beings whose Being is an issue for us: that is, whose behavior and expression manifest a stance toward the sort of being we resolve to be within the constraints of our "thrownness." Whatever significance the rest of nature has, then, it possesses through the lens of our care.

Jonas charges that Heidegger's fundamental ontology does not fulfill its promise to delineate *transhistorical* structures of human existence. Instead, it bears testimony to the particular, historically fated situation of modern humanity: a situation defined by "the spiritual denudation of [the concept of nature] at the hands of physical science" since the Copernican revolution.[4] In Heidegger's analysis of existence, there is no room for nature as possessing intrinsic value, for nature is assumed to be a realm of "un-meaning" (*sinnlos*), taking on significance only in relation to our workaday world. As Karl Löwith puts it, "The world which is concretely analyzed by contemporary existentialism . . . is only our historical world of selfhood and interhuman relations. . . . Nature, says Heidegger, cannot elucidate the ontological character of world and of our being because it is only a kind of being *within* our world, and we encounter it therefore *within* the analysis of man's being-there."[5]

Heidegger's supposedly ontological account of temporality presupposes the modern, materialistic understanding of nature in which there is no eternity, only the flux of time. This flux lacks a genuine present, according to Jonas, because at the moment of decision, authentic *Dasein* stands unguided by any eternal measure. *Dasein* is not seen as being part of "an objective order of essences in the totality of nature," but only as a "transessential, freely projecting existence" who must create values on the basis of nothing but the shifting soil of history. The loss of eternity, Jonas contends, "accounts for the loss of a genuine present. . . . If values are not beheld in vision *as being* (like the Good and the Beautiful of Plato), but are *posited* by the will as projects, then indeed existence is committed to constant futurity with death as its goal; and a merely formal resolution to be, without a *nomos* for that resolution, becomes a project from nothingness to nothingness. . . . Will replaces vision; temporality of the act ousts the eternity of the "good in itself."[6]

On Jonas's reading, Heidegger's existentialism uncritically accepts the metaphysical background of the nihilistic situation: the dualism between humanity and nature. The idea that nature has no ends and is indifferent to human purposes throws us back on ourselves in our quest for meaning. No longer able to find our place in a sacred order of creation or an objective order of essences comprising the totality of nature, we have lost not only the grounds for cosmic piety, but also a stable image of our own nature, even the conviction that we have a nature. Jonas writes, "That nature does not care, one way or another, is the true abyss. That only man cares, in his finitude facing nothing but death, alone with his contingency and the objective meaninglessness of his projecting meanings, is a truly unprecedented situation. . . . As the product of the indifferent, his being, too, must be indifferent. Then, the facing of his mortality would simply warrant the reaction: 'Let us eat and drink. For tomorrow we must die.' There is no point in caring for what has no sanction behind it in any creative intention."[7] In other words, Heidegger's existentialism gives us no good reason to care about future generations or the long-term fate of planet Earth.

But existentialism, as we have seen, is not an idiosyncracy of modern thought. It is rather the most complete expression of "the ethical vacuum" caused by the two key assumptions of the modern credo: (1) that the idea of obligation is a human invention, not a discovery based on the objective being of the good-in-itself, and (2) that the rest of Being is indifferent to our experience of obligation.[8] Jonas's whole philosophy aims at providing a reasonable account, consistent with modern science, of why it does make an objective difference how we relate to nature because living nature, from which our own caring selves emerge and on which we depend, is essentially good, is worth being cared for, and even cares that we care for her so that we, her most magnificent creation, can continue to be.

What is so strikingly subversive about the first stage of Jonas's reply to nihilism is that he uses Heidegger's own existential categories as a point of departure for undermining the modern credo that human being is the source of all value in nature. By attributing "existence" to all organisms, not only humans, Jonas challenges the metaphysical prejudice of materialism and begins to ground a heteronomy that supersedes all authenticity. Though only we humans can take stock of our lives as a whole, reflect on the ontological structure of existence, and be thematically aware of death, all organisms show concern for their own being and reach out to the world in order to fend off not-being. The materialist, of course, claims to remain metaphysically neutral and accuses those who impute purposes to nature of being all too metaphysical. But Jonas insists that one cannot offer a phenomenological description of an organism's way of being without finding purposes in "the things themselves." Even the simplest organism's way of being is utterly different in kind from that of a lifeless object.

Organisms are not mere destructible chunks of matter, but possess an inward relation to their own being. Their being is their own doing. Metabolism is the most basic expression of an organism's struggle for life. Each organism exhibits what Jonas calls "needful freedom." It is free with respect to its own substance: a dynamic unity, not identical with any simultaneous totality of its parts. Yet it remains forever needful: dependent on exchange with an environment which it must use to avoid dying. By clinging to itself, Jonas remarks, life says "Yes" to itself.[9] Only in confrontation with the ever-present potentiality of not-being can Being come to feel itself, affirm itself, make itself its own purpose. Through negated not-being, "to be" turns into "existence": a constant choosing of itself. Because all life is relation and relation implies transcendence, polarities that we find in ourselves—being/not-being, self/world, form/matter, and freedom/necessity—have traces in even the most primitive forms of life. Existential categories—concern, transcendence, freedom, possibility, world, not-being—are necessary to describe powers of mind (*psyche*) that are "objectively discernible" in all life. Some measure of "mind" (or "subjective inwardness") and "freedom" are present at all levels of the life-world.

Each new level of mind—metabolism, moving and desiring, sensing and perceiving, imagining and thinking—brings with it a new dimension of freedom and peril alike. Though all organisms share the vegetative functions of nutrition and reproduction, the evolution of sensitive capacities in animals marks a major advance in the quality of freedom in the life-world. Whereas plants, moved by need but not desire, are immersed immediately in their surroundings and are at the mercy of adjacent matter and impinging forces, animals have a mediated relationship to their environment in virtue of their perceptual awareness of discrete things at a distance from them and their ability to move toward or away from these things in response to their passions. The "secret of animal life" is the gap they are able maintain between their immediate desires and mediate satisfactions. They participate in existential space and time because, as Jonas puts it, "motility guided by perception and driven by desire turns 'there' into 'here' and 'not yet' into 'now.' "[10] Among animals, an inner dimension blooms forth and externalizes itself in behavior and communication, but at the same time there emerges a new liability: the capacity to suffer pain, fear, and even abandonment.

Only with humans, and the interposition of an *eidos* between subject and object, does mediacy become reflective: an explicit relation between a self-conscious subject and objects identified and classified as such. In our imaginations, we can make present what is materially absent. By comparing past and present perceptions, we can discern between truth and falsehood. Furthermore, we can reflect upon essences or "thought-objects" in their own right. The eidetic capacity present in our power to recognize and create images as representations of objects is at the root of our ability to live in light of an image of who we are and ought to be. And we are fundamentally metaphysical organisms for we can, and even must, try to comprehend our place in the whole of which we are a part. Though the twofoldness of self-reflection is the condition of having a "self" and of possessing the special kind of freedom that Heidegger delineates when he calls us the only beings "whose Being is an issue for us," this split within the self is the source of perils peculiar to human existence: the threat of anxiety when we are deeply uncertain about what our standards should be, and the sting of unhappiness, guilt and even despair when we judge ourselves unfavorably from the distance of our wishes, aspirations and approvals.

Once we recognize mechanistic materialism for what it is—a metaphysical prejudice, not a neutral description of the physical world, we are free to interpret the biological facts existentially and to appreciate the reality of value independent of us, for all organisms are "ends-in-themselves" who value whatever contributes to their existence and welfare. This is not the case simply from our point of view, for we have no reason to doubt that they flourish or suffer in their own right. A crucial step in Jonas's reply to nihilism has been achieved, and one might think that there is little, if anything, more to be done. Once we see that humanity is not the sole locus, much less the creator of all value in nature, we would seem to be in a position to accept our duty as guardians of living nature. But if our technological incursions into nature are destined to disrupt the ecosystem and destroy whole species at an unprecedented rate, why would we not be doing nature a favor by taking ourselves out of the picture? Even if there are centers of purpose in nature outside of us that command our respect, it has not been shown that there is an overarching purposiveness in the evolutionary process as a whole in which we play a crucial role and by virtue of which we have an obligation to ensure that human beings remain among the Earth's citizens.

Darwin's theory of evolution does not seem to be of much help in this regard, for it explains natural history as a wholly mechanistic process in which higher and more complex species result from utterly contingent alterations in lower elements. The official Darwinian view is resolutely antiteleological in holding that life first came into being through spontaneous generation from inorganic matter and evolved by chance through the joint processes of random genetic variation and natural selection. But Darwin's materialist explanation of evolution, according to Jonas, "contains the germ of its own overcoming"[11] for, against the spirit of mind/matter dualism, evolution displays a continuity between nonhuman and human organisms. Rather than interpreting mind as an utterly novel and sudden emergence coinciding with humans, Darwin makes room for the idea that the whole life-world is a chain of psychophysical organisms whose minds and bodies coevolve so as to allow for greater freedom and individuality. Being has achieved in us a being who can reflect her back to herself in thought. We can interpret ourselves as the outcome of a teleological process whose immanent purposes are self-knowledge and freedom: an evolution in which mind ultimately answers to the qualities that nature shows forth and individuality, rooted in each organism's concern for its own being, reaches its maximal intensiveness in the capacity of each person to speak for himself and forge his own unique story.

Against the reductionist tendency of modern thought which boils the complex down to its simplest parts, Jonas, in a synthesis of Aristotle, Hegel and Darwin, finds the germ of what is higher in the lower forms from which it evolves. "Reality, or nature, is one and testifies to itself in what it *allows* to come forth from it."[12] The extension of mind to the entire organic world enables

Jonas in *The Phenomenon of Life* to make several speculative, metaphysical claims that cannot be proven but are consistent with the biological facts, existentially interpreted: (1) matter's feat of organizing itself for life attests to latent organic tendencies in the depths of being, and (2) the emergence of the human mind does not mark a great divide within nature but elaborates what is prefigured in all organic existence.[13] These two points allow for a third speculation with dramatic ethical consequences: insofar as we see ourselves, with our capacity for reflecting Being in knowledge, as "a 'coming to itself' of original substance," we should understand ourselves as being called by nature, our own source, to be her guardian.[14] By extending the category of existence to all organisms, Jonas makes possible a radical conversion of modern thought: "a principle of ethics which is ultimately grounded neither in the authority of the self nor the needs of the community, but in an objective assignment by the nature of things."[15]

The Imperative of Responsibility: Living Nature as a Good-in-Itself

The Phenomenon of Life is only preliminary to the ethical task of *The Imperative of Responsibility*; the disclosure of value in nature is not yet sufficient to ground a principle of responsibility for the future. It is still possible, Jonas suggests, for a "nihilist" to acknowledge the presence of subjective value in being, of organisms willing their own existence, while doubting "whether the whole toilsome and terrible drama is worth the trouble."[16] Jonas worries that "our showing up to now that Nature harbors values because it harbors ends and is thus anything but value-free, has not yet answered the question of whether we are at pleasure or duty-bound to join in her "value-decisions," though he admits that once the immanence of purpose in nature has been shown, "the decisive battle for ethical theory has already been won."[17]

What must be established in the second stage of Jonas's reply to nihilism is the *objective* reality of value, a *good-in-itself*, because only from it can a binding responsibility to guard Being be derived. Jonas seeks to establish that the good is not relative to already existing subjective purposes, but rather that it is good that there is purposiveness in nature in the first place: "the very capacity to have any purposes at all is a good-in-itself."[18] He admits that, empirically speaking, the quantity of suffering in life may well outweigh the sum of enjoyment, but nonetheless, suffering rarely destroys the sentient self's will to live. "The very record of suffering mankind," writes Jonas, "teaches us that the partisanship of inwardness for itself invincibly withstands the balancing of pains and pleasures and rebuffs our judging it by this standard."[19] The metaphysical judgment of life's essential goodness cannot be made on hedonistic grounds.

On the basis of the "intuitive certainty" that purposiveness is absolutely and infinitely superior to purposelessness in Being, Jonas derives the "ontological axiom" that "purpose as such is its own accreditation within Being."[20] Being is "for itself," and the facts bear testimony to the idea that being favors the maximization and intensification of purposiveness, ultimately the freedom and peril that come with the human ability to think about, and deeply affect, nature as a whole. From these ontological premises, filled as they are with axiological significance, he draws the ethical conclusion that purposive nature, being good-in-itself, addresses an "ought" whenever it comes under the custody of a will. Though such a will must be infused with a feeling of responsibility in order to be moved by the object that obligates it, the will must experience itself as responding to a transcendent summons in order for the moral sentiment "to be in its own eyes more than a mere impulse."[21]

Just as Jonas grounds the ontological goodness of *Being as such* prior to the ontic goods that are relative to the purposes of particular living beings, so he grounds the ontological goodness of *human being* prior to the goods that are relative to existing human individuals and to the rights that belong to them. Though from an ontic point of view our first commitment is to members of the present generation, especially our kin (and nature guarantees it), this proves inadequate to the task of demonstrating a *principle* of responsibility for future mankind. For if the not-yet-existent have no rights, on what basis can there be a duty to ensure the possibility of their existence and the quality of their lives? Jonas argues that our first duty is ontological; future human individuals matter because the idea of Humanity matters. "Since in [man] the principle of purposiveness has reached its highest and most dangerous peak through the freedom to set himself ends and the power to carry them out, he himself becomes, in the name of that principle, the first object of his obligation, which we expressed in our "first imperative": not to ruin (as he well can do) what nature has achieved in him by the way of his using it."[22]

Essential to the idea of Humanity is the capacity for responsibility. The duty to ensure the future existence of mankind includes the duty to preserve his essence by not undermining the conditions in which man can show himself to be "the executor of a trust which only he can see, but did not create."[23] Here we are reminded of a Heideggerian theme: a command of Being arouses our *conscience* and calls us to take responsibility. But instead of Heidegger's silent call of conscience commanding us to own up to our own possibilities in a moment of resolve in the face of nothingness, Jonas's call has moral substance because it emanates from the *plenitude of being*. The expansion of existential interpretation to include the biological facts allows Jonas "to ground an ethics in the depths of Being": to find value in nature and so to conceive of our freedom as subject to a heteronomous source of authority. The goodness of Being, reality, or nature opens up a "*genuine present*" because it gives us a future worth caring for. This is the meaning of Jonas's pointedly anti-Heideggerian motto: "Responsibility is the moral complement to the ontological constitution of our temporality."[24]

The present, according to Jonas, is governed by an imperative to preserve the human essence forever, and this because our essence is the outcome of a natural process that is good. Our fundamental responsibility is to that which allowed us to come into being, nature herself, but this responsibility is exercised first of all in our relationships to other human beings. The archetype of responsibility is the care of parent for child, where the goal of parenting is the perpetuation of the capacity for responsibility itself. "[The] highest fulfillment [of parenting], which it must be able to dare, is its abdication before the right of the never-anticipated which emerges as the outcome of its care. Its highest duty, therefore, is to see that responsibility itself is not stifled, whether from its source within or from constraints without."[25] The ultimate ground for our duty to our children, however, is not our ontic relationship to them in particular, but our duty to humankind as such: to the idea of Humanity which is part of the idea of purposive nature.

Has Jonas provided the glimpse of eternity that he claims is necessary to overcome the loss of a genuine present in Heidegger's nihilism? Jonas remains a modern in that he cannot rely on the vertical orientation of Platonic ontology which found the eternal beyond the transient: pure Being apart from becoming. Today, Jonas reminds us, it must be becoming rather than abiding nature that holds out the promise of a reunion between ontology and ethics. But nature evolves and can be irreversibly altered by our behavior. If all becoming is simply change *within* nature, and not the destruction of nature, then how can our spoiling of the environment or even collective suicide be a violation *of* nature? "We must seek the essential in transience itself. It is in this context that responsibility can become dominant in morality. . . . Only for the changeable and perishable can one be responsible, for what is threatened by corruption."[26] The good-in-itself is living nature, including humanity as the highest expression of nature's purposiveness. Unlike the permanent and indestructible Good of Platonic ontology, Jonas's Good—our privileged, but delicate place within the totality of nature—is at the mercy of our actions. Insofar as our coming-to-be was "of cosmic importance," an "event *for* Being itself," it matters *to* Being that we preserve our existence and essence.[27]

Jonas hopes to have disclosed an ontological foundation for a principle of responsibility for the future. His *ontological axioms*—(1) that the very presence of purposiveness in Being implies that being is better than not-being, and (2) that the idea of Humanity matters to Being itself as the maximal actualization of its potentiality for purposiveness—ground his *ethical axiom* that "never must the existence or essence of man as a whole be made a stake in the hazards of action." Given scientists' uncertain ability to predict the long-term effects of our technological incursions into nature, Jonas's ethical axiom yields the "*pragmatic rule*" that we be cautious and pursue only modest goals, paying heed to prophets of doom before being seduced by prophets of bliss. Jonas chides utopians for being, in a sense, more pessimistic than prophets of doom, because the promise to perfect humanity presupposes that our nature is not essentially good enough already. For the sake of the perfect, they would risk the good. The rhetoric of hope animating their "starry-eyed ethics of perfection" conceals contempt for the essential condition nature grants us. Against this, Jonas advocates the "sterner" ethic of responsibility, born of a veneration for "the image of man" and fear for what we might lose in the name of progress: a meliorism that aims at modest improvements in our lot. The link between nihilism and utopianism is the absence of veneration for the image of Humanity, and so the loss of a "genuine present" oriented by the "plain truth" that "genuine man is always already there and was there throughout known history: in his heights and in his depths, his greatness and wretchedness, his bliss and torment, his justice and guilt—in short, all the ambiguity that is inseparable from his humanity."[28]

Beginning from an existential interpretation of the biological facts, Jonas accomplishes precisely the *turn toward Being* that Heidegger hoped to initiate by way of fundamental ontology, but could not accomplish so long as existential categories were confined to the human way of being. Having so restricted his existential analysis, Heidegger could see nature only as a mode of being *within* our world, and so could not appreciate its intrinsic value or our continuity with it. For Heidegger, the turn toward Being spells "the *end of metaphysics*" because it reminds us that Being hides itself and that our approach to any supposed unconditioned ground is a conceit that necessarily bears the trace of our finitude. For Jonas, on the other hand, existential analysis provides the thread by which we can find our way beyond ourselves to what is good-in-itself. His expansion of "the ontological locus of purpose from what is apparent at the subjective peak to what is hidden in the breadth of being"[29] permits him to make the *metaphysical* claim, though on admittedly speculative grounds, that Being reveals itself to contemplation *if* thought gives itself the right to transcend materialist assumptions about nature and hence to see the human essence as part of a unified, continuous, intelligible order (what in older times was called a "great chain of Being"). Existentialism, far from presaging the end of metaphysics, proves to be the beginning of a metaphysics that is compatible with the modern idea that Being is dynamic and unfolds in time.

Theology's Impotence in Grounding the Goodness of Life

We have now passed through the first two stages of Jonas's reply to the nihilist who asks, "Why should I care about the distant future of humankind and the planet?" The first, or existential, stage establishes the presence of "subjective inwardness"—mind, freedom, value, and purpose—throughout the entire organic realm. We are not aliens caught up in a lifeless machine, but citizens of a biotic community teeming with life. But the nihilist still may ask how we know that the presence of subjective value throughout the earthly environment is objectively good, and so really worth protecting: not simply "a tale told by an idiot, full of sound and fury, signifying nothing."[30] The second, or metaphysical, stage offers a grounding of the objective goodness of life and of our ethical obligation to safeguard the existence and essence of evolution's most sublime outcome: humanity itself. If Jonas has succeeded, what more could the nihilist want?

There is one last condition that the nihilist may believe is necessary to ensure that human life is worth caring for, namely, that we are creatures of a benevolent God. For if God does not exist, the nihilist may cry out, then we are but a flicker in the darkness and our prayers ultimately go unanswered in the silence of the universe. Jonas invites this worry when he asserts that "there is no point in caring for what has no sanction behind it in any creative intention."[31] One might assume, therefore, that theology comprises a necessary third stage for the overcoming of nihilism. Although Jonas does provide us with a theology that is compatible with his existentialism and metaphysics, he calls theology "a luxury of reason," not a necessity, and explicitly denies that nature needs to be created by God in order to ground an imperative of responsibility. If rational metaphysics proves insufficient on its own, then no appeal to a Creator will satisfy the nihilist. For if life can be shown to be worthwhile without reference to God, then whether God exists is superfluous to ethics, but if life cannot be shown to be worthwhile without reference to God, then appealing to a benevolent Creator will provide only hollow consolation for our experience of life's worthlessness.

In reaction against theological voluntarism, whose conceptual formulation dates back to Plato's *Euthyphro*, Jonas insists that the cause of life's coming-to-be is irrelevant to the question whether it ought to be, for if the world was created by a God worthy of our reverence, then the intrinsic goodness of the world must have been a prior and independent reason for His creating it. The presupposition of a Creator offers us no reason for judging the world to be good if the world does not justify our perception of its value in its own right. Theological voluntarism runs the risk of lapsing into Gnostic dualism wherein faith is owed to a divine will that is wholly absent from the perceptible testimony of His creation. Yet, because theological rationalism posits the priority of the good-in-itself over God's will, it renders the question of life's authorship irrelevant to the issue of its goodness, and so cedes to metaphysics the task of providing a foundation for ethics.[32]

If the nihilist is to be defeated, reason must be able to replace revelation in the office of guiding our ultimate choices, but reason triumphant through modern natural science is precisely the root of nihilism. It undermines the cosmic piety that is supported by the Old Testament tradition, a tradition that Jonas identifies with four key propositions: (1) God created heaven and earth,

(2) God saw that His creation was good, (3) God created man in His own image, and (4) God makes known to man what is good because His word is inscribed in our hearts.[33] It may seem, then, that theology provides the only alternative to the nihilism of modern reason, but I think that Jonas's metaphysics should be read as an attempt to preserve the meaning of the latter three biblical propositions on rational grounds without relying on theology at all. Once the metaphysical prejudice of modern reason—materialism—has been challenged by Jonas's "existential interpretation of the biological facts," he can defend the three latter propositions without recourse to the premise, at the heart of the biblical tradition, that God created heaven and earth.

In each case, we must be able to translate a theological proposition into a naturalistic one. First, that "God saw that His creation was good" gets reinterpreted in Jonas's metaphysics as the good-in-itself of living nature whose very being imposes an "ought-to-be" whenever a responsible agent is there to appreciate it. Second, that "God created man in His own image" gets recast as the notion that the idea of Humanity is an event of cosmic importance because our unprecedented power to reflect Being in knowledge and to recreate nature in our own image is constrained by the good-in-itself: the uncreated, ethical measure of our cognitive and technical powers. Finally, that "God makes known to man what is good because His word is written in our hearts" gets translated as the idea that the objective imperative of responsibility is answered by our subjective capacity to feel responsible for the totality, continuity, and futurity of the object that commands our respect, namely, ourselves as, in Hegelian language, the "coming to itself" of original substance.

Jonas's metaphysics provides a wholly naturalistic interpretation of these Judaic ideas based on an internal teleology: the view that nature is purposive even if there is no purposer, that it makes sense to speak of a creative intention in nature without reference to a Creator, and that life is a gift to be received with gratitude even if there is no giver. Though external teleology (the view that nature is God's creation) may be grafted onto an internal teleology, there is no need to ground metaphysics in theology. Rational metaphysics must stand on its own. If it fails, Jonas contends, theology cannot rescue it from the nihilist's protests. It may be gratifying to those of a religious temper that faith complements Jonas's post-Kantian *Grundlegung*, but theocentric commitments are not necessary to address nihilism and so to meet the emergency of ecological ethics today.

Beyond the Anthropocentrism/ Nonanthropocentrism Debate

Even if an ecological ethic need not be theocentric, it is a commonplace among environmentalists that any ethic adequate to the task of saving our habitat must be *non*anthropocentric. We often hear that by according moral worth only to persons, traditional ethics is anthropocentric and so cannot do justice to our duties to extrahuman nature. If nature has no moral worth for its own sake, then our environment has value only insofar as it is a means to satisfying our human ends. Environmentalists often decry this instrumental attitude as the philosophical core of our ecological crisis.

Two alternatives to anthropocentrism are usually proposed. Some thinkers follow Albert Schweitzer down the path of biotic egalitarianism, claiming that all living beings have equal moral worth. Others (like the American preservationist, John Muir, and more recently deep ecologists like Arne Naess) pursue the way of ecocentric holism, asserting that the environment as a whole has intrinsic moral worth independent of its relationship to us. From both perspectives, Jonas is open to the accusation that his philosophy is too anthropocentric, for he accords us a privileged place in evolution and insists that our primary obligation is to protect the existence and essence of human life.

I believe, however, that Jonas's metaphysics undercuts the very distinction between anthropocentrism and nonanthropocentrism. He thinks we can, and indeed must have it *both* ways. While living nature is a good-in-itself commanding our reverence, and while all organisms participating in this goodness are vulnerable ends-in-themselves who exhibit concern for their own being, humans have special dignity as moral agents, for our will is responsive to ends beyond our own vital ones. "Only human freedom," Jonas writes, "permits the setting and choosing of ends and thereby the willing inclusion of the ends of others in one's immediate own, to the point of fully and devotedly making them one's own."[34] Our first duty is to preserve the noble presence of moral responsibility in nature: of a being who is able to recognize the good-in-itself as such.

This does not mean, however, that we must adopt an instrumental stance toward the rest of nature, viewing it as a mere means or resource to be mastered for our sake. We must come instead to respect ourselves not in virtue of our utter difference from nature, but for "what nature has achieved in us."[35] Our difference in kind is rooted in our continuity with the rest of the biotic community, with others who share in life's goodness, regardless whether they serve our vital needs. Our self-respect requires "cosmic piety," or reverence for the whole of which we are a part. Our obligation to the future of humanity is not based simply on our natural prejudice toward ourselves, but on an objective "assignment" by Being to take care of the delicate web of life that has allowed us to come forth from it. According to Jonas, "the threatened plenitude of the living world" issues a "silent plea for sparing its integrity."[36] This claim upon us is "not a mere sentiment which we may indulge as far as we wish or can afford to," but a categorical imperative emanating from Being when it is "disclosed to a sight not blocked by selfishness or dimmed by dullness."[37] There need be no important divide between anthropocentrism and nonanthropocentrism, between self-respect for our unique title as nature's stewards and reverence for "mother nature" who gave birth to us as her guardian.

Jonas is not a biotic egalitarian because the amount of moral regard organisms deserve as individuals depends upon the quality of their individuality in the psychophysical hierarchy. There is a natural distinction between nutritive and sensitive souls and, among sensitive souls, between animals who experience primitive pleasure/pain responses and those who suffer the passions of desire and fear. Finally, there are humans, metaphysical and moral beings who can take the interests of others to heart beyond our own vital needs. It is no anthropocentric conceit that our first obligation is toward ourselves, for we are the necessary condition of there being obligation in the world at all. Though we should feel awed by the diverse web of species that lets us evolve and sustains us, we are often justified in letting our nonvital interests (like desires for cultural institutions) override even the vital needs of other organisms. There should be no guilt attached to swatting a mosquito and we do not owe mammals in the wild the sort of protection from suffering that they merit when we have domesticated them. Our primary focus should not be on each and every individual organism, but in the words of Aldo Leopold, the American advocate of the land ethics, on protecting "the integrity, stability and beauty" of our environment as a whole.[38]

Having rejected biotic egalitarianism, however, Jonas would not join the camp of radical ecocentric holists either, for they contend that nature has moral worth regardless whether human beings are on the scene. Although Jonas believes that nature carries value independent of us because Being is "for itself" from the inception of life, the *moral* worth of life only comes into being with the phenomenon of obligation, and obligation requires the evolution of a being capable of moral responsibility. We are "an event for Being" precisely because our arrival marks the transition from vital goodness to moral rightness, from desire to responsibility. It might seem that we would do the greatest justice to the ecosystem by removing ourselves from it in an act of supreme impartiality so that other species might flourish, but collective suicide would annihilate the phenomena of justice and injustice alike and so deprive Being of the metaphysical and moral dimensions it took so long to produce. We not only must, but should—out of respect for what nature has achieved in us—appreciate the ecosystem from the perspective of its suitability for our well-being.

Concluding Critical Reflections

I have traced Jonas's reply to nihilism through three stages—existential, metaphysical and theological—and have argued that his answer to the question, "Why should we care about the distant future of humankind and the planet?" does not require an appeal to either theology or radical nonanthropocentrism. At the same time, his answer is far more metaphysically substantial than that of simple anthropocentrists who appeal only to our fear for ourselves or disgust at handing down a degraded habitat to our children. The need to move beyond pragmatism to metaphysics arose because a "heuristics of fear," and prudence for our childrens' sake proved inadequate to the task of grounding a principle of responsibility on behalf of the not-yet-existent who have, Jonas contends, no individual rights. But if Occam's razor permits us to answer the nihilist without theological commitments, might we also, in this allegedly postmetaphysical time, meet the emergency of our ecological plight without relying on a metaphysics that is the functional equivalent of theology?

I am persuaded by Jonas's extension of existential interpretation to the biological life-world. It helps us to acknowledge the continuity and kinship among life forms, and to appreciate what we lose when we cut ourselves off from them and replace nature with artifice. Even without the metaphysical leap to the idea of nature as a good-in-itself, Jonas's critique of dualism lets us "see" differently, and in such a way that we may be moved to act less violently and instrumentally. This monistic revisioning of life opens up new existential choices for us. However, I wonder whether we need to turn the purposiveness embedded in nature into an *imperative* in order to establish limits. Correlatively, I wonder whether we need such a weighty *metaphysics*—one which turns the presence of value throughout nature into a *command* proceeding from Being itself to us—in order to ground our responsibility toward future generations.

This worry takes several forms: pedagogical, speculative, and ethical. First, the pedagogical. If a nihilist is not moved to appreciate nature by an existential interpretation of the biological facts combined with his own experience of the world around him, I doubt there is any hope of convincing him by way of the metaphysical claim that we are the self-realization of nature's purposiveness and are called upon to be her guardian (any more than an atheist can really be converted by the ontological proof of God's existence or an immoralist can be shown the errors of his ways by an argument that persons are, after all, ends-in-themselves). It is not clear to me how or whether the metaphysical arguments that Jonas thinks *need* to be added to his existential interpretations improve the possibilities of moral *education*. I worry that insisting on metaphysical support as being necessary to ground a principle of responsibility may have the negative effect of deflecting attention from the concrete appreciation of nature to abstract argumentation that may be understood by only a few.

The response to this point may well be that metaphysical argument is not meant primarily to convert the unconverted but, like Kant's *Grundlegung*, to articulate the background assumptions that are necessary to support a feeling of responsibility that many feel prereflectively, but that run counter to the presuppositions of the offical philosophical culture. This leads to my second, speculative point. Jonas claims that an ethics rooted in metaphysics does not *require* theology, though his metaphysics remains *compatible* with the theology laid out in his lecture "The Concept of God After Auschwitz."[39] It remains unclear to me what it can mean to speak of human life as "an event for Being" or "an event of cosmic importance" or as being the outcome of "a creative intention" sanctioning our care *without* there being a conscious perspective outside our own *for* whom our destiny matters. If human life were somehow to be extinguished, would this be a disaster for nature herself? One can hold that it would be a tragedy for *us*, and that we ought to do everything we can to avoid it, without assuming that it would be a subversion of nature's final cause or purpose. The death of our star will occur eventually by nature. In the meantime, the biosphere itself might be better off without us; certainly many species would. Can we concur with Jonas that "purpose is in general indigenous to nature"[40] *without* concluding that *our* way of being purposive is special from a larger perspective than our own?

This raises a third, psychological point. One might wonder, along with all advocates of "the hermeneutics of suspicion," whether the insistence that we be subject to an imperative coming from beyond us lest things fall apart is not driven by a psychological need to be at the center, linked to eternity, in order to avoid the bitter truth of our contingency. There is no hint of the terror of nature in Jonas's writing. Though he rejects Hegelian dialectic as a way of understanding human history, he finds a kind of reason at work in cosmic history. It is one thing to say that teleology is a more coherent way of understanding our condition than absurdity, but it is another to suppose that unless cosmic teleology is true, there are no grounds for substantive responsibility. There are many thinkers (I have in mind, for example, the evolutionary biologist Stephen Jay Gould) who reject the idea that nature exhibits an overarching purpose, but who surely feel that we are lucky to be alive on earth and argue forcefully that we ought to respect the delicate natural environment into which we happen to be thrown. Living within a nature that does not care about us or to which we do not make a difference need not be an abyss so long as *we* care—perhaps for its sake as well as ours, but not necessarily because we take ourselves to be special from a cosmic point of view.

Against Jonas's Kantian account of responsibility, which insists that ethics requires a metaphysical good-in-itself sanctioning obligation beyond the web of our natural sentiments, I can imagine a more Humean story that would build on feelings (like those most parents have for their children or field ecologists often have for the niches they come to know) in order to motivate a desire to care about the future habitability of our planet. An empiricist ethic of this sort does not have the systematic force of Jonas's cosmic deontology, but it may be more concrete and generally persuasive.

Finally, a question about the application of Jonas's primary ethical commandment, the duty to perpetuate the existence and essence of human life. What would it mean to undermine our condition in such a fundamental way that our essence had been violated? How do we draw the line between meliorism, which seeks to improve our lot without risking our essence, and utopianism, which would wager "the genuine present" for a "new and improved" humankind? When do proposed improvements violate the legitimate demands of justice, charity, and reason? Few would disagree with Jonas's general commandment. Instead, they would plead innocent to the charge of being utopians, and would call themselves meliorists and their critics neo-Luddites. As ecologist Garret Hardin points out, "Even if the scenario ending in doom is correct, at every point short of the ultimate step, the technological optimist wins out in competition with the technological pessimist."[41]

One worries that in spite of Jonas's critique of formalism, his commandment remains as formal as Kant's categorical imperative, and so just as subject to conflicting contents. It would be an unfortunate outcome if the power and glory of Jonas's ontological arguments succeeded in commanding a consensus, while concealing deep disagreements over the fundamental existential and policy choices we face as we approach the twenty-first century. Even if we need an ontological grounding of ethics, there is no shortcut around paying the closest attention to the ontic domain in which real conflicts persist, for this is where each of us will figure out what it means to *live* like a good Jonasian today.

Notes

1. Hans Jonas, *The Imperative of Responsibility* (Chicago: University of Chicago Press, 1984), p. 22. Hereafter cited as *IR*.
2. Jonas, *The Gnostic Religion* (Boston: Beacon Press, 1963).
3. Jonas, *The Phenomenon of Life* (New York: Delta, 1968), p. 229. Hereafter cited as *PL*.
4. Ibid., p. 232
5. Karl Löwith, "Nature, History and Existentialism," in *Nature, History and Existentialism*, trans. Arnold Levinson (Evanston, Ill.: Northwestern University Press, 1965), pp. 28, 37.
6. *PL*, p. 215
7. Ibid., p. 233.
8. Ibid., p. 283
9. *IR*, p. 81.
10. *PL*, p. 101.
11. Ibid., p. 53.
12. *IR*, p. 69.
13. *PL*, p. 4.
14. Ibid., p. 284.
15. Ibid., p. 283.
16. *IR*, p. 49.
17. Ibid., p. 78.
18. Ibid., p. 80.
19. Jonas, "The Burden and Blessing of Mortality," *Hastings Center Report*, January–February 1992.

20. *IR*, p. 80.
21. Ibid., p. 86.
22. Ibid., p. 130.
23. *PL*, p. 283.
24. *IR*, p. 107.
25. Ibid., p. 107.
26. Ibid., p. 125.
27. *PL*, p. 284.
28. *IR*, p. 200.
29. Ibid., p. 71.
30. Ibid., p. 50.
31. *PL*, p. 234.
32. *IR*, p. 47.
33. Jonas, *Philosophical Essays* (Chicago: University of Chicago Press, 1974), p. 169.
34. *IR*, p. 235.
35. Ibid., p. 129.
36. Ibid., p. 8.
37. Ibid., p. 90.
38. Aldo Leopold, *A Sand County Almanac and Sketches Here and There* (New York: Oxford University Press, 1968), pp. 224–255.
39. "The Concept of God After Auschwitz: A Jewish Voice," in *The Journal of Religion* (1987): 1–13. Also in *Mortality and Morality: A Search for the Good After Auschwitz*, ed. Lawrence Vogel (Evanston, Ill.: Northwestern University Press, in press).
40. *IR*, p. 74.
41. Garrett Hardin, "Review of *The Imperative of Responsibility*," *Hastings Center Report*, December 1984.

Chapter Eight

Domination and Utopia: Marcuse's Discourse on Nature, Psyche, and Culture

Henry T. Blanke

Introduction

During the era encompassing the First and Second World Wars, Marxism suffered a series of historical developments which shook the foundations of its theory and practice. Among the vicissitudes of the Left were the capitulation of the European working classes to militaristic nationalism, the failure of revolutionary activity in Germany followed by the rise of Nazism, and the degeneration of the Russian Revolution into a bureaucratic and repressive regime. After the Second World War, Western capitalism entered a period of unprecedented stability and affluence as prominent social theorists proclaimed "the end of ideology." It was becoming increasingly clear that Marxism was in a state of crisis and, if it was to survive at all as a philosophy of human emancipation, much of its theoretical structure was in need of revision and rejuvenation.

Among those who attempted such a revision was the German-American philosopher, Herbert Marcuse. Along with his colleagues in the group of critical Marxists known as the Frankfurt School, Marcuse became immersed in far-

First published in *Capitalism, Nature, Socialism* 5, no. 3 (September 1994): 99–123; reprinted here with minor corrections.

ranging, interdisciplinary examination of patterns of social domination and their reproduction in personality structure, the family, bureaucracy, and cultural practices. Especially in his writings from the 1950s on, Marcuse explored areas which had been ignored or denigrated by mainstream Marxist theory. Sexuality, art, play, and psychic structure were all investigated as he developed a cultural radicalism which was to have a considerable impact on the New Left and the counterculture of the 1960s. Among the most compelling and unorthodox aspects of Marcuse's thought during this period was his unabashed utopian speculation. Contrary to the traditional Marxist disdain for utopian theorizing, he felt that it was now fruitful and necessary to elaborate a vision of a truly nonrepressive society.

Central to Marcuse's analysis of advanced capitalism, and to his utopian vision, was his critique of the domination of nature and of the instrumental, objectifying epistemology which underlies it. He posited the possibility of a qualitatively new relationship with the environment (and new social relations) based on profound changes in psychic structure and modes of perception. A particular strength of Marcuse's theoretical project was his effort to trace social and ecological domination to the roots of bourgeois civilization while also theorizing the possibility of radical alternatives in psyche and society. Marcuse's treatment of the interconnections between environmental depredation, psychic repression, and unfree social relations will be the focus of this essay.

It is my contention that in light of the ecological crisis confronting contemporary society, Marcuse's thought remains relevant as theoretical inspiration for a new, noninstrumental world-view. When first articulated, Marcuse's critique of technological domination was almost alone on the Left. Over the last twenty years or so, with the emergence of environmentalism as a prominent social movement, themes which Marcuse was among the first to theorize have come to the fore in critical discourse. Much of the latest crisis in Marxist theory concerns its apparent inability to account adequately for ecological devastation and the repressive nature of technocratic culture. I propose that Marcuse's vision of a reconciliation with nature based on a new, nonobjectifying sensibility represents an alternative not only to the technocratic capitalist world-view, but also to Marx's instrumental conception of nature. At the same time Marcuse's critical theory provides a corrective to the tendencies toward nature romanticism, irrationalism, and technophobia into which ecological theorists sometimes lapse.

Marcuse's Appropriation of Freud

Unsatisfied with orthodox Marxist accounts of psychic and social reality, Marcuse undertook a critical appropriation of Freud which culminated in his most seminal work, *Eros and Civilization*. The later Freudian metapsychology, in particular, provided Marcuse with the concepts necessary to uncover the psychosocial roots of both human domination and the drive to dominate nature. However, in order to posit the possibility of a nonrepressive society and a more harmonious relationship with nature, it was necessary to refute Freud's equation of civilized life and social progress with instinctual repression. Marcuse did this by introducing a distinction between "basic repression," the instinctual modifications allowing for work and cooperation which any society must demand of its members, and "surplus repression," the considerable additional controls imposed by a social elite in order to dominate others and maintain its privileged position.[1] Most instinctual repression suffered by humanity is not necessitated by the brute fact of material scarcity, as Freud had thought, but by particular historical distributions of scarcity imposed by ruling classes that support the exploitation of labor. Rather than being organized around equitable provision for human needs and the gratification of desire, societies have exacted a harsh toll from the libidinal energy of their members to ensure systems of hierarchy and control.

By arguing that the rigid controls imposed on deep human impulses toward the erotic, playful, and aesthetic are not universally necessitated by the imperatives of material survival, but rather have been imposed by particular sociohistoric institutions which are subject to change, Marcuse opens the possibility of a society based on gratification and fulfillment. In fact, the very achievements of bourgeois society, in terms of its abundance of social wealth and resources, undermine the traditional rationalizations for repression based on scarcity. Technology and automation offer the promise of minimizing the amount of labor-time required to produce material necessities, thus allowing for the free expression of instinctual energy. Although the ruling order fights to maintain itself, the increasing abundance of resources and the development of technology provide the material foundations for a society beyond scarcity, toil, and domination. Further, Marcuse contends that only liberated erotic instincts can effectively counteract the aggressive and destructive impulses in human nature which violate the natural environment and threaten to destroy civilization.

Marcuse roots humanity's violent degradation of nature in an extroversion of the Freudian death instinct. The increasing surplus in libidinal repression characteristic of bourgeois civilization has so weakened Eros that it is unable to perform its function of neutralizing the death instinct. While aggressive impulses are diverted from the ego toward the external world in the form of technology, since they are not sufficiently stabilized by Eros, their essentially destructive character remains unmodified. "Then, through constructive technological destruction, through the constructive violation of nature, the instincts would still operate toward the annihilation of life."[2] While providing the impetus for technological development, the destructive instincts increasingly threaten the future of both nature and humanity. Only a revivified Eros, freed from the demands of surplus repression, can effectively neutralize its great adversary Thanatos. It might be objected here that Marcuse is misreading Freud, who contended that it is not Eros per se that counteracts the aggressive, destructive impulses, but sublimated sexuality (e.g., bonds of affection and friendship). However, Marcuse is not rejecting sublimation in toto, but only the repressive, externally imposed forms of sublimation characteristic of a culture based on competitive economic performance. A different reality principle (the constellation of introjected and institutionalized norms and values governing behavior in a given society) would permit different forms of sublimation based on a gentle diffusion of the libido throughout the body.

Marcuse argues that a liberation of the life instincts is possible because current levels of technological development and material productivity undermine the rationale for alienated labor. He emphasizes that the automation of work offers the possibility that material needs could be met with a minimum of labor. The possibility of reversing the relationship between labor-time and free time challenges the essence of the established reality principle, which is based on the imperatives of hard work, productivity, and acquisitiveness. The reduction of necessary labor-time to a minimum would "release time and energy for the free play of human faculties outside the realm of alienated labor."[3] This implies a profound reorientation of human existence since the instinctual energy subordinated to the demands of alienated labor would become available for the satisfaction of desire and the fulfillment of human capacities.

Central to Marcuse's vision of a new reality principle is a radical reorientation of humanity's relationship with nature. The classical Western ego "which undertook the rational transformation of the . . . environment revealed itself an essentially aggressive, offensive subject, whose thoughts and actions were designed for mastering objects. . . . Nature (its own as well as the external world) was 'given' to the ego as something that had to be fought, conquered, and even violated."[4] For his symbols of a new orientation toward nature, Marcuse delves into the mythopoeic imagination where, beneath Promethean images of mastery, productivity, and progress through repression, he finds Orpheus and Narcissus, representatives of an entirely different sensibility. "Theirs is the image of joy and fulfillment; the voice which does not command but sings; the gesture which offers and receives; the deed which is peace and ends the labor of conquest; the liberation from time which unites man with god, man with nature."[5]

Whereas Prometheus, the archetypal hero of Western civilization, embodies relentless striving and unceasing labor, Orpheus and Narcissus represent peaceful passivity, play, sensual languor, and aesthetic contemplation. Prometheus stands for a reality principle based on instinctual repression and the domination of nature. Interestingly, Marcuse's selection of Marx's favorite culture hero as the symbol of the repressive reality principle reveals his intent to critique not only bourgeois civilization but implicitly, the centrality of work, production, and purposeful activity in Marx's view of human nature.

As we shall see, Marcuse proposes an existential attitude and relationship with the environment far different from the instrumentalism at the core of Marx's attitude toward nature. The Western self, suppressing its own erotic urges and bent on control, confronts nature across an unbridgeable psychic gulf and perceives only manipulable objects. Orpheus and Narcissus symbolize a liberation of erotic energy which flows outward, forming a bridge between humanity and nature and revealing nature as a subject in its own right. Once liberated, the erotic instincts would generate a new existential attitude in which "being is experienced as gratification, which unites man and nature so that the fulfillment of man is at the same time the fulfillment . . . of nature." Through the experience of a loving union with nature, "flowers and springs and animals appear as what they are—beautiful, not only for those who regard them, but for themselves."[6]

Marcuse supports his interpretation of the Orphic-Narcissistic symbols as representing the erotic union of the self and nature by referring to Freud's concept of primary narcissism. However, Marcuse's interpretation of narcissism differs from the usual one of an autoerotic or egoistic withdrawal from reality, consisting instead in a fundamental experience of oneness with the universe. He quotes Freud to the effect that the normal feeling of separation between the self and the world is only a later development of an original "feeling which embraced the universe and expressed an inseparable connection of the ego with the external world." Marcuse sees this conception as denoting "a fundamental relatedness to reality which may generate a comprehensive existential order" based on "a reunion of what has become separated" (subject and object, humanity and nature).[7]

As the basis for man's reconciliation with his environment, Marcuse advocates a revolution in consciousness and perception that is unabashedly mystical. Even for so unorthodox a Marxist as Marcuse, this is striking and points to the radical nature of his thinking on these issues. The transcendence of the subject-object division, or the temporary dissolution of ego boundaries in an ecstatic encounter with the environment are definitive characteristics of the mystical experience which have some claim to universality.[8] In fact, the term "oceanic feeling," which Freud used to describe the state of primary narcissism, was supplied by his friend Romain Rolland in describing to Freud his own mystical experience.[9] Marcuse anticipates a leitmotif of those ecological theorists who call for a radically new relationship with the environment grounded in a mystical consciousness.

Critics of Marcuse, with Jürgen Habermas leading the way, inevitably accuse him of mysticism as if the word itself—connoting in their minds quietism, anachronism, and irrationalism—were enough to condemn him. However, if to be radical is to go to the root of things, then by proposing a fundamental reorientation of Western consciousness as a prerequisite for overcoming the domination of nature, Marcuse is uncompromisingly radical.

Central to the Western Marxist project is the attempt to account for the failure of socialist revolution in Europe despite the presence of the objective factors which Marx believed necessary for such an event. Consciousness, subjectivity, and depth psychology were employed to explain the reproduction of social exploitation.

Marcuse deepened and extended that project by locating the flaws in the development of Western rationality, meaning the ascendance of purely instrumental, objectifying modes of perception and knowledge which have stunted human nature and violated the environment. He proposed alternative modes of perception which transcend the rigid subject-object dichotomy and allow for a more empathic, sensuous relationship with the environment and between people. Evidence for the possibility of this kind of transformed perception can be found in the mystical literature of almost all times and places. If Marcuse's advocacy of a revolution in consciousness (grounded in observable historical developments such as material abundance) is mystical, then so be it. But far from being quietistic, it is the kind of revolutionary mysticism which animated the utopianism of Fourier or the visionary anarchism of radical Protestant sects in the seventeenth century. It may be the only kind of radicalism sufficiently deep and far-reaching to save the planet from ecological disaster.

For all its power and originality, Marcuse's analysis in *Eros and Civilization* is not without problems. Chief among these is his failure to account for the origins of domination. Marcuse claims that the prevailing repressive reality principle ("performance principle") is only the bourgeois version in a series of hierarchical "organizations of scarcity" that have characterized human history. He asserts that "the gradual conquest of scarcity was inextricably bound up with . . . domination," but fails to explain why this necessarily should have occurred.[10] Marcuse assumes that a harsh and recalcitrant natural environment forced "primitive" people to labor ceaselessly under the threat of starvation. In fact, anthropological literature suggests that many of our Paleolithic and Neolithic ancestors lived reasonably comfortable lives in a close, symbiotic relationship with their environment. Because their demands on nature were minimal, their means were more than sufficient to meet their material needs, leaving time for leisure and communal cultural pursuits. Furthermore, many of these early cultures were characterized by cooperative, egalitarian forms of social organization.[11]

The point is not to idealize primal societies. Marcuse might well point to their relative poverty of needs and failure to develop humanity's potential for great literary traditions, science, and universal concepts of justice and rights based on a sense of shared humanity. However, given the relatively nonrepressive, nonhierarchical conditions of life in some primal societies, the point is to provide an adequate account of the origin of the "logic of domination." Other than flirting with Freud's apocryphal tale of the primal horde, Marcuse fails to do this.

Marcuse grounds his theory of liberation in what he sees as the unprecedented potential for creative leisure made possible by advanced technology. He sees technology as overcoming scarcity and thus undermining both the rationale for alienated labor and the imputed necessity of erotic repression. A strengthened and liberated Eros would then be capable of pacifying the death instincts which have assumed the form of the technological destruction of nature. Marcuse's argument moves in a circle here; as a cure for the disease of domination and repression, he proposes increased doses of the very technological rationality which is a symptom of that selfsame disease. In a subsequent major work, *One-Dimensional Man*, Marcuse addressed this contradiction by providing a trenchant critique of scientific and technological reason and by calling for new forms of science and technology.

New Science and the Pacification of Existence

Marcuse seemed originally to believe that science and technology were neutral. Depending on the social context of their application, they could be used for liberation as easily as for domination. Later he contended that scientific rationality is historically and inextricably bound up with a specific world-view which "experiences, comprehends and shapes the world in terms of calculable, predictable relationships among exactly identifiable units. In this project, universal quantifiability is a prerequisite for the domination of nature," which is reduced to "potential instrumentality, the stuff of control and organization." Science cannot transcend the "lifeworld" in which it is forged; it has an "internal instrumentalist character" and is thus inherently repressive. By virtue of its instrumentalism, science is "*a priori* technology, and the *a priori* of a specific technology—namely, technology as form of social control and domination."[12] As scientific-technological rationality proved increasingly efficient in the control and manipulation of nature, the possibility opened up for the same logic and method to be applied to the social world. In his analysis of "one-dimensionality," Marcuse sees technological rationality as a prime force in the development of a society in which all segments of the population, including potentially oppositional elements, have been manipulated and integrated into a total system of administration, efficiency, and relatively comfortable domination.

In its emphasis on the threat to human autonomy posed by social rationalization and the triumph of a purely means-oriented instrumental reason over a rationality concerned with social goals and values, *One-Dimensional Man* bears the influence of Max Weber. But while Weber saw no antidote to the "disenchantment" of the world, Marcuse posits the possibility (and necessity) of a radically reconstructed science and technology based on a "qualitatively new mode of 'seeing' and qualitatively new relations between men and between man and nature."[13] He envisions the transcendence of the historical separation of means and ends and facts and values, and the integration of philosophy, art and science. Ideals such as the "pacification of existence" would infuse scientific concepts and be translated into concrete technical possibilities. As previously utopian speculations about "the Good Life, the Good Society, Permanent Peace" become material possibilities, they might inform and guide science so that reconstructed "scientific concepts could project and define the possible realities of a free and pacified existence." Through such a reconstructed science of liberation, "the free play of thought and imagination assumes a rational and directing function in the realization of a pacified existence of man and nature."[14]

The goal of pacification involves the reduction of misery, violence and cruelty in society and nature. Humanity's relationship with the environment would be based on gratification rather than domination, and would involve cultivation and careful extraction of resources instead of destruction and exploitation. Marcuse calls for a radical change in a standard of living based on the wasteful production of superfluous consumer goods, and advocates a reduction of overdevelopment which squanders natural resources.[15] Interestingly, however, pacification for Marcuse also presupposes a "liberating mastery" of nature which protects humanity from the blindness and ferocity of natural forces. He sees the relationship between humanity and nature in terms of the former's historical transformation of the latter so that the qualities of nature become, in some sense, socially mediated historical qualities. In this, Marcuse distinguishes himself from those ecological theorists who advocate a submissive adoration of the environment and a return to a more "natural" condition. For him, an environment unmediated by human intervention (if this were possible) would be harsh and brutish, and ideologies which idealize nature tend to rationalize oppressive social relations (such as racial hierarchies) as being natural. "Civilization produces the means for freeing Nature from its own brutality . . . by virtue of . . . the transforming power of Reason."[16] To fulfill this function, reason itself (and technology) must be transformed so as to bring out its affinities with art.

Marcuse emphasizes that a liberating science and technology must be infused with aesthetic values. "The rationality of art, its ability . . . to define yet unrealized possibilities could then be envisaged as validated by and functioning in the scientific-technological transformation of the world."[17] Through a "technology of pacification," the productive machinery of society would reflect aesthetic categories and be organized to allow for the free play of human faculties. For Marcuse, science and technology must be oriented toward the reduction of disease, the elimination of poverty, and the provision of a decent standard of living for all. Beyond these goals, he envisions a society in which automation has reduced socially necessary labor-time to a minimum, not only allowing workers an unprecedented amount of leisure, but also giving them the opportunity to exercise their creative and imaginative faculties within the technological realm. They would be free to experiment and play with technology in order to fulfill genuine human needs, promoting an aesthetically gratifying, life-enhancing environment.

At times, Marcuse seems to be interested in nature only as a medium which can stimulate man's own aesthetic and sensuous nature, a subtle version of the instrumentalism he critiques elsewhere. The stronger trend, though, seems to be toward a persistent regard for nature as a subjectivity in its own right. However, the question remains how nature can be perceived in and for itself ("in its own right") if it is, as Marcuse contends, always mediated by human history and social institutions. It is one of the strengths of Marcuse's analysis that he rejects the romantic ideal of a return to a pure and pristine natural state. Humanity is, after all, part of nature; therefore, its historical transformation of the environment cannot be considered "unnatural." Marcuse posits a distinction between a repressive and a liberating mastery of nature, with the latter involving "the reduction of misery, violence, and cruelty."[18] Still, one wonders to what extent nature can be subsumed under human history, even via a liberating mastery, without losing its "inherent" qualities. Further, if instrumental rationality is as inherently oppressive as Marcuse believes, it is not clear to what extent the technological mastery of nature could be liberating (for both nature and humanity) in the radical sense Marcuse intends. Would not even a liberating mastery contain symptoms of the disease he wishes to eradicate? Marcuse, perhaps, would argue that a certain minimum threshold of technological development must be reached before qualitative changes in human sensibility, technology, and the relationship with nature can occur.

Ambiguities such as these have resulted in some curious misinterpretations. C. Fred Alford reads Marcuse's call for reconciliation with nature as mere rhetoric softening his real interest in technological conquest as the precondition for human freedom. He interprets Marcuse as an advocate of "the complete subordination of nature to human purposes." Only the technological domination of nature can provide the material abundance, and freedom from work, necessary for the liberation of humanity's erotic-aesthetic capacities. For Alford, Marcuse's new science is merely "an ideology. It grants the aura of reconciliation with nature to what is actually projected to be humanity's final victory over it."[19]

Alford's contention that Marcuse was only interested in expanding human freedom and pleasure ignores the extent to which Marcuse believed that the liberation of human instincts depends on respecting and promoting the intrinsic potential of nature. Just as ecological domination entails the repression of human desire and instinct, so human and ecological liberation are deeply intertwined. This is explicitly set forth in the chapter on nature in *Counterrevolution and Revolt*, as is Marcuse's rejection of the anthropocentric paradigm which reduces nature to "matter, raw material . . . only a productive force," and which cannot recognize that nature "also exist[s] for its own sake."[20] Remarkably, Alford does not discuss this crucial chapter.

For Alford to reduce Marcuse's ideas on reconciliation with nature, which I have been arguing are central to his *oeuvre*, to the status of "rhetoric" and "ideology" seems extreme, if not perverse. However, such an interpretation is perhaps understandable in light of the contradictions in Marcuse's view of nature between the Marxist elements (which Alford emphasizes) and the ecological-mystical elements (which he does not take seriously). The tension between these two elements (which would never be completely resolved) would lead Marcuse to argue, on the one hand, that "pacification presupposes mastery of Nature, which is and remains the object opposed to the developing subject" and, on the other hand, that through a new sensibility "the opposition between man and nature, subject and object, is overcome."[21] Ultimately, Marcuse's last word on the subject, as we shall see, is that the liberation of human and nonhuman nature are inextricably bound and that the nonhuman nature is a subject in its own right which must be honored.

Alford does not believe that Marcuse sincerely envisions a reconciliation with nature because he does not take into account sufficiently the radical nature of Marcuse's new science and technology, upon which such a reconciliation depends. Indeed, there is some question as to how radical Marcuse intends these concepts to be. In the face of the critique of Marcuse's ideas on a noninstrumental relationship with nature based on a qualitatively new science and technology (most prominently Habermas's), William Leiss has put forth an odd defense of Marcuse by claiming that these ideas are minor "inconsistencies" found in "isolated" passages.[22] According to Leiss, Marcuse is really talking about changing the social context (from capitalist to socialist) of an otherwise neutral science and the social applications of technology, not a transformation of scientific concepts and technological rationality. Granted, there are inconsistencies in Marcuse's writings which might lend credence to this interpretation: "Science and technology are the great vehicles of liberation, and...it is only their use and restriction in the repressive society which makes them into vehicles of domination." Elsewhere he contradicts himself: "Not only the application of technology but technology itself is domination (of nature and men)."[23] The overwhelming thrust of Marcuse's thought, however, is toward a radically reconstituted scientific project, based on profound changes in human sensibility, which "would arrive at essentially different concepts . . . different facts . . . new mode[s] of 'seeing.'" Not merely a more enlightened use of existing technologies is at issue here, but a decisively different technology of liberation, "conceivable only . . . as expressive of the needs of a new type of man."[24]

Leiss imputes to Marcuse his own belief that "the fault lies not with technological rationality itself, but with the repressive social institutions which exploit the achievements of that rationality."[25] Scientific rationality is somehow external to and independent of social exploitation. However, as Robert

D'Amico points out in a position much closer to Marcuse's, "exploitation is not an afterthought, it does not precede and appropriate a fully formed science ready at hand, since it is first a lived human relationship in the very production of knowledge." Social exploitation determines scientific concepts; "it is embedded in the intentional structure that defines operative reason."[26] Marcuse would be fully in accord with this statement. In fact, it is among his most representative positions that the "principles of modern science were *a priori* structured in such a way that they could serve as conceptual instruments for the universe of self-propelling productive control."[27] Leiss shies away from coming to terms with Marcuse's most radical and provocative ideas on science and technology by simply denying that they are meant to be as radical as they seem.

Habermas, on the other hand, is correct in attributing to Marcuse a serious belief in the possibility of an alternative science grounded in a quasi-mystical perception of nature. However, he is wrong in categorically foreclosing such a possibility; in so doing, he capitulates to current forms of destructive technology and ecological devastation. Habermas roots science and technology as we know them in the structure of human purposive-rational action and maintains that "as long as the organization of human nature does not change" and social labor remains a necessity, a qualitatively different technology is impossible.[28] The instrumental control of nature is unavoidable and is the price we must pay for the possibility of free, noninstrumental human relations. Such a view precludes, in principal, the possibility of deep changes, not only in humanity's relation to nature but within human nature itself. It is not only "the resurrection of fallen nature" which Marcuse seeks, but the transformation of human nature, including its relation with the environment and the nature of work.

Habermas rejects Marcuse's contention that science and technology are historically contingent projects, bound up with the emergence and development of capitalism, and thus subject to transformation under different historical conditions. For Habermas, technology as we know it cannot be historically transcended because it is a project of the human species as a whole. It is intrinsically involved with the structure of purposive-rational action which is, in turn, rooted in the function of the human body and therefore cannot be altered.[29] Further, science and technology are expressions of the fundamental "human interest" in the control and manipulation of nature necessary for survival. However, given the mounting environmental crisis, the question must be raised whether there is also an imperative human interest in preserving the ecosystem which demands a relationship with nature other than one based on control. The anthropological literature provides evidence of peoples whose attitude toward their environment was based on reciprocity and reverence and who developed technologies consistent with this attitude.[30] The anthropological evidence belies Habermas's contention that Western technology represents a universal human project. Marcuse's idea of a new science based on a "liberating mastery" of nature is an attempt to incorporate such alternative conceptions of nature within a higher stage of historical and technological development.

Habermas posits a fundamental distinction between purposive-rational action (work) oriented toward control of the environment and symbolic interaction (communication) oriented toward mutual linguistic understanding. Because work and communication are basic to human nature and because the logics governing these two modes are categorically distinct, nature cannot be known to humanity as anything other than an object of "possible technical control."[31] The possibility of an empathic understanding of the environment and awareness of humanity's embeddedness in it is categorically foreclosed. Human communication can be made freer, less distorted, and more enlightened, but the human interaction with nature must remain instrumental. A thoroughly anthropocentric ethic follows from this perspective by which intrinsic purpose, value, and meaning are attributed solely to communicating humans and denied to nature.

By rooting work in the unalterable constitution of the human organism and rigidly separating it from interaction, Habermas rules out a priori the possibility of radical transformations of work, technology, and the human relationship with nature. Treating these distinctions as dialectical rather than absolute enables Marcuse to envision the possibility that work can become more playful, technology more aesthetic, and our relationship with nature more harmonious. Because he denies any intimate connection between human and nonhuman nature, Habermas does not appreciate the ways in which ecological domination rebounds on the human subject. Yet, Theodor Adorno, Max Horkheimer, and Marcuse interpreted the dialectic of enlightenment in terms of how the progressive control of nature necessitates the repression of the erotic and playful in human nature. As Marcuse recognized toward the end of his life, now the situation is such that not just certain human qualities, but the very survival of humanity is threatened. We must ask, as Joel Whitebook does with regard to Habermas, whether we can "continue to deny all worth to nature and treat it as a mere means without destroying the natural preconditions for the existence of subjects?"[32]

When first formulated, few disagreed with Habermas's rejection of Marcuse's call for a substantively transformed science and technology. However, Whitebook is representative of those recent critics who, motivated by the escalating environmental crisis, have questioned Habermas's instrumentalist view of nature and his anthropocentric ethics. Henning Ottman equates Habermas's technical cognitive interest with a boundless "will-to-control" unable to recognize nature as a "purpose-for-itself." Such a recognition of "the dimension of nature beyond control and exploitation" is now imperative if we are to resolve the paradox that the technical interest in control (for Habermas the only possible cognitive orientation toward nature), the legitimacy of which is based on human survival, increasingly threatens survival.[33] In a similar vein, Vincent Di Norcia argues that the technical interest in control suggests a rigid demarcation between the natural "object controlled and its [ecological] surroundings, and between controller and controlled. Such boundaries are in reality fluid and relative."[34]

Ultimately, what separates Marcuse from Habermas on these issues is their respective standpoints toward modernity. Habermas seeks to complete the Enlightenment project by holding bourgeois society accountable to its own highest ideals. Contrary to Marcuse, "he takes rational autonomony as a perfectly adequate idea of selfhood, and feels no need [for] . . . a new sensibility."[35] In agreement with the classical Enlightenment perspective, Habermas sees the ideal of rational autonomony as being realized through the progressive conquest and mastery of nonhuman and human nature.

By contrast, the first generation of Frankfurt School theorists traced man's inhumanity to man (exemplified by fascism) to an intrinsic and irremedial flaw in the Enlightenment conception of reason and selfhood. Nothing less than an epochal transformation would rescue the hope of human emancipation. Marcuse lived long enough to see the degradation of the environment assume a status comparable to fascism, as evidence of the inadequacy of the project of modernity.

In a judicious appraisal of these respective positions, Whitebook notes how Habermas's view avoids the reactionary potential of chiliastic and total rejections of modernity. However, "it may be that the scope and depth of the social and ecological crisis are so great that nothing short of an epochal transformation will be commensurate with them."[36] It is in the possibility, indeed, the necessity of such a radical transformation that Marcuse's relevance lies.

On the Reconciliation with Nature

When *One-Dimensional Man* was written, Marcuse had not moved fully beyond an objectifying, instrumentalist conception of nature. There remained a tension, perhaps never fully resolved, between the Marxist and mystical elements in his thought. As Martin Jay has pointed out, as early as 1932, Marcuse had argued that Marx had sought the unity of man and nature while, in the same article, he implied that nature had to be dominated.[37] These same contradictions would recur over subsequent decades with Marcuse maintaining, on the one hand, that "pacification presupposes mastery of Nature, which is and remains the object opposed to the developing subject" and, on the other hand, that through a new sensibility "the opposition between man and nature, subject and object, is overcome."[38]

However, Marcuse's last significant writings on the subject, in *Counter-*

revolution and Revolt, finally render interpretations such as Alford's and Leiss's untenable. Here he moves beyond the view, still lingering in *One-Dimensional Man*, that nature is an object opposed to a human subject. He is concerned with the recognition of nature as a "subject in its own right . . . as a *cosmos* with its own potentialities," which humanity should strive to enhance. Marcuse no longer refers to the "liberating mastery" of nature, but to the "liberation of nature" itself.[39] He stresses that nature is part of history and, under the historical horizon of the instrumentalist-technological world-view, appears as value-free raw material to be controlled and manipulated. A different historical world-view could generate a reciprocal relationship with the environment through which human beings would recognize that the fulfillment of their own inner nature requires a respect for the needs of outer nature.

Marcuse emphasizes the interconnections between the repression of human nature and the violation of the environment, on the one hand, and a new sensual-aesthetic epistemology and the liberation of nature, on the other. The homology between human nature and external nature (both animated by erotic energy) is central to Marcuse's thought, as is his view that both are historical entities subject to social transformation. In fact, in a lecture delivered shortly before his death in 1979, Marcuse reiterates the necessity of transforming inner nature in order to preserve external nature. Reciprocally, the project to protect the environment works to pacify our psychic nature. Continuing along the lines developed almost twenty-five years earlier in *Eros and Civilization*, Marcuse speculates whether striving for a state of freedom from pain is characteristic of the life instinct rather than of the death instinct, as Freud believed. If so,

> the drive for painlessness, for the pacification of existence, would then seek fulfillment in protective care for living things . . . in the restoration of nature, both external and within human beings. . . . The protection of the life-environment will also pacify nature within men and women. A successful environmentalism will, within individuals, subordinate destructive energy to erotic energy.[40]

Just as industrial-capitalist society has fostered an aggressive, competitive personality structure, it has produced a polluted, commercialized environment. Individuals in this society are blocked from seeing their own creative activity reflected in the environment, and they cannot experience nature as a subject in its own right. However, Marcuse stresses that since nature is historical, its liberation would not mean regressing to a pretechnological situation, but advancing to a stage where the achievements of industrialism would be preserved while its destructiveness would be eliminated.[41]

Marcuse says little on exactly how this remarkable development would occur, but locates the possibility of a nonexploitative relationship with nature in the absolute necessity of a profound transformation in human perception. The rigid epistemological dichotomy through which an atomized subject confronts a world of inert objects must be overcome in favor of an empathic, receptive mode of perception, "the ability to see things in their own right, to experience . . . the erotic energy of nature." Marcuse is not speaking of merely a more relaxed, contemplative appreciation of nature, but of an "erotic cathexis," a deep sympathetic vibration of the life-energy animating both human and nonhuman nature. Such a mode of perception would affect a "synthesis, reassembling the bits and fragments [of] distorted humanity and distorted nature" into an experience of wholeness, integration, and fulfillment. "The liberated consciousness would promote the development of a science and technology free to discover and realize the possibilities of things and men in the protection and gratification of life, playing with the potentialities of form and matter for the attainment of this goal."[42]

Marcuse bases his analysis on the young Marx's ideas on the emancipation of the senses and the human appropriation of nature. Humans transform nature according to their needs and capacities, and the environment comes to reflect human sensibility. This "humanized" environment provides a medium for the further unfolding of human faculties and abilities. Freed from the distorting economic and technical imperatives of capitalism, this process would allow for the limitless expansion of human creative capacities. According to Marcuse's interpretation, this conception comprehends nature as a universe which becomes the congenial medium for human gratification in the degree to which nature's own gratifying forces and qualities are recovered and released. In sharp contrast to the capitalist exploitation of nature, its "'human appropriation' would be nonviolent, nondestructive: oriented to the life-enhancing, sensuous, aesthetic qualities inherent in nature."[43]

While it is clear that Marx regarded nature as a medium for the liberation of human sensibility, it is not apparent how, through this same process of appropriation, nature itself would be emancipated and perceived as subject in its own right. The latter concern is more Marcuse's than Marx's. In fact, Marcuse admits that "Marx's notion of appropriation of nature retains something of the hubris of domination. 'Appropriation,' no matter how human . . . offends that which is essentially other than the appropriating subject" and which exists as a subject in and of itself. The idea of appropriation retains the implication of humanity's struggle with nature, but Marcuse suggests that "the struggle may also subside and make room for peace, tranquillity, fulfillment. In this case, not appropriation but rather its negation would be the nonexploitative relation: surrender, 'letting-be,' acceptance."[44]

Marcuse's Contemporary Relevance

Marcuse's unyielding radicalism and his relentless efforts to uncover the roots of the repression of human nature and the violation of nonhuman nature in the development of Western rationality and bourgeois civilization provides a clear alternative to the shallow liberalism characteristic of mainstream "environmentalism." Murray Bookchin uses the latter term, as opposed to "ecology," to designate "a mechanistic, instrumental outlook that sees nature as a passive habitat" which must be protected so as to ensure continued human use. This approach, which dominates mainstream discourse, does not begin to challenge the assumptions of industrial capitalism, but merely seeks less crude, more efficient means of extracting natural resources. "Environmentalism does not question the most basic premise of the present society, notably, that humanity must dominate nature; rather it seeks to *facilitate* that notion by developing techniques for diminishing the hazards caused by the reckless despoliation of the environment."[45]

Environmentalism argues for a more rational science and technology without challenging, as did Marcuse, their normative and conceptual foundations. According to Stanley Aronowitz, "the reason for this omission is that liberal ecology, like much of Marxism, separates the domination of *nature* from *human* domination." It does not recognize that in capitalism, the same logic which reduces nature to its abstract, measurable features is extended to all spheres of economic and social life. However, our external environment, like our psychic nature, "can neither be reduced to their quantitative aspects for the purposes of control, nor exploited instrumentally, without dire consequences for us."[46] This is a central thesis of *One-Dimensional Man*. That one of the most astute of current radical theorists felt compelled to reiterate the point in his reassessment of Marxist theory in the early 1980s speaks to Marcuse's continued relevance. The latter's critique of instrumental rationality and his positing of the necessity of nonobjectifying modes of perception remain fruitful alternatives to the instrumentalism at the core of Marxist and liberal environmentalist theory.

At the same time, Marcuse's views on the domination of nature have advantages over some of the more radical philosophies and movements that have emerged in critique of reformist environmentalism. Prominent among these is the philosophy of deep ecology, which gives ontological and ethical primacy to the nonhuman biosphere and assumes the inviolability of natural processes.[47] In an often implicit, sometimes overt antihumanistic position called "biocentrism," deep ecology opposes an abstract, idealized nature to an equally abstract, undifferentiated concept of humanity, valuing the former over the latter. Without considering racial, class, or gender differences, humanity as a whole is seen as the enemy of the biosphere. This view reduces people from complex social beings to a simple species and ignores the social origins of the environmental crisis in the emergence of social hierarchy, instrumental rationality, and the imperatives of capital.

Deep ecology, in common with other currents of radical ecological thought, often lapses into a crude dualism in which human culture and nature are seen as antithetical. For Marcuse, human society is part of natural history just as nature is mediated by society. While he emphasized nature's autonomy as a corrective to both bourgeois and Marxian instrumentalism (in Marx the "humanization of nature" takes precedence over the "naturalization of man" with the result that nature is reduced to "raw material" for labor), Marcuse never lost sight of the two-sidedness of humanity's relationship with the environment. He would have critiqued biocentrism as he did anthropocentrism (or any other centrism) as being undialectical and hierarchical. Marcuse did not oppose the sociohistoric mediation of the environment, but rather the instrumental subsumption of nature under the specific historical imperatives of technocratic capitalism. Bookchin, greatly influenced by the Frankfurt School, puts it well when he writes:

> natural evolution has conferred on human beings the capacity to form a "second" or cultural nature out of "first" or primeval nature. Natural evolution has not only provided humans with the ability, but also the necessity to be purposive interveners into "first nature." . . . [Human society] can be placed at the service of natural evolution to consciously increase biotic diversity, diminish suffering, foster the further development of new and ecologically valuable life-forms.[48]

Marcuse used the concept of second nature to analyze human psychic nature as well. In particular, he felt that women, by virtue of their forced exclusion from the dehumanizing competition and aggressiveness of capitalist work relations, had retained the receptive, empathic qualities essential to the sensibility capable of forging a nonexploitative relationship with nature. Marcuse saw that environmental domination was deeply implicated in patriarchal social relations and came to believe that individual, social, and ecological liberation required the ascendance and institutionalization of those qualities (tenderness, nurturance, nonviolence) historically attributed to women. In this he anticipated ecofeminism, one of the more interesting developments in contemporary ecological theory because of its attempt to unite two of the most important progressive movements of the last two decades.

Briefly stated, ecofeminism seeks to challenge the premises and practices (in all spheres of human life) of patriarchal societies (especially Western industrial society) which have historically identified women and nature as fundamentally "other" and thus subject to manipulation, control, and exploitation.[49] Some versions of ecofeminism, in arguing that women are in a unique position to protect the environment due to the closeness of female nature to nonhuman nature, fall victim to a biological determinism which reifies gender differences.[50] At worst, this undifferentiated and ahistorical valorization of an essential female nature as the basis for a moral alternative to male culture degenerates into a politically quietistic, romantic celebration of nature and feminine spirituality.

Marcuse is careful not to ascribe the female qualities he sees as essential to a new sensibility to any intrinsic, biological, female nature. The obvious physiological characteristics specific to women have been translated historically into social factors used to justify their oppression (e.g., the identification of woman as mother or sex object). However, in true dialectical fashion, this same oppression, specifically the relegation of women to the home and their exclusion from the workplace, has allowed women to remain closer to more sensuous, life-enhancing, human qualities, while men have been subject to the harshness and dehumanizing impact of the capitalist workplace. These life-enhancing qualities have become, over the course of centuries, female "second nature": not innate characteristics exclusive to women, but qualities deeper than those resulting from socialization alone.[51] While the liberation of women from isolation in domestic life and complete social and economic equality are essential, the true subversive potential of feminism lies in a condition "beyond equality." Marcuse fears that equality, within a capitalist framework, would mean the capitulation of women to male values of aggression, competition, and domination of nature. Patriarchal values and capitalist institutions must be transcended by a psychic, political, and cultural revolution grounded in "the ascent of Eros over aggression . . . and this means, in a male-dominated civilization, the 'femalization' of the male."[52]

The view that women possess unique, tender qualities has, of course, been used historically to justify their marginalization. Marcuse recognizes the deeply conservative implications of the idea of an innate female essence associated with motherhood and a bond with nature. Marcuse's concept of "second nature" (historically sedimented psychosocial conditioning) allows him to avoid the ahistoricism of both patriarchal and feminist essentialism. The qualities he identifies as historically embodied in women are fundamental human qualities, equally accessible to men under different social conditions. At the same time, the idea of second nature provides for an appreciation of female difference and rejects the formal equality of liberal (and much of socialist) feminism, which tends to consider positive regard for psychocultural female differences as misguided. Marcuse recognizes that facile egalitarianism under present conditions "would be regressive: it would be a new form of female acceptance of a male principle."[53] Owing to their unique historical experience, women are positioned to develop the qualities of empathy, mutuality, and peacefulness into an ecological sensibility. This sensibility may then be brought into the arena of political struggle as the foundation for a universal project of psychosocial and ecological transformation.

Marcuse's examination of the epistemological foundations of modern science anticipated the ecofeminist scholarship of the last decade which has questioned "the Cartesian masculinization of thought" characteristic of the Scientific Revolution.[54] Beginning with Carolyn Merchant's *Death of Nature*, feminist thinkers have argued that the modern subjugation of nature and women has been implicated with the historical ascendance of ways of knowing which privilege detachment, abstraction, and control over more organic, participatory modes of cognition.[55] Similarly, Marcuse forcefully criticized the disembodied dualism and will to power underlying Western modes of being and knowing.

Recently Ynestra King, one of the first and most nuanced ecofeminist theorists, has developed a position remarkably congruent with Marcuse's (in fact, she cites the Frankfurt School in one of her early essays) in her attempt to forge an "epistemology based on a noninstrumental way of knowing." King rejects, as did Marcuse, the nature-culture dualism of both essentialist ecofeminism (with its ahistorical romantic naturalism) or liberal and socialist feminism (which emphasize gender equality and deny that women are uniquely positioned to develop an ecological, noninstrumental sensibility). King advocates "a new dialectical way of thinking about our relationship to nature. . . ." For her, "the domination of nature originates in society and therefore must be resolved in society." In a passage redolent of Marcuse, she writes,

> Both feminism and ecology embody the revolt of nature against human domination. They demand that we rethink the relationship between humanity and the rest of nature, including our natural embodied selves. . . . An analysis of the interrelated dominations of nature—psyche and sexuality, human oppression, and non-human nature—and the historic position of women in relation to those forms of domination is the starting point of ecofeminist theory.[56]

It is precisely the analysis of the interrelated dominations of both human and nonhuman nature that is the particular virtue of Marcuse's project. At a time when the so-called new social movements (ecology, feminism, gay rights) suffer from fragmentation and the lack of a theoretical basis to form coalitions among each other or with socialist, labor and other older, progressive movements, Marcuse's critical theory provides a unified analysis of the complex mediations and interrelationships structuring domination in the psychosexual, cultural, political, economic, and ecological realms. At a time when issues of sexual, racial, and ecological "otherness" are high on the progressive theoretical agenda, and when particularity and difference are celebrated, Marcuse provides a perspective from which we can trace the domination of the Other to the roots of bourgeois civilization while, at the same time, advancing a *universal* project for human emancipation.

Marcuse's critical theory is not without flaws. His formulations are too often vague and lacking in empirical detail. For instance, his ideas on the need for a radically reconstituted science and technology respecting and enhancing the potentiality of nature are richly suggestive, but bereft of concrete specificity. For example, what would the conceptual structure of a new physics (presumably one which does not reduce nature to abstract, quantifiable elements) look like?

What forms would a new technology take? Such lacunae in Marcuse's writings provide an opportunity to develop the insights of the Frankfurt School. Murray Bookchin, in "Towards a Liberatory Technology"[57] and later writings, has concretized Marcuse's ideas on alternative technologies and ecologically sound social structures, and has influenced younger thinkers in the small but vital social ecology movement.

Andrew Light has recently emphasized the similarities between Marcuse's views and Bookchin's social ecology by describing both in terms of "environmental materialism," which he counterposes with "environmental ontology."[58] For the former, environmental depredation is rooted in the specific technical-economic system for the reproduction of material life characteristic of capitalism. Solutions to both ecological and social domination are located in a radical restructuring of the organization and control of the means of material production. Environmental ontology, on the other hand, stresses the need for a fundamental change in consciousness toward a perception of the symbiotic embeddedness of humanity within the larger ecological system. While locating Marcuse in the materialist camp, Light correctly maintains that Marcuse's perspective is actually an amalgam of the two positions. The values and imperatives of capitalism become introjected into the instinctual structure of the individual and it is from these depths, Marcuse insists, that any genuine rebellion must emanate. At the same time, changes in the material conditions of social life provide the necessary impetus for transformations in consciousness. What Light is alluding to in terms of Marcuse's "combined materialist-ontological . . . framework"[59] is the same tension in Marcuse's thought that I have described between its Marxist and mystical elements.

Objecting to these types of ambiguities, Bookchin overstates his differences with Marcuse. In responding to Light, he criticizes Marcuse's "often romantic view of 'Nature' " which contradicts "his call to 'master' this 'Nature' socially and technologically in order to 'liberate her.'" Bookchin also separates himself from Marcuse's instinct theory, which he finds "mythopoeic and personalistic."[60] He goes too far, however, in implying that Marcuse's thought is compatible with the misanthropy, primitivistic antirationalism, and social quietism of deep ecology and other strands of ecomysticism. Marcuse's thought is especially relevant today precisely because it offers a corrective to these types of problems within the ecology movement. When Bookchin says that "without in any way eschewing the need for a new spirituality, character structure, individuality, or sensibility, I have always grounded my ideas of ecology and anarchism in *social* relationships," he is not as out-of-sync with Marcuse as he believes.[61] Marcuse was always concerned with the material and social conditions necessary for psychic transformation. Analyzing the dialectical interplay between individual consciousness and social structure, between nature (both internal and external) and culture, was a hallmark of his approach. Furthermore, Marcuse's thought fosters the utopian tendencies whose disappearance from today's Left Bookchin laments.

While Marcuse's uncompromising utopianism is to be admired, one has to wonder if he does not sacrifice careful consideration of practical political strategies for addressing the problems of human and ecological domination. His insistence on a total revolution in consciousness and culture overlooks the more immediate and realistic possibilities for social change, and perhaps obscures the extent to which gradual, incremental reforms can do much to reduce the suffering of people and nature here and now.

Yet, utopian theory is valuable in that it provides normative standards by which to measure and critique the quality of life in present-day society. The utopian imagination produces images of peace, freedom, and fulfillment that can inspire efforts to approximate them in reality. Postmodernist fashion currently hails the collapse of the Marxian metanarrative, yet the transnational, corporate-capitalist grand narrative of market integration, cultural commodification, and social and ecological homogenization (the latter signaled by the perilous reduction of biospheric diversity) continues unabated. This is the social reality we capitulate to when we give up the impulse to critique what is by imagining what could be. Marcuse's vision of a society beyond domination, where the free development of nonhuman nature is the condition for the free development of human nature and vice versa, challenges us to keep the utopian imagination alive. That is his continuing legacy.

Notes

1. Herbert Marcuse, *Eros and Civilization* (Boston: Beacon Press, 1966), pp. 35, 37.
2. Ibid., pp. 86–87.
3. Ibid., p. 156.
4. Ibid., pp. 109–110.
5. Ibid., p. 162.
6. Ibid., p. 166.
7. Ibid., pp. 168–170.
8. Agehananda Bharati, *The Light at the Center: Context and Pretext of Modern Mysticism* (Santa Barbara, Calif.: Ross-Erikson, 1976), pp. 15–61 and passim; W.T. Stace, *Mysticism and Philosophy* (London: Macmillan, 1960), pp. 41–133 passim.
9. Sigmund Freud, *Civilization and Its Discontents*, trans. and ed. James Strachey (New York: W. W. Norton and Company, 1961), pp. 11–12, 15.
10. Marcuse, *Eros and Civilization*, p. 36.
11. Paul Radin, *The World of Primitive Man* (New York: Grove Press, 1960); Marshall Sahlins, *Stone-Age Economics* (New York: Adine-Atherton, 1972).
12. Herbert Marcuse, *One-Dimensional Man* (Boston: Beacon Press, 1964), pp. 164, 153, 157–158.
13. Ibid., p. 165.
14. Ibid., pp. 230–231, 234.

15. Ibid., pp. 240–242.
16. Ibid., p. 238.
17. Ibid., p. 239.
18. Ibid., p. 236.
19. C. Fred Alford, *Science and the Revenge of Nature* (Gainesville: University of Florida Press, 1985), p. 64.
20. Herbert Marcuse, *Counterrevolution and Revolt* (Boston: Beacon Press, 1972), p. 62.
21. Marcuse, *One-Dimensional Man*, p. 236; *Eros and Civilization*, p. 166.
22. William Leiss, *The Domination of Nature* (New York: George Braziller, 1972), pp. 199–212.
23. Herbert Marcuse, *An Essay on Liberation* (Boston: Beacon Press, 1969), p. 12; Marcuse, *Negations: Essays in Critical Theory* (Boston: Beacon Press, 1968), p. 223.
24. Marcuse, *One-Dimensional Man*, pp. 167, 165; *An Essay on Liberation*, p. 19.
25. Leiss, *The Domination of Nature*, p. 203.
26. Robert D'Amico, "Review of *The Domination of Nature* by William Leiss," *Telos* 15 (Spring 1973): 147.
27. Marcuse, *One-Dimensional Man*, p. 158.
28. Jürgen Habermas, *Toward a Rational Society*, trans. Jeremy J. Shapiro (Boston: Beacon Press, 1970), p. 87.
29. Ibid.
30. Dorothy Lee, *Freedom and Culture* (Englewood Cliffs, N.J.: Prentice-Hall, 1959); Colin Turnbull, *The Forest People: A Study of the Pygmies of the Congo* (New York: Simon and Schuster, 1961).
31. Habermas, *Toward a Rational Society*, pp. 88–92.
32. Joel Whitebook, "The Problem of Nature in Habermas," Chapter Twelve, this volume, p. 297.
33. Hennig Ottmann, "Cognitive Interests and Self-Reflection," in *Habermas: Critical Debates*, ed. John B. Thompson and David Held (Cambridge, Mass.: The MIT Press, 1982), p. 89.
34. Vincent Di Norcia, "From Critical Theory to Critical Ecology," *Telos* 22 (Winter 1974–1975): 74, 89.
35. Joel Whitebook, "The Problem of Nature in Habermas," Chapter Twelve, this volume, p. 310.
36. Ibid., p. 69.
37. Martin Jay, *The Dialectical Imagination* (Boston: Little, Brown, 1973), p. 75.
38. Marcuse, *One-Dimensional Man*, p. 236; *Eros and Civilization*, p. 166.
39. Marcuse, *Counterrevolution and Revolt*, pp. 61, 69, 60.
40. Herbert Marcuse, "Ecology and the Critique of Modern Society," *Capitalism, Nature, Socialism* 3, no. 3 (September 1992): 36.
41. Marcuse, *Counterrevolution and Revolt*, p. 60.
42. Ibid., pp. 74, 60, 70; *An Essay on Liberation*, p. 24.
43. Marcuse, *Counterrevolution and Revolt*, p. 67.
44. Ibid., pp. 68–69.
45. Murray Bookchin, *The Ecology of Freedom* (Palo Alto, Calif.: Cheshire Books, 1982), pp. 21–22.
46. Stanley Aronowitz, *The Crisis in Historical Materialism* (New York: Praeger, 1981), p. 56.
47. Bill Devall and George Sessions, *Deep Ecology: Living As If Nature Mattered* (Salt Lake City, Utah: Gibbs M. Smith, 1985); Arne Naess, *Ecology, Community and*

Lifestyle: Outline of an Ecosophy (Cambridge, England: Cambridge University Press, 1989).

48. Bookchin, "Social Ecology versus Deep Ecology," *Socialist Review* 18 (July–September 1988): 28.

49. Ynestra King, "The Ecology of Feminism and the Feminism of Ecology," in *Healing the Wounds: The Promise of Ecofeminism*, ed. Judith Plant (Philadelphia: New Society Publishers, 1989).

50. Mary Daly, *Gyn/Ecology: The Meta-Ethics of Radical Feminism* (Boston: Beacon Press, 1978); Susan Griffin, *Women and Nature* (New York: Harper and Row, 1978).

51. Marcuse, *Counterrevolution and Revolt*, pp. 77–78; Marcuse, "Marxism and Feminism," *Women's Studies* 2 (1974): 280.

52. *Counterrevolution and Revolt*, p. 75.

53. Ibid., p. 78.

54. Susan Bordo, "The Cartesian Masculinization of Thought," *Signs* 11, no. 3 (1986): 439–456.

55. Carolyn Merchant, *The Death of Nature: Women, Ecology and the Scientific Revolution* (New York: Harper and Row, 1980); Evelyn Fox Keller, *Reflections on Gender and Science* (New Haven, Conn.: Yale University Press, 1985); Ruth Bleier, ed., *Feminist Approaches to Science* (New York: Pergamon Press, 1986).

56. Ynestra King, "Healing the Wounds: Feminism, Ecology and Nature/Culture Dualism," in *Gender/Body/Knowledge*, ed. Alison M. Jagger and Susan R. Bordo (New Brunswick, N.J.: Rutgers University Press, 1989), pp. 131–132.

57. Murray Bookchin, *Post-Scarcity Anarchism* (San Francisco: Ramparts Press, 1971).

58. Andrew Light, "Rereading Bookchin and Marcuse as Environmental Materialists," *Capitalism, Nature, Socialism* 4, no. 1 (March 1993).

59. Ibid., p. 77.

60. Bookchin, "Response to Andrew Light's 'Rereading Bookchin and Marcuse as Environmental Materialists,' " *Capitalism, Nature, Socialism* 4, no. 2 (June 1993): 112.

61. Ibid.

Chapter Nine

Lewis Mumford, the Forgotten American Environmentalist: An Essay in Rehabilitation

Ramachandra Guha

Introduction

When the Western environmental movement broke out in the early seventies, a young British journalist wrote a book profiling scientists whose work had direct bearing on the ecological predicament. Not surprisingly, her roster was dominated by university dons with impeccable scholarly credentials, including René Dubos, Raymond Dasmann, Estella Leopold, and Kenneth Boulding. Yet she chose to begin her celebration of ecological pioneers with a man without any formal training in ecology—indeed, without any formal intellectual training whatsoever (the man's only university, as he was to recall in his autobiography, was the city of Mannahatta, or New York). But for Anne Chisholm, this man had the most visible influence on contemporary environmental thought; as she wrote, "Of all the wise men whose thinking and writing over the years has prepared the ground for the environmental revolution, Lewis Mumford, the American philosopher and writer, must be preeminent."[1]

Anne Chisholm's judgment would have found strong support in the

First published in *Capitalism, Nature, Socialism* 2, no. 3 (October 1991): 67–91; reprinted here with minor corrections.

scientific community, for Lewis Mumford had already been chosen to sum up the deliberations of two seminal scientific symposia on ecological change.[2] Yet, in the two decades since Chisholm wrote her book, Lewis Mumford's reputation as an ecological thinker has suffered an extraordinary eclipse. Meanwhile, the environmental movement has grown enormously and, in the manner of any mature and self-confident social movement, has begun to construct its own genealogy and pantheon of heroes. The prehistory of environmentalism has been documented most abundantly for Mumford's own country, the United States of America, yet nowhere is the ignorance of Mumford's environmental writings more acute. That, at any rate, is the conclusion which follows from a reading of the most authoritative histories of American environmentalism, including those by Roderick Nash, Stephen Fox, and Samuel Hays.[3]

The commonly acknowledged patron saints of American environmentalism, this reading tells us, are the naturalist and nature lover, John Muir, and the forester and biologist, Aldo Leopold. Why U.S. environmentalists make Muir and Leopold into cultural icons while neglecting Mumford is a fascinating question, to which we will return toward the end of this essay. For the moment, I can only say that I find Mumford's ecological thought as congenial as did Chisholm, also an outsider to U.S. environmentalism. This study, then, is primarily an essay in rehabilitation. Its analysis of Mumford's ecological ideas is aimed especially at U.S. environmentalists, who have failed to recognize or even acknowledge one of their own most authentic voices.

Mumford's Ecological Histories

Mumford's own appreciation of nature started with boyhood summers spent in Vermont. Toward the end of his life, he remembered those early encounters in the wild with skunks, woodchucks, deer, and river trout as having "deepened my native American roots."[4] Mumford's experience was perfectly in tune with a long line of environmentalists, from Henry Thoreau to Edward Abbey, whose love of nature followed directly from their experience of the diversity and beauty of the North American wilderness.

But had Mumford's ecological horizons remained confined to the wild, he would merit no more than a footnote in the history of environmental ideas. What distinguishes Mumford from the pantheon of American wilderness heroes—and the reason I am writing about him in the first place—is his fundamentally ecological understanding of the ebb and flow of human history. Unlike Muir, Leopold, and a dozen other cultural icons, he refused to disembed individual attitudes to nature from their social, cultural, and historical contexts. The range and richness of Mumford's thought mark him as the pioneer American social ecologist and environmental historian.

For the origins of his ecological approach, we must turn briefly to the only man Mumford acknowledged as his teacher, the maverick Scotsman, Patrick Geddes. Like Mumford he was a polymath who ranged widely over the humanities and life sciences, but unlike his disciple, Geddes was a maddeningly obscure writer. As a longtime professor of botany and activist city planner in Scotland, he inspired students primarily through the spoken word and by force of example. For those with patience, however, there are veritable nuggets to be found in his writings.

The centrality of nature to Geddes's theory of city planning is evident in the one general treatise he wrote on the subject as well as in the several dozen town plans he wrote on assignment in India between 1915 and 1919, all of which reveal a subtle understanding of ecological processes in the formation, functioning, rise, and decline of cities.[5] Apart from his groundbreaking work in the theory and practice of town planning, Geddes also made a more general contribution to ecological thinking. No less a person than A. G. Tansley, one of the premier ecologists of this century, noted Geddes's influence on early ecological studies of the Scottish Highlands, while the American ecologist Paul Sears hailed his impact on the geographer Dudley Stamp, the ecologist C. C. Adams, and our own subject, Lewis Mumford.[6] Mumford wrote that by "both training and general habit of mind Geddes was an ecologist long before that branch of biology had obtained the status of a special discipline. . . . And it is not as a bold innovator in urban planning, but as an ecologist, the patient investigator of historic filiations and dynamic biological and social relationships that Geddes's most important work in cities was done."[7] At a more philosophical level, Geddes was an early harbinger of that "general revolution in science now in rapid progress, the change from a mechanocentric view and treatment of nature and her processes to a more and more fully biocentric one."[8]

"Biocentric" is, of course, a term much favored by radical environmentalists of the present time. Whereas self-styled "deep ecologists" use biocentricism only as a standard by which to judge the alleged moral feelings of those they call "shallow ecologists," Geddes (and in time, Mumford) used a biocentric approach to more constructive ends, viz., the patient investigation and understanding of "historic filiations and dynamic biological and social relationships." Mumford inherited from Geddes both a fundamentally ecological approach and a repertoire of neologisms (paleotechnic/neotechnic, conurbation, megalopolis) that he put to remarkably innovative use, especially in his authoritative histories of technology and the city. Mumford also owed to Geddes his respect for premodern technologies and patterns of resource use.

Notably, it was Geddes who drew the disciple's attention to the work of that forgotten American conservationist, George Perkins Marsh. And as Mumford noted in an early appreciation, it was Marsh who first treated man as an "active geological agent" who could "upbuild or degrade," but who was, one way or another, a "disturbing agent, who upset the harmonies of nature and overthrew the stability of existing arrangements and accommodations, extirpating indigenous vegetables and animal species, introducing foreign varieties, restricting spontaneous growth, and covering the earth with 'new and reluctant vegetable forms and with alien tribes of animal life.' "[9]

Marsh focused on the destruction of forest cover. In Mumford's assessment, deforestation served as but one example of the many ways in which Americans, in "the very act of seizing all the habitable parts of the earth," had "systematically misused and neglected our possessions."[10]

The ecological implications of early American economic development were fleshed out by Mumford in a remarkable (and unjustly forgotten) series of essays on regionalism published in *The Sociological Review*, a journal edited by Patrick Geddes's close associate, Victor Branford. Those essays constitute Mumford's first systematic attempt to apply the Geddesian ecological framework to historical phenomena.[11] This regional approach to social analysis, which the Indian sociologist Radhakamal Mukerjee was developing at the same time,[12] took off from Geddes's conceptual trinity of folk/work/place (borrowed from the work of the French sociologist, Frédéric Le Play).

In his *Sociological Review* essays, Mumford used the regional framework to analyze the ecological crimes of American pioneer civilization (the epitome of "irregionalism") and to outline the prospects for a more sustainable economy and culture (which he termed "regionalism").[13] The refusal to base industry and institutions on regional ecological endowments had led, on the one side, to enormous ecological devastation and, on the other, to a parasitical relationship between the city and the hinterland. "In America during the last century," wrote Mumford, "we mined soils, gutted forests, misplaced industry, wasted vast sums in needless transportation, congested population and lowered the physical vitality of the community without immediately feeling the consequences of our actions." During this period, "it has suited us to ignore the basic realities of the land: its contours and landscape, its vegetation areas, its power [and] minerals resources, its industry, its types of community. . . . " This was a "miner's kind of civilization," exalting the miner's cut-and-run attitude to nature, as exemplified by timber, mining, and the relentless skimming of the soils. This civilization's cities were likewise unmindful of ecological realities: bloated in their proportions, they became "prime offenders in their misuse of regional resources." Mumford did not fail to notice, either, the proliferation of slums and slag heaps *within* city boundaries.

Following Geddes, Mumford characterized the processes of what he called the paleotechnic age as "doubly ruinous: they impoverish the earth by hastily removing, for the benefit of a few generations, the common resources which, once expended and dissipated, can never be restored; and second, in its technique, its habits, its processes, the paleotechnic period is equally inimical to the earth considered as a human habitat, by its destruction of the beauty of the landscape, its ruining of streams, its pollution of drinking water, its filling the air with a finely divided carboniferous deposit, which chokes both life and vegetation."[14]

But, warned Mumford, the day of the pioneer had passed. No longer could American economic development afford to neglect regional realities. If one thought not discretely of products and resources but of the region as a whole, it would become clear "that in each geographic area a certain balance of natural resources and human institutions is possible, for the finest development of the land and the people." In America, the regionalist movement (notably the Regional Planning Association which Mumford helped to initiate) emphasized the conservation of natural resources, but in a more inclusive framework. Thus, regionalism "must not merely, through conservation, prevent waste: it must also provide the economic foundations for a continuous and flourishing life." In particular, regionalism sought to harmonize urban living with the countryside by making the city an integral part of the region. Here, Mumford drew attention to Ebenezer Howard's Garden City movement (also strongly influenced by Geddes): the creation of cities, limited in size, surrounded by farmland, with easy access to natural areas, and in other ways in organic unity with their hinterland.[15]

These early and penetrating essays illustrate Mumford's deepening interest in the ecological infrastructure of human life. Shortly afterward, he wrote that the three main threats to civilization were the continuing destruction of forest cover and soil erosion, the depletion of irreplaceable mineral resources, and the destructive potential of modern warfare.[16] The *Sociological Review* series acted as a trailer, as it were, for his masterly histories, *Technics and Civilization* (1934) and *The Culture of Cities* (1938). These—Mumford's most celebrated books, written at the height of his powers—should be read as essentially *ecological* histories of the rise of modern Western civilization.

Both outline a three-stage interpretation of the development of industrial civilization. Mumford termed these successive but overlapping and interpenetrating phases "eotechnic," "paleotechnic," and "neotechnic." The last two terms he owed to Geddes, while he added the first to designate the preparatory stage in which, he argued, most of the technical and social innovations of the modern world had been anticipated.[17]

Most, if not all treatments of Mumford's histories neglect their ecological underpinnings, but in fact, he owed his understanding of society to his three-stage model. Thus, "each of the three phases of machine civilization has left its deposits in society. Each has changed the landscape, altered the physical layout of cities, used certain resources and spurned others, favored certain types of commodity and certain paths of activity and modified the common technical heritage." Viewed from the point of view of characteristic inputs of energy and materials, "the eotechnic is a water-and-wood complex, the paleotechnic phase is a coal-and-iron complex, and the neotechnic phase is an electricity-and-iron complex."[18]

In a strictly ecological sense, the ecotechnic phase was largely benign. The resources it most heavily relied on (wood, water, and wind) were all renewable; it created exquisite landscapes and it did not lead to pollution. The "energy of the eotechnic phase did not vanish in smoke nor were its products thrown quickly in junkheaps: by the seventeenth century it had transformed the woods and swamps of northern Europe into a continuous vista of wood and field, village and garden." Its ecological impact could be regarded even more favorably when set against the record of its succeeding phase, the paleotechnic era of "carboniferous capitalism."[19]

After 1750, industrial development "passed into a new phase, with a different source of power, different materials, different social objectives." The new source of energy was coal; the dominant new material, iron; the overriding social objectives, power, profit and efficiency. The widespread dependence on coal and iron meant that for the first time in human history, societies were living not on current income from nature, but on nature's capital. At the same time, the characteristic by-products of carboniferous capitalism were polluted air, water, and homes; creating abominable living conditions made worse by the concentration and congestion brought about by factory production and modern urban living. The newer chemical industries introduced dangerous substances into the air and water and that handmaiden of industrial capitalism, the railroad, "distributed smut and dirt." Indeed, the "reek of coal was the very incense of the new industrialism," and the rare sight of a "clear sky in an industrial district was the sign of a strike or a lock-out or an industrial depression." These varied and often deadly forms of environmental degradation were a consequence of the values of the money economy, in which the environment was treated as an abstraction while air and sunlight, "because of their deplorable lack of value in exchange, had no reality at all."[20]

Despite all this, Mumford remained hopeful that the paleotechnic phase was but "a period of transition, a busy, congested, rubbish strewn avenue between the eotechnic and neotechnic economies."[21] The neotechnic phase which Mumford saw emerging would rely on a new and nonpolluting source of energy—hydroelectricity—and devise long-lasting materials (alloys) and synthetic chemical compounds. Mumford was also hopeful, in the 1930s, for a push in the direction of solar energy. As water was readily available in Africa, South America, and Asia, the arrival of electricity would also tend to displace Europe and North America from their position of industrial dominance. As far as pollution was concerned, the "smoke pall of paleotechnic industry begins to lift: with electricity the clear sky and the clean waters of the eotechnic phase comes back again." Meanwhile, the renewed utilization of human excrement and the development of nitrogen-fixing fertilizers would arrest the soil erosion caused by the miner's civilization of the earlier phase.[22] The neotechnic phase, as and when it came fully into its own, would restore three vital equilibria: that between humans and nature; that between industry and agriculture; and that in population, through the balancing of birth and death rates.[23]

Mumford's magisterial history of the city also follows a three-stage interpretation of environmental use, abuse, and renewal. He begins with the medieval city (corresponding to the eotechnic phase) against which, he claimed, modern writers had developed a violent but largely unfounded prejudice. In Mumford's reconstruction, the premodern city blended easily with its rural surroundings, while the extent of usable open space within its bounds contrasted sharply with the "notorious fact of *post*-medieval overcrowding." Again, the waste materials of city life were largely organic, hence easily decomposable. In essence, the medieval city was more than adequate "on the biological side," with its sights, smells, and sounds infinitely more pleasurable than those of its modern successor. Indeed, architecturally speaking, "the town itself was an omnipresent work of art."[24]

Once again, Mumford's evocation of a harmonious and organic past was preparatory to his condemnation of the living present, the "insensate industrial town" of the paleotechnic era. In the urban complex which superseded the medieval city, the factory and the slum were the two main elements. Whereas the effluents of a single factory often could be absorbed by the surrounding landscape, the characteristic massing of industries in the paleotechnic city polluted "air and water beyond remedy." Meanwhile, in the congested living quarters of the slum, "a pitch of foulness and filth was reached that the lowest serf's cottage scarcely achieved in medieval Europe." Sanitation and waste disposal also fell far short of minimal hygienic standards. As "night spread over the coal-town," wrote Mumford dramatically, "its prevailing color was black. Black clouds of smoke rolled out of the factory chimneys, and the railroad yards, which often cut clean into the town, mangling the very organism, spread [dust] and cinders everywhere." To the historian of the paleotechnic city, it was "plain that never before in recorded history had such vast masses of people lived in such a savagely deteriorated environment."[25]

The way out lay in the growing movement for regionalism. With the epoch of land colonization coming to a close, Mumford thought he discerned a change in attitudes towards the earth, with the parasitic and predatory attitudes of the pioneer being supplanted by the more caring values of the emerging biotechnic regime. In European countries, the regionalist movement had fought against excessive centralization, reclaimed the folk heritage, and fostered the growth of cooperatives. In the United States, the conservation movement, under the romantic impulse, had helped set aside large areas of wilderness; now, under a more scientific guise, it was actively promoting the conservation of raw materials. Meanwhile, Ebenezer Howard's Garden City movement, which stressed the creation of balanced urban communities within balanced regions, was growing in influence.[26]

The common analytical framework of Mumford's two great ecological histories has a markedly Hegelian ring: the ecotechnic, paleotechnic, and neotechnic stages being analogous to the dialectic of thesis, antithesis, and synthesis.[27] While his philosophical frame may have been inherited, his ecological sophistication is, for its time and place, quite remarkable. The major organizing principles of his histories are truly ecological in nature: the use of energy and materials as indices of technical and environmental change; the mapping of resource flows within and between regions at different stages; the forms of environmental degradation and movements of environmental redress typical of different epochs; and the role of values in creating the "money economy" of destruction and the (future) "life economy" of renewal. Underlying it all is a commitment to environmental conservation as a positive force, in contrast to the negativism with which environmentalism, then as now, was beset. In a passage which is strikingly contemporary, Mumford wrote in 1938 that

> originating in the spectacle of waste and defilement, the conservation movement has tended to have a negative influence: it has sought to isolate wilderness areas from encroachment and it has endeavored to diminish waste and prevent damage. The present task of regional planning is a more positive one: it seeks to bring the earth as a whole up to the highest pitch of perfection and appropriate use—not merely preserving the primeval, but extending the range of the garden, and introducing the deliberate culture of the landscape into every part of the open country.[28]

Mumford's Environmental Philosophy

The optimism of Mumford's ecological histories of the 1930s would surprise anyone acquainted only with his later writings. At the time, he was hopeful that the emerging values of the neotechnic economy would humanize and domesticate the machine. From the standpoint of democracy, too, neotechnic technology (in particular, hydroelectricity) worked in favor of decentralization and the human scale, in direct contrast to the giantism and concentration of the paleotechnic epoch. Mumford even has something positive to say for the automobile. Although he deplored its reliance on gasoline, he believed its growing displacement of the railroad meant that humans would no longer crowd around railheads, pitheads, and ports.[29]

Mumford's early ecological philosophy was, therefore, deeply *historicist*. He believed that the forces of history were moving in the direction of a cleaner environment, a more benign technology, and a more democratic social order. Meanwhile, his association with the regionalist movement—probably the only time in his long career as a public intellectual that he participated in collective action—also favored a more optimistic outlook on social change.

All this sits oddly with Mumford's more common reputation, based on his later writings, as a prophet of doom. Locating this transition in time, it appears that the aftermath of the Second World War fundamentally altered Mumford's faith in the forward movement of history. The carpet bombing of German cities, the dropping of atomic bombs on Japan, and the paranoia of the cold war all deeply affected Mumford. No longer could history be relied on to usher in the neotechnic age, for technology, and the "gentlemen" who controlled its development, had gone mad. This change in outlook is captured in a preface Mumford wrote in 1973, for the reprinting of a book first published nearly thirty years before. He defended the book's support of John Stuart Mill's theory of the "steady state" as opposed to the Victorian belief in progress and the expansionary thrust of modern Western civilization, thus continuing the call in *Technics and Civilization* for a dynamic equilibrium between man and nature and industry and agriculture. But, he noted significantly, "the chief effect of the regressive transformations that have taken place in the last quarter of a century [i.e., since the end of World War II] has been to change my conclusions from the indicative to the imperative mood; not 'we shall' achieve a dynamic equilibrium but '*we must*'—if we are not to destroy the delicate ecological balance upon which all life depends."[30]

In this more somber, reflective phase, Mumford's social and environmental values remained steadfast, yet he was considerably less sanguine about their wider acceptance. Nonetheless, scattered through his writings one can see the elements of an ecological philosophy that is at once analytic and programmatic. No doubt it is difficult to find a compact or authorized statement of his views in the postwar period; no canonical text exists comparable to *Technics and Civilization* or *The Culture of Cities*. Rather, his perspective on ecology, culture, and politics must be reconstructed from his diverse writings, particularly his neglected periodical essays and articles.

Let us first consider Mumford's reconsideration of modern technology, beginning with his criticisms of atomic energy and culminating in the full-scale attack in *The Pentagon of Power* (1970). Abandoning the hope that modern technology would develop in a benign direction, he now believed that modern science and technology bore the impress of capitalism, with "the capitalist's interest in quantity—his belief that there are no natural limits to acquisition" being "supplemented in technology, by the notion that quantitative production had no natural limits either."[31] Where "the machine takes precedence of the man," he wrote elsewhere, "and where all activities and values that sustain the human spirit are subordinated to making money and privately devouring only such goods as money will buy, even the physical environment tends to become degraded and inefficient."[32] Mumford reserved his strongest strictures on technology for atomic energy, which to him exemplified the one-sided, life-denying development of modern technics. He argued for this technology to be put on "strict probation," refusing to accept the "sedative explanations" of the Atomic Energy Commission that pollution would be negligible and easy to control. Such reassurances gave no confidence, for the history of industrial pollution was one where "our childish shortsightedness under the excitement of novelty, our contempt for health when profits are at stake, our lack of reverence for life, even our own life, continue to poison the atmosphere in every industrial area, and to make the streams and rivers, as well as the air we breathe, unfit for organic life."[33]

Mumford's faith in science and technology was also shaken by their role in World War II and the arms race that followed its conclusion. He was an early and percipient critic of the atomic bomb, and urged America to share its nuclear knowledge with the Soviet Union rather than embark on a meaningless and costly competition. Both the development of atomic energy and the perfection of weapons of mass destruction, he argued, undermined democracy by fostering secrecy by and within the state.[34] The military-industrial complex was itself only one part of a wider denial of democracy; large areas of central government had passed out "of all popular surveillance and control, operating in secret, defiantly withholding or adulterating the information needed by democracy in order to pass judgement on the work of its officers."[35] He unfavorably compared the state of democracy to that of the United States a century before, when there had been a great diffusion of property, wealth and political power. In calling for a renewal of democracy, Mumford was putting forward a cyclical theory of political structures curiously similar to his (by now modified) cyclical theory of technical development: a harmonious but irretrievable past, an abominable present, and a future that had yet to be claimed.[36]

Two elements of Mumford's democratic vision bear highlighting. First, he stressed that citizens must have control over public programs that vitally affect their lives. For Mumford, high among Patrick Geddes's revolutionary contributions to planning—that which set him apart from the archetypal administrator, bureaucrat, or businessman—was his "willingness to leave an essential part of the process to those who are most intimately connected with it—the ultimate users, consumers, or citizens."[37] Mumford also inherited Geddes's high regard for folk (or premodern) knowledge. In the early days of the most savage war in human history, he hoped for the time when the "mechanically more primitive cultures . . . may influence and civilize their European conquerors; may restore to them some of that deep organic sense of unity with the environment, some of that sensuous enrichment and playful enjoyment that Western man has so often forfeited in his aggressive conquest of the environment."[38]

These sentiments were perfectly consistent with Mumford's larger plea for what we would now call cultural and biological diversity. The machine world, he complained, "has insulated its occupants from every form of reality except the machine process itself: heat and cold, day and night, the earth and the stars, woodland, crop land, vine land, garden land—all forms of organic partnership between the millions of species that add to the vitality and wealth of the earth—are either suppressed entirely from the mind or homogenized into a uniform mixture which can be fed into the machine." Against this deadly uniformity, Mumford called for us to cherish our own history by "promoting character and variety and beauty wherever we find it, whether in landscapes or in people."[39]

I have argued elsewhere that the three generic environmental philosophies are wilderness thinking (or primitivism), agrarianism, and scientific industrialism. Mumford is rare (and possibly unique) among environmental philosophers in his ability to synthesize and transcend partisan stances on behalf of the wilderness, countryside, or city.[40] As his close associate Benton Mckaye pointed out, the primeval, the rural, and the urban were all environments necessary for man's full development. Consequently, a regionalist program had to incorporate all three elements: the preservation of the primeval wilderness, the restoration of the stable rural landscape, and the salvaging of the true urban.[41]

The humanizing of technology and the protection of diversity were both contingent on a fundamental change in values. As Mumford's biographer has perceptively noted, while other radicals "expected such a value change to occur after the revolution, for Mumford this value change *was* the revolution."[42] In the machine age, the disintegration of the human personality had reached an advanced stage, as the pathologies of the civilized world bore witness. So, as Mumford told a gathering of international scientists in 1955, "if we are to achieve some degree of ecological balance . . . we must aim at human balance too."[43] In an address delivered at the centenary meeting of the American Association for the Advancement of Science and published (coincidentally, but appositely) on Mahatma Gandhi's birthday, he called for a greater human element in technics, for technics to fully engage the human personality. But at a deeper level, he went on to call for a dethroning of technics from its superior place in modern society. In this larger task of cultural renewal:

> Not the Power Man, not the Profit Man, not the Mechanical Man, but the Whole Man, must be the central actor in the new drama of civilization. This means that we must reverse the order of development which first produced the machine; we must now explore the world of history, culture, organic life, human development, as we once explored the non-living world of nature. We must understand the organics of personality as we first understood the statics and mechanics of physical processes; we must center attention on quality, value, pattern, and purpose, as we once centered attention on quantity, on physical relationships, on mass and motion.[44]

Even more than values, individuals and societies need viable myths. Mumford hoped for the overthrow of the myth of the machine which had, for such an extended period, held Western man in its thrall. For sanity, stability, and survival, the myth of the machine had to be replaced with "a new myth of life, a myth based upon a richer understanding of all organic processes, a sharper insight into man's positive role in changing the face of the earth . . . and above all a deeply religious faith in man's own capacity to transform and perfect his own self and his own institutions in cooperative relation with all the forces of nature, and above all, with his fellow men."[45]

In contemporary terms, Mumford's philosophy can fairly be characterized as ecological socialist, but unlike radical socialists and radical environmentalists, he did not pin his faith on a chosen agent of history (whether proletarian or deep ecologist). In one sense, the refusal to project one's aspirations onto an agent is wholly laudable.[46] Looked at another way, it exemplifies a curious silence in Mumford's work regarding the role and place of purposive social action. Mumford frequently invokes paradigmatic individuals, values, and ways of life, but never social movements.

Remembering Lewis Mumford

If the second section of this essay closely followed Mumford's ecological histories, this section reconstructs his environmental philosophy from the social commentaries he contributed to periodicals. I now explore, more tentatively, the outlines of how Mumford might have chosen to be remembered as a social and ecological thinker. To do so, I use three appreciations Mumford wrote of other people: his son, an ecological pioneer, and a nineteenth-century polymath whom he greatly admired. None of these works is well known, but it is in these apparently ephemeral writings that we can glimpse hints of Mumford's self-image and uncertainty regarding his place in history.[47]

Mumford's only son, named for Patrick Geddes, died in action at the age of nineteen in World War II. The loss of his son shattered the writer, and it contributed significantly to his deepening pessimism about the direction of Western civilization. Yet, in the brave memoir he wrote after Geddes Mumford's death, the father was able to celebrate in his son attitudes toward the land and people Mumford himself had so long cherished.

In a strongly pastoral chapter, "The Land and the Seasons," Mumford called his son a "true countryman." Through his deep feelings for the countryside, Geddes Mumford "was renewing the spirit Thoreau had brought to the American landscape. . . . Geddes responded in every fiber to Thoreau's question: 'who would not rise to meet the expectations of the land?' " In a later chapter, "Country Ways and Country Neighbors," also written in a pastoral vein, Mumford remembered his son liking to work with his hands and notably, having an intense dislike for the machine. An illustration of his "antipathy toward the machine" was Geddes's preference, expressed to his mother when he was a child, for a horse to plow the fields. When his mother suggested that a tractor could do the job just as well, Geddes replied in amazement, "But, Mommie, you'd never use a tractor on the *ground* would you? Have you ever seen the fields after a tractor has gone over them? A tractor doesn't care what it does: it digs right into the earth and hacks it up. A horse goes gently. I'd never use a tractor."

Moreover, Geddes's "feelings for the country included country people." Here, Mumford wrote evocatively of a neighbor of theirs, Sam Honour, a small holder and farmer. Of English stock, Sam was "full of country love . . . and nearer to the peasant than any other American I have ever met." In his person, Sam was "a living specimen of an older and homelier America, which was closer to Geddes's ideal than the one he was part of."[48]

If Mumford's own love of the land, like his son's, was derived from a youthful exposure to the country, his mature ecological consciousness was indebted to the work of Patrick Geddes and the great American geographer, George Perkins Marsh. In *The Brown Decades* (1931), Mumford had alerted the American public to the significance of this forgotten writer, so when a full-length biography appeared a quarter of a century later, Mumford was uniquely placed to write a further appreciation of Marsh. Marsh's *Man and Nature* was unquestionably a "comprehensive ecological study before the very word ecology had been invented." Mumford was also quite justified in claiming that Marsh would have opposed "the vast program of pollution and extermination that has been engineered in our country in the name of scientific progress" and, in particular, that he would have spoken out against the production of nuclear energy with its potential for "permanently crippling" the human race and making the planet unfit for habitation.

It is in setting Marsh's thought in perspective that Mumford truly reveals himself. It was not that "Marsh undervalued science or the products of science," he wrote, "but he valued the integrity of life even more." For the Vermonter's "unique contribution was his combination of the naturalist's approach with that of the moralist and the humanist; he supplied both the intellectual tools and the moral direction necessary." Here, his "type of mind was the exact opposite of the German-trained specialists who began to dominate America in the 1880s, when Marsh died; for the latter narrowed their life experience and segregated their specialized interests." Ironically, it was Marsh's very ability to transcend narrow spheres of thought which "made his work suspect to the following generation, who dodged the task of evaluating his genius by ignoring it."[49]

This uncharacteristically defensive tone also crept into Mumford's appreciation of William Morris, a nineteenth-century genius whose achievements were even more wide-ranging than the versatile Vermonter's. Morris, wrote Mumford, was not merely a "dreamer of dreams" but also a "resolute realist, who refused to take the sordid Victorian triumphs of mechanical progress as the ultimate achievements of the human spirit." He was not, as commonly supposed, a revivalist, but rather what Henry Russel Hitchcock had called a "New Traditionalist, seeking not to revive the past but to nourish and develop what was still alive in it." Morris had undoubtedly devoted immense time and energy to the recovery of traditional techniques that were being rendered superfluous in the machine age. Indeed Morris, "a whole generation before the anthropologists began their belated work of salvage with surviving stone age and tribal communities, performed a similar task for the arts and crafts of the Old World past," but "if he had been more sympathetic with the peculiar triumphs of his own age, he might not have had the copious, concentrated energies to perform this necessary salvage operation."

Nor did Morris want to abolish all machines. He thought they could do the necessary work and leave other, more joyous tasks to be done by human beings themselves. Morris was in effect an early appropriate technologist who, instead "of accepting either megatechnics or monotechnics as inevitable . . . sought to keep alive or if necessary to restore those forms of art and craft whose continued existence would enrich human life and even keep the way open for fresh technical achievements."

In challenging the stereotype of Morris as an impractical dreamer, Mumford also instanced his engagement with socialism. Although it came rather late, Morris's lifework was strengthened immeasurably by the socialist vision, which "bestowed a fuller social content and a larger human purpose on all his private achievements as an artist, and gave him the confidence to work for a future in which all men might know the joys of creative labor that he himself had experienced."[50]

Written nearly ten years apart and on two very different nineteenth-century giants, there are nonetheless striking similarities in Mumford's tributes to George Perkins Marsh and William Morris. In either case, he appears to have projected himself and, more notably, *society's preferred evaluation of himself*, onto a kinsman in ideas and action, albeit of an earlier generation. He anticipates and contests the criticism that Marsh was against science, and goes on to attribute his neglect to his refusal to be trapped within narrow specialisms. He had faced the first criticism himself, and was coming to terms with a neglect of his counsel in the intellectual and political forums where he might have expected to be given a hearing. His defense of Morris can likewise be understood as the product of a close personal identification. Morris, he argues, was not blindly against technology, knew how to use the past without being a revivalist, was as much a "realist" as a "dreamer," and was guided to a deeper social vision by his engagement with socialism. In writing his tribute, Mumford may or may not have been conscious how his defense of Morris was at once a defense of his own life and work. He was seventy-three years old at the time, almost at the end of his active career and profoundly unsure how history would judge him.

Conclusion

We return to where we began—with the reception, or more accurately the nonreception of Mumford's environmental writings in his own country. Illustrative in this regard is a major round table on environmental history recently organized by the prestigious *Journal of American History*. In the keynote essay, Donald Worster recalls Aldo Leopold's call in *A Sand County Almanac* for an "ecological interpretation of history," commenting that it has "taken awhile for historians to heed Leopold's advice," but that at long last the field of environmental history "has begun to take shape and its practitioners are trying to build on his (Leopold's) initiative."[51]

Now I have nothing against Aldo Leopold, and considerable admiration for Donald Worster. Not only is Worster the most brilliant of American environmental historians, he is also among the most wide-ranging. More, he once seriously contemplated writing a doctoral dissertation on the subject of this essay. Why, then, would he invoke Leopold's call for an ecological history, made in passing and in a wholly different context, rather than the work of the man who may justly be regarded as having founded the field in the United States? Mumford used the term "ecological history" as early as 1917,[52] outlined an ecological theory of history in his brilliant essays of the 1920s on regionalism, wrote two full-blown ecological histories in the next decade, and continued to write on environmental themes till the end of his life.

Worster's preference must be explained positively rather than negatively, that is, through a positive identification with Aldo Leopold in which Mumford's more weighty contributions are obscured. It is nonetheless emblematic of a far wider neglect of Mumford's ecological writings by American environmental historians, philosophers, and activists in the United States. Aldo Leopold and John Muir appear to be far more congenial to the mind and heart of the American environmentalist. Both were remarkable human beings, acute observers of the natural world, and powerful moralists. Neither had Mumford's historical sweep, sociological sensibilities, or philosophical depth.

This neglect needs to be explained, and the following paragraphs offer a preliminary interpretation. A primary influence has been the dominance of wilderness thinking in the American environmental movement. Like Muir and Leopold, Mumford valued primeval nature and biological diversity, but unlike them, he focused simultaneously on cultural diversity and relations of power *within* human society. Moreover, his subtle, nuanced and complex philosophy cannot be reduced to the Manichean oppositions of black and white, good and evil, to which environmentalism (like other social movements) has so often succumbed. Wrenching Muir and Leopold's thought out of context, radical environmentalists can reduce it to the polar opposition of biocentric/anthropocentric, an option foreclosed in Mumford's thought.[53] There is little hint in his *oeuvre* of a scapegoat "out there" (whether capitalist or shallow ecologist). Rather, the burden of his work is toward internal social reform, or recognizing that the enemy is "us."

Second, Mumford is not a narrow nationalist. Several writers have noted the historical interpenetration of environmentalism and nationalism in the United States. The wilderness movement began as a nationalist crusade to preserve "monuments" of nature not found in Europe, and it has been closely identified with the need to challenge the world's identification of American culture with materialism.[54] Politically, Mumford opposed American nationalism and its most egregious expression, isolationism.[55] Intellectually, the internationalism of his outlook is indisputable. Deploring the "false tribal god of nationalism," he was clear that "cultural advances usually work by cross-fertilization," a credo to which his own thought bore eloquent testimony.[56]

Mumford's political beliefs must have worked against acceptance of his thought in the United States. An early critic of Stalinism, he was nonetheless a lifelong socialist. He deplored the tyranny and continuing worship of technology in Soviet Russia, but recognized that the promise of equality underlying communism, though perverted in practice, was wholly in keeping with the spirit of the age. The task of democracy, he pointed out, was to show that there were better ways of promoting economic and political equality than tyranny and thought control.[57]

Finally, Mumford was indeed too much the polymath. He made fundamental contributions in so many fields that it is easy to overlook the ecological underpinnings of his work. Two recent books on Mumford, Donald Miller's authorized biography and the collection of essays edited by Thomas and Agatha Hughes, are models of sympathetic and rigorous scholarship. They carefully appraise Mumford's contributions to architecture, technology, urban history, regional planning, and literature, but contain little awareness of his ecologically oriented writings.[58]

Although Mumford continues to be neglected by American environmentalists, he continues to speak powerfully to environmentalists from other cultures. The noted green thinker (and former German Green) Rudolf Bahro came upon Mumford's thought rather late, but immediately recognized that his work "has the same significance for the ecological movement as the achievement of Marx once had for the labor movement."[59] What the visionary Bahro knew intuitively, I have tried painfully to demonstrate with the apparatus of scholarship. But then, I should have been more than ordinarily prepared for the phenomenon of a prophet with little honor in his own culture, but deeply respected elsewhere. For what else has been the fate of the greatest Indian of this century, Mahatma Gandhi, and the greatest Indian who ever lived, Gautama Buddha?[60]

Acknowledgments

I am grateful to Mike Bell, Bill Burch, and James O'Connor for encouraging me to write this essay. My research was greatly aided by E. Newman's indispensable bibliography of Lewis Mumford's writings, published by Harcourt Brace Jovanovich in 1970. This essay is dedicated to the memory of the writer Richard J. Margolis.

Notes

1. Anne Chisholm, *Philosophers of the Earth: Conversations with Ecologists* (London: Sidgwick and Jackson, 1972). Mumford's invocation of the city as his university is in his *Sketches from a Life: The Autobiography of Lewis Mumford—The Early Years* (New York: The Dial Press, 1982), esp. chap. 11.

2. See W. L. Thomas, ed., *Man's Role in Changing the Face of the Earth* (Chicago: University of Chicago Press, 1956); F. Fraser Darling and John P. Milton, eds., *Future Environments of North America* (Garden City, N.Y.: Natural History Press, 1966).

3. See Roderick Nash, *Wilderness and the American Mind*, 3d ed. (New Haven, Conn.: Yale University Press, 1982); Stephen Fox, *The American Conservation Movement: John Muir and His Legacy*, 2nd ed. (Madison: University of Wisconsin Press, 1985); and Samuel P. Hays, *Beauty, Health and Permanence: Environmental Politics in the United States, 1955–1985* (New York: Cambridge University Press, 1987).

4. Mumford, *Sketches from a Life*, p. 90.

5. Patrick Geddes, *Cities in Evolution* (1915; rev. ed., London: William and Norgate, 1949), p. 51; Jacqueline Tyrwhitt, ed., *Patrick Geddes in India* (London: Lund Humphries, 1947), pp. 57–58, 78–83. The latter work is an edited compilation of extracts from some of Geddes's reports on Indian towns. He wrote some fifty reports in all, as a guest of the colonial government and several princely states.

6. See R. P. McIntosh, *The Background of Ecology: Concept and Theory* (Cambridge, England: Cambridge University Press, 1985), pp. 293–294. Geddes was also one of the first to criticize conventional economics from the perspective of ecological energetics. See Juan Martinez-Alier, *Ecological Economics: Energy, Economics, Society* (Oxford: Basil Blackwell, 1987), pp. 89–98.

7. Essay first published in the *Architectural Review* (1950) and excerpted in Mumford, *My Works and Days: A Personal Chronicle* (New York: Harcourt, Brace, Jovanovich, 1979). See also Mumford, *Sketches from a Life*, p. 147, and "Patrick Geddes, Insurgent," *The New Republic*, 30 October 1929.

8. Patrick Geddes, *Report on Town Planning, Dacca* (Calcutta: Bengal Secretariat Book Depot, 1917), p. 17, emphasis in original.

9. Mumford, *The Brown Decades: A Study of the Arts in America* (1931; reprint, New York: Dover Publications, 1955), pp. 76–77. Years later, Mumford speculated that he was invited to cochair the 1955 Werner-Gren Conference on "Man's Role in Changing the Face of the Earth" because of his memoir on George Perkins Marsh in *The Brown Decades*; as Geddes had introduced him to Marsh, in effect he owed that invitation to his master (see *Sketches From a Life*, p. 408). The proceedings were dedicated to Marsh (see Thomas, *Man's Role in Changing the Face of the Earth*).

10. Mumford, "Regionalism and Irregionalism," *The Sociological Review* 19, no. 4 (1927): 277.

11. Ibid.; Mumford, "The Theory and Practice of Regionalism," *The Sociological Review* 20, nos. 1 and 2 (1928).

12. See Radhakamal Mukerjee, *Regional Sociology* (New York: Century Company, 1926). Mukerjee was also powerfully influenced by Patrick Geddes, with whom he came in close contact during the latter's stay in Calcutta, c. 1915–1916.

13. The following paragraphs are based on Mumford, "The Theory and Practice of Regionalism."

14. Mumford rarely missed the opportunity to berate the pioneer for his crimes against nature. As late as 1962, he was still complaining that "even when the pioneer didn't rape Nature, he divorced her a little too easily: he missed the great lesson that both ecology and medicine teach—that man's great mission is not to conquer nature by main force but to cooperate with her intelligently but lovingly for his own purposes." Mumford, "California and the Human Prospect," *Sierra Club Bulletin* 47, no. 9 (1962): 45–46.

15. In the event, American economic development has continued to ignore regional realities. But ecological disaster has been forestalled by drawing in natural resources from all over the globe. At the time of Mumford's *Sociological Review* series, America was still relying largely on its own resources, but, especially since World War II, the development of consumer society has rested on a fundamentally exploitative relationship with the rest of the world. Consumers in the high centers of industrial civilization can take for granted the continued supply of mink from the Arctic, teakwood from India, and ivory from Africa, without being in the slightest degree responsible for the environmental implications of their lifestyles.

16 Mumford, "Science on the Loose" (review of Robert Millikan's *Science and the New Civilization*), *The New Republic*, 6 August 1930. See also the section entitled "Pre-1970 Ecology," in *My Works and Days*, pp. 29–32.

17. Mumford, *Technics and Civilization* (New York: Harcourt, Brace and Company, 1934), p. 109.

18. Ibid., pp. 268, 110.

19. Ibid., pp. 111, 118, 147.

20. Ibid., pp. 151, 168–169.

21. Ibid., p. 211.

22. Ibid., p. 255.

23. Ibid., pp. 429–431.

24. Mumford, *The Culture of Cities* (New York: Harcourt Brace and Company, 1938), pp. 49, 51.

25. Ibid., pp. 162, 164, 191, 195.

26. Ibid., especially chaps. 5–6.

27. Hegel is mentioned only once in *Technics and Civilization* and not at all in *The Culture of Cities*. However, Mumford had read Karl Marx closely, and perhaps his stages approach unconsciously drew upon Marx's interpretation of the Hegelian dialectic. Marx's theory of history is open to both an evolutionist and a cyclical reading; while Marxists have usually preferred the former, Mumford undoubtedly would have been more comfortable with the latter.

28. Mumford, *The Culture of Cities*, p. 331.

29. Mumford, *Technics and Civilization*, pp. 221–223, 247, 250, 267; "The Theory and Practice of Regionalism," p. 19.

30. Mumford, Preface to *The Condition of Man* (1944; reprint, New York: Harcourt Brace Jovanovich, 1973), p. viii.

31. Mumford, "Technics and the Future of Western Civilization," in *In the Name of Sanity* (New York: Harcourt Brace Jovanovich, 1954), p. 47.

32. Mumford, "California and the Human Prospect," p. 43.

33. Mumford, "Prospect," in Thomas, *Man's Role in Changing the Face of the Earth*, pp. 1147–1148.

34. See, for example, Mumford, "Gentlemen! You Are Mad!" *Saturday Review of Literature*, 2 March 1946; "The Morals of Extermination," *Atlantic Monthly*, October 1959; and the collection, *In the Name of Sanity*. See Paul Boyer, *By the Bomb's Early Light: American Thought and Culture at the Dawn of the Atomic Age* (New York: Pantheon, 1985), especially pp. 284–287.

35. Mumford, "The Moral Challenge to Democracy," *Virginia Quarterly Review* 35, no. 4 (1959): 565.

36. Ibid., pp. 562–567.

37. Mumford, *My Works and Days*, pp. 115–116.

38. Mumford, "Looking Forward," *Proceedings of the American Philosophical Society* 83, no. 4 (1940): 541.

39. Mumford, "California and the Human Prospect," pp. 45–47.

40. Ramachandra Guha, "Toward a Cross Cultural Environment Ethic," *Alternatives* 16, no. 3 (1990).

41. Benton Mckaye to Lewis Mumford, 3 December 1926, quoted in John L. Thomas, "Lewis Mumford, Benton Mckaye and the Regional Vision," in *Lewis Mumford: Public Intellectual*, ed. Thomas P. Hughes and Agatha C. Hughes (New York: Oxford University Press, 1990).

42. Donald L. Miller, *Lewis Mumford: A Life* (New York: Weidenfeld and Nicholson, 1989), p. 166.

43. Mumford, "Prospect," p. 1146.

44. Mumford, "Let Man Take Command," *The Saturday Review of Literature*, 2 October 1948, 35.

45. Mumford, "California and the Human Prospect," pp. 58–59.

46. Cf. Alvin Gouldner, *Against Fragmentation: The Origins of Marxism and the Sociology of Intellectuals* (New York: Oxford University Press, 1985).

47. It is, of course, not at all uncommon for writers to project their hopes, prejudices, and aspirations in their tributes to other people. On Orwell's identification with Dickens, see John Rodden's suggestive study, *The Politics of Literary Reputation: The Making and Claiming of 'St. George' Orwell* (New York: Oxford University Press, 1989), especially pp. 181–182, 238–239.

48. Mumford, *Green Memories: The Story of Geddes Mumford* (New York: Harcourt, Brace and Company, 1947), pp. 114–115, 126–128.

49. Mumford, "Marsh's Naturalist-Moralist-Humanist Approach" (Review of David Lowenthal's *George Perkins Marsh: Versatile Vermonter*), *Living Wilderness* 71 (Winter 1959–1960): 11–13.

50. Mumford, "A Universal Man," *New York Review of Books*, 23 May 1968, 8, 10, 12, 15.

51. Donald Worster, "Transformations of the Earth: Toward an Agroecological Perspective in History," *Journal of American History* 76, no. 4 (1990): 1087.

52. Miller, *Lewis Mumford*, p. 87.

53. In this connection, see my polemic, "Radical American Environmentalism and Wilderness Preservation: A Third World Critique," *Environmental Ethics* 11 (Spring 1989).

54. See Alfred Runte, *National Parks: The American Experience* (Lincoln: University of Nebraska Press, 1979). Mumford anticipated Runte's critique of monumentalism when he deplored the tendency of American planners in the past to "single out the most striking forms of landscape." He went on, "If the culture of the environment had yet

entered deeply into our consciousness, our aesthetic appreciations would not stop short with stupendous geological formations like the Grand Canyon of Arizona: we should have equal regard for every nook and corner of the earth, and we should not be indifferent to the fate of less romantic areas" (*The Culture of Cities*, p. 332). This is yet another striking example of Mumford's prescience. It is only in the last decade or so that the wilderness movement has begun to shift its priorities toward the protection of biological diversity and away from a narrow emphasis on the spectacular.

55. See, for instance, his controversial essay, "The Corruption of Liberalism," *The New Republic*, 29 April 1940. For a conservative critique of Mumford as antipatriotic, see Edward Shils, "Lewis Mumford: On the Way to the New Jerusalem," *The New Criterion* 9, no. 1 (1983).

56. Mumford, "Let Man Take Command," p. 8; "Looking Forward," p. 545.

57. Mumford, "The Bolshevist Religion," *The New Republic*, 1 April 1928; "Alternatives to the H-Bomb," *The New Leader*, 28 June 1954.

58. See Miller, *Lewis Mumford*; and Hughes and Hughes, eds. *Lewis Mumford*. One reason is probably the writers' lack of interest in ecology. Another is methodological; they rely on Mumford's books and private papers, ignoring his essays in periodicals, on which I have drawn so heavily here. I may note that my own reconstruction of Mumford's ecological philosophy is itself based on a deliberately restricted focus; I have completely ignored his writings in American studies and architecture, two fields in which his reputation is assured.

59. Quoted in Kirkpatrick Sale, "Lewis Mumford" (obituary), *The Nation*, 19 February 1990. Although he has not written at length on Mumford, Sale is rare among American environmentalists in his frequently expressed admiration for the man. See his *Dwellers in the Land: The Bioregional Vision* (San Francisco: Sierra Club Books, 1985).

60. See Ramachandru Guha, "Mahatma Gandhi and the Environmental Movement in India," *Capitalism, Nature, Socialism* 6, no. 3 (September 1995): 47–61.

Chapter Ten

Change and Continuity in Environmental World-View: The Politics of Nature in Rachel Carson's *Silent Spring*

Yaakov Garb

Introduction

The publication of Rachel Carson's *Silent Spring* is often marked as a watershed in the history of the American environmental movement.[1] Within a year of its publication, the book's critique of pesticides had prompted scientific research into their hazards, brought significant changes in their regulation, spurred public debate on environmental practices, inspired a younger generation of environmental activists, and made ecology a household word. Arguably the century's most consequential book on an environmental topic, *Silent Spring* achieved its impact largely through the balance it achieved between questioning the prevailing environmental beliefs and practices, on the one hand, and remaining continous with them, on the other. While sufficiently disturbing to galvanize change, the book, on the *New York Times* bestseller list for almost three years, was nevertheless ideologically acceptable to a large audience. In this essay I read Carson's text to better understand this balance between challenge and compromise, and to evaluate its political opportunities and hazards.[2]

Let me begin my account with a brief biographical sketch of Carson, highlighting those aspects of her background that brought her to a seemingly unlikely yet perfect position from which to initiate this reworking of the environmental consciousness of middle-class Americans. Carson's childhood on a family farm in Pennsylvania helped establish her visceral and lifelong connection to the outdoors and wildlife. Her talents as a writer had emerged by her early teens and, in 1925, she entered a small women's college in Pennsylvania as an English major, switching to biology at the end of her sophomore year. Academic achievements earned Carson a summer fellowship at Woods Hole Marine Biological Laboratory, and later support for completing her M.A. in zoology at Johns Hopkins University. However, her studies were constantly accompanied by financial insecurity and, with the death of her father in 1935, she abandoned doctoral work to become her family's breadwinner, working for the federal government primarily in the Fish and Wildlife Service as a writer for service publications in a position that drew on both her scientific expertise and writing talent.

Working slowly after hours on her own writing, in 1941 she produced a natural history of the ocean, *Under the Sea-Wind*, and a decade later, *The Sea Around Us*, whose bestseller status allowed her to buy a home on the Maine coast and retire to pursue her own writing. After completing a third work on the sea in 1955, she became increasingly preoccupied with the dangers of pesticides, which she had first encountered in her work for the government. Drawing on a range of written materials and her connections to agricultural and medical researchers and individuals in government agencies and wildlife societies, Carson put together the picture that appeared as *Silent Spring* in the summer of 1962. The book created a storm of media and government attention and elicited angry rebuttal and attacks, including a well-funded campaign by the chemical industry.

Several factors made Carson ideally suited to synthesize and present a case against indiscriminate pesticide use. She felt a deep passion for the natural world that, combined with the sensibilities absorbed from the nature writers whose works she had grown up with and her own considerable writing skill, made her a familiar and uncontroversial writer within an established tradition. This standing served her well with the publication of her more controversial book. Carson's training as a biologist and familiarity with government bureaucracies enabled her to locate and assimilate technical material, and her income as a writer allowed her to pursue her topic independently. Paradoxically, her relatively marginalized status assisted her in her task; because she never had the opportunity for narrow specialization, as did many of her male colleagues, and because she did not share their allegiances to specific research programs, she could range freely across disciplinary lines and was less constrained in her conclusions. Furthermore, her gender and lack of "professional standing," which her detractors later turned against her,[3] probably made her a less threatening figure as she gathered what was often quite sensitive information from a wide range of sources. After the furor of her book's publication, her quiet, dignified bearing helped to stabilize her position and made the rather frenzied attacks of her opponents seem ridiculous.

Carson's intermediate position—involved in yet removed from scientific orthodoxy and bureaucratic structures—enabled the writing of *Silent Spring*. The book itself occupied a similar middle ground, challenging yet remaining continuous with prevailing beliefs about the environment, science, and society. On the one hand, it introduced its readers to the new and frightening hazards of an increasingly toxic landscape, described insidious and far-reaching disruptions of "the natural," and questioned prevailing images of science and the technological mastery of the natural world. On the other, by drawing on familiar cultural themes and notions of nature, and by not extending hints about the role of greed into direct and specific accusations or a substantial challenge to existing economic and political structures, the book remained safely within the bounds of the American mainstream.

Was *Silent Spring* an extraordinary balancing act or a disappointing compromise? If Carson's critique of pesticides had diverged greatly from common understandings of nature and society, it is unlikely it would have been so popular and thereby effective. I will explore this claim by comparing *Silent Spring* to Murray Bookchin's *Our Synthetic Environment*,[4] a scarcely remembered work published a half-year earlier, which described the same pesticide problems as Carson's book. Several circumstances contributed, no doubt, to his book's slight impact. Bookchin lacked Carson's standing and skills as an accomplished nature writer as well as the platform of a *New Yorker* series that placed *Silent Spring* in the center of public attention. Probably the most serious impediment to *Our Synthetic Environment*'s broad acceptance, however, was its content. Bookchin's account of the dangers of pesticides was part of a comprehensive and politically forthright chronicle of the many assaults on the environment and human well-being that he claimed were inevitable in an industrial capitalist society. This pill was too big, bitter, and unfamiliar for most Americans to countenance, much less swallow. But Carson's moderation, I will argue, also had its costs. There are places where her account can barely sustain the logical strain of not pursuing its own implications and, more seriously, structures and assumptions she left unchallenged continue to underlie pesticide problems today.

Nature: Whole, Harmed, and Regained

Silent Spring issues its challenge to environmental practices from within deeply conventional conceptions of nature. It is energized by a familiar, simple, and powerful narrative: nature disturbed, wholeness undone. Nature's integrity is centered on a concept of natural balance that has a long heritage in Western thought. Carson's time-honored pastoral vision of human reconciliation with the natural world, her ease with mechanical and productivist metaphors of natural process, and the conscious downplaying of her biocentric leanings in favor of more acceptable anthropocentric and utilitarian grounds of persuasion, further contribute to a book that offers its readers a remarkably familiar framework through which to understand the potentially disturbing phenomenon of pesticide use.

NATURE WHOLE

"A fanatic defender of the cult of the balance of nature" was the way that the president of Monsanto characterized Carson in a cover letter to the distributors of an industry hit piece on the virtues of chemicals, commissioned to counteract the growing support for *Silent Spring*.[5] He was reacting to the book's central metaphor, the balance Carson defined as "a complex, precise, and highly integrated system of relationships between living things which cannot be safely ignored any more than the law of gravity can be defied with impunity. . . . The balance of nature is not a status quo; it is fluid, ever shifting, in a constant state of adjustment (218).[6]

This "economy of nature"—a "precise and delicate" system in which everything has its place, where relations of "interdependence and mutual benefit" and checks and balances prevail—provides the norm against which human interference can be assessed and challenged (72, 73, 77). In her use of these concepts of nature's balance or economy, Carson invokes an element of Western thought that can be traced back to antiquity. In the eighteenth century, it achieved quite explicit form in theological beliefs that regarded the world as underlain by harmony and order; God's providence ensured a system of perpetual balance among all living things, in which each creature had its allotted place.[7]

A second strong guiding metaphor in the book is the notion of an "ecological web of life" whose "threads" "bind" together organisms and their environment so that even minute changes in one area reverberate over space and time throughout the natural world (75, 57, 170). "The earth's vegetation," for example, "is part of a web of life in which there are intimate and essential relations between plants and the earth, between plants and other plants, between plants and animals" (64).

Much of the power and novelty of Carson's warnings about the hazards of pesticides to human health is achieved through extending these ecological notions of balance and interconnectedness to include the internal realm of physiology: "there is also an ecology of the world within our bodies" (170). The seamlessness of inner and outer landscapes is beautiful, but its implications are chilling. It is a "web of life—or death" (170). The complex and precisely regulated nuances of physiological functioning exist within a system that is now seen to include far-flung and unsuspected agents, linking chemicals and radiation to the fragile processes of mitosis.[8]

A tremendous amount of the persuasive work in her book is achieved through these well-established notions: the balance of nature, the ecological web, "the natural." They provide her with a versatile conceptual framework and evocative imagery with which to underscore the durability or fragility of natural systems, and to evoke wonder at the intricacy of connections or dismay at the consequences of their disruption. Nature whole is the basis for *Silent Spring*'s most unsettling tidings of balance lost, and the implicit ground of critique in her portrait of an emerging toxic landscape. The unnaturalness and damage of pesticides lies in the ways they "disturb," "upset," "alter," and "do violence to" nature's lawlike integrity (17, 60, 231, 17).

BALANCE LOST

Like her previous books and many others on natural history, *Silent Spring* presents the unseen currents, mechanisms, and balances that animate the natural world. Now, however, discovery is simultaneous with a realization of disruption.[9] Nature's wonders are known through their loss: we learn of the dark hidden sea of groundwater through its contamination; of the "aerial highways" because of the culling of the birds that normally fly through them (111); of the "unseen and intangible" paths from river mouths to far offshore because the fish that have followed them for thousands of years are now being attracted to their deaths (120). "The soil exists in a state of constant change," we are told, "taking part in cycles that have no beginning and no end" (57), even as we learn of the sterilizing effect of pesticides on soil life. This convergence of natural history and elegy continues today: the public discovered the miraculous protective layer of ozone through news of its breaching, the critical functions of the rain forest through its disappearance, and the significance of genetic diversity through its loss.

The density of Carson's imagery of disruption strikes me as new to natural-history writing. She speaks of the penetration and shattering of the living world, a "breaking [of] the threads that bind life to life," the "bludgeoning of the landscape," the "ripping apart" of "the whole closely knit fabric of life," the "uncoupling" of the phosphorylation process by radiation or chemicals, the disruption of checks and balances and "age-old patterns" (73, 64, 67, 203, 209). She reaches for jarring metaphors to convey the crudity of the human rupture of natural process: "a crowbar between the spokes of a wheel" (183); "a cave man's club" hurled against the delicate fabric of life (261).

THE TOXIC LANDSCAPE

Perhaps the greatest novelty of *Silent Spring* was its sustained presentation of what it might mean to live in a world of unfamiliar vulnerabilities and hazards: the increasingly toxic landscape of postwar America. To do this,

Carson sketched a new phenomenology of hazard, for intuitions learned from encounters with macroscopic physical threats or pathogenically borne disease were no longer adequate for grasping the effects of exposure to radiation and toxins. Though we have grown more accustomed to them today, these hazards were still quite new in the late 1950s and early 1960s, and much of the impact of *Silent Spring* stemmed from its reworking of popular perceptions of self-in-environment.[10]

In this new and alien toxic landscape, what was once life-giving has become poisonous. The title of chapter 3, "Elixirs of Death," conveys this reversal. Molecules based on carbon, the foundation of life, have become agents of death (27). Exposure to "air and sunlight," formerly principles of wholesomeness, can spontaneously transform chemicals into poisonous substances" in a way that is not only unpredictable but beyond control" (48–49). For birds in a sprayed area, "the once beneficial rain had been changed, through the evil power of the poison introduced into their world, into an agent of destruction" (90). This world we are coming to inhabit, suggests Carson, is like a scary fairy tale where shiny red apples poison and chemists have become evil sorcerers, tricking nature into becoming death-dealing with their systemic pesticides that offer poison to pests in place of the nourishment they expect (39).

Unlike many dangers, we encounter these chemicals regularly and unsuspectingly in our everyday life. This results in the "innumerable small-scale exposures to which we are subjected day by day, year after year. . . . Each of these recurrent exposures . . . contributes to the progressive buildup of chemicals in our bodies ands so to cumulative poisoning. . . . Lulled by the soft sell and the hidden persuader, the average citizen is seldom aware of the deadly materials with which he is surrounding himself; indeed, he may not realize he is using them at all" (157).

Carson emphasizes the everydayness and "intimacy" of our contact with these threats (197), juxtaposing mundane occasions, such as taking home fish to fry for dinner (53, 130), with their terrible consequences. When an average diet can produce liver damage (31), normalcy has become hazardous—surely engendering a new stance towards the world.

Our usual indications of hazard are useless in this new landscape, for these novel threats are invisible:[11] "Whether detected or not, the pesticides are there" (46). Even in the absence of apparent bodily harm in one generation, reproduction may be affected (113) or damage may appear only in subsequent generations.

These chemicals also disrupt our usual judgments of scale and sense of causality in assessing danger. At the outset of the book Carson establishes the potency of minuscule amounts, which make everyday estimations of hazard meaningless (29–30). By the book's end, the reader will come to read concentrations measured in parts per million with shock rather than dismissal. Exposures to small amounts ("no matter how slight" [157]) can have huge consequences and cause shifts that are "subtle but far-reaching" (78). "Minute causes produce mighty effects" (170). Carson repeats variations on this theme to drive home an alternative notion of causality: "a little twist" in structure makes a molecule "five times as poisonous" (34); "seemingly moderate applications of insecticides . . . may build up fantastic quantities in soil" (60), and over the course of several pages in her chapter on cancer, Carson explicates how tiny distortions of chromosomal material lead to horrendous consequences (such as Mongolism, leukemia, Klinefelter's syndrome, and Turner's syndrome). When "a change at one point in one molecule even may reverberate throughout the entire system to initiate changes in seemingly unrelated organs and tissues," our habit of "looking for gross and immediate effect and . . . ignoring all else" no longer applies. In this new landscape, "cause and effect are seldom simple and easily demonstrated relationships. They may be widely separated in space and time" (170).

These toxic threats are also new in the degree to which they penetrate into the innermost levels of our physiology, the very core of the self. Synthetic chemicals, as opposed to naturally occurring ones, "enter into the most vital processes of the body and change them in . . . often deadly ways" (25). Because they "make use of all available portals to enter the body" (31) and easily "penetrate [its] inadequate defenses" (196), pesticides affect the "chemical conversions and transformations that lie *at the very heart* of the living world" (60, emphasis added). They compromise reproduction and "strike at the genetic material of the race" (41); they lodge in the "very marrow of our bones" (25) and affect the nervous system that defines higher life forms (178); they undermine even the innermost lines of defense, "the very [liver] enzymes whose function is to protect" (25), and pass across the "traditional protective shield" of the placenta (31), a profoundly symbolic affront.

In addition to the traits already mentioned, the impact of these new chemicals is cross-generational (52), cumulative, synergistic, and often delayed (chap. 11, pp. 169, 201). It is probably because they defy conventional mental schemata in so many ways that these chemicals readily acquire their aura of eeriness, evil, and unnaturalness, which Carson reinforces. The book's title *Silent Spring*, sets the tone, evoking a disruption of annual seasons, the largest and most reliable natural cycle. The regular return of the robin (99), the "early mornings . . . filled with song" (97)—"all [this] is changed, and not even the return of the birds may be taken for granted" (99), says Carson, their absence leaving a silence she calls "eerie, terrifying" and "weird" (98). She describes as "sinister" (45) the chemicals that can do such strange things as speeding up animal metabolism to the point of death, or causing "the leaves of the oaks [to begin] to curl and turn brown, although it was the season for spring growth" (71). Her image of the sorcerer's cauldron reinforces these connotations.

IS NATURE OURS?

Carson's ethic of nature in *Silent Spring* was calculatedly moderate. She was careful not to express in the book her own sense that nature's rights exist apart from human needs, though she argued for the political representation of those who enjoy a noninstrumental (i.e., recreational and aesthetic)[12] relationship with nature. She claimed, for instance, that the suburban bird watcher's enjoyment of nature, or that of the hunter, fisherman, or explorer of wilderness, is just as legitimate as the interests of those ostensibly safeguarded by chemical spraying (84) and should be represented in decision making. "Who," she asks,

> has made the decision that sets in motion these chains of poisonings . . . who has placed in one pan of the scales the leaves that might have been eaten by the beetles and in the other the pitiful heaps of many-hued feathers . . . of the birds that fell before the unselective bludgeon of insecticidal poisons? Who has decided—who has the *right* to decide—for the countless legions of *people* who were not consulted that the supreme value is a world without insects, even though it be also a sterile world ungraced by the curving wing of a bird in flight? (emphasis added)

"The decision," she answers, "is that of the authoritarian temporarily entrusted with power." And through a typically indirect kind of formulation, whose significance in *Silent Spring* I will discuss later, she inserts a veiled call for greater public oversight: "He has made it during a moment of inattention by millions to whom beauty and the ordered world of nature still have a meaning that is deep and imperative" (119).

Ultimately, however, the nature of *Silent Spring* remains a resource, albeit for the relatively benign ends of the hunter and wildlife enthusiast. I agree with Roderick Nash's and Donald Fleming's assessments of why Carson chose not to raise the more biocentric views, influenced in part by Transcendentalist thinking,[13] evident in her three earlier books[14] and unpublished works.[15] "The ethical philosophy displayed in *Silent Spring*," says Nash,

> is a blend of old ideas and new ones. Carson intended the book to shock Americans into awareness and action. She was angry. Her objective was to outlaw pesticides or at least greatly constrain their use. Like Muir and Leopold she wanted to be effective in the political arena, and she knew she would lose her audience if she stepped too far ahead of public opinion. As a result there is no direct mention in *Silent Spring* of the rights of insects, birds, fish, and other victims of the poisons.[16]

More than any of her previous works, says Fleming, *Silent Spring*

> was written to persuade and arouse. To this end Rachel Carson had to calculate her effects as never before, trim her sails to catch the prevailing winds. That was the great contrast between her last testament and Aldo

> Leopold's. He thought of himself as writing *A Sand County Almanac* for a minority. He was almost despairing at the changes that had to be wrought in the world to make it livable again, and [was] correspondingly reckless and intransigent, ready to throw off all disguises and defiantly insist on his own vision on his own terms, with no concession to the timid or temporizing. He would save absolutely all the species as a matter of 'biotic right,' and that was that, take it or leave it. Rachel Carson could not strike any such posture in *Silent Spring*. She could not defy the majority as if she despaired of reaching them, for she meant to get around them, however skeptical or hostile to begin with, and force them to capitulate. She certainly could not afford to lose them straight off by taking a religious tone about biotic rights.[17]

Instead, in a voice that was "studiously reasonable,"[18] Carson rallied a huge amount of scientific information on the impact of pesticides and appealed in her argument primarily to considerations of human health, prudence and, to a lesser degree, appreciation of nature as aesthetic and recreational resource. Keenly aware, no doubt, that human concern for the nonhuman is not distributed evenly over all of Creation, she presented images of destruction most likely to elicit her audience's empathy: the impact of chemicals on *young* animals or humans;[19] the poisoning of the emblematic Bald Eagle (111); and a page-long description of how pesticides undermine male virility (weakening bull sperm, for example, and causing smaller testicles in animals [186]).

Despite her vision of radical interconnectedness, Carson employs familiarly mechanistic, utilitarian, and industrial imagery in describing the natural world. At the cellular level, she images mitochondria as "powerhouses"; ATP and ADP as "a storage battery" or "a common currency of energy" that is circulated and exchanged; and cellular processes as a "rotating wheel" or "racing engine" (185). The soil life she describes has a sensible division of labor (58), and the plants are busy little factories: "harnessing" the sun's energy, "manufacturing" products (64), and "extracting" nutrients (59). Indeed, her argument for the conservation of soil microbiota and for the use of natural predators rather than chemicals to control insect populations is presented in sound (one might say ruthless) managerial and economic terms. Whereas chemicals are costly, soil microbes, a "horde of . . . ceaselessly toiling creatures" (58), provide free labor to "improve" the soil (59). And the industriousness of insects—"working," she marvels at them, "working in sun and rain, during the hours of darkness, even when the winter's grip has damped down the fires of life to mere embers" (222)—invites their enlistment as rather short-shrifted "allies" in our war against pests. One might argue that Carson could scarcely have described biological process otherwise in the early sixties—no other language was available. That is precisely my point: *Silent Spring*'s challenge to accepted thinking about nature was fused with quite conventional conceptions of it.

Hints of Carson's broader conception of natural rights emerge only subtly at two points in the book: its dedication to Albert Schweitzer, with whose biocentric reverence for life she sympathized, and her complaint that innocent animals were "doomed by a judge and jury who neither knew of their existence nor cared" (118), where she seems implicitly (if only in metaphor) to extend a juridical model to include the rights of other species and to charge that due process was not observed.[20]

THE CONTRADICTIONS OF "NATURE"

Terms like "nature," the "natural," and the "balance of nature" have great discursive force not in spite of but because of their fuzziness. Their multiple connotations and self-evident (thus unexamined) definition within the community that shares them, enable protean versatility. We add great force to any argument by adducing the "natural" to it, so long as no one asks too carefully what we mean by the term. If they do, it will often turn out that nature (and its cognates) are not preexisting, ontologically firm objects or conditions in the natural world, but a reification of human criteria and definitions.

The socially defined nature of "nature" sometimes becomes apparent when the concept is used as a guide for concrete actions in the world, such as the appropriate way to "control" insect pests. Carson's approval of certain forms of biological over chemical methods, for example, ostensibly rests on their noninterference with "the balance of nature." The actual operative criteria, however, lie largely elsewhere, in judgments about the benefits and costs—to humans—of these methods.

The social criteria smuggled in through talk of "the balance of nature" emerge as internal contradictions in Carson's account. Why, for instance, is the importation of an exotic pathogen (a bacteria) to kill the Japanese beetle (93–95) a "natural" means of control? Is this intervention which, as Carson notes in passing in another context (94), kills not only the target species but at least forty other species in the scarabaeid family, more respectful of the balance of nature than certain pesticides?

Similar questions could be asked of each of the methods for "biological control" that Carson celebrates: juvenile hormones, chemical attractants, repellent sounds, microbial and viral infection of insects, introduced predators and parasites. Carson enthusiastically endorses, for instance, the dispersal of X-ray-sterilized males, and heralds the "complete extinction of the screw-worm in the Southeast" (247) as a "brilliant success" (248) and "a triumphant demonstration of the worth of scientific creativity" (247). Slipping into the same militaristic imagery to which she objects in the mentality of proponents of pesticide spraying, she writes approvingly of research by Dr. Knipling that turned "insect sterilization into a weapon that would wipe out a major insect enemy" (246).[21]

Surely the difference between this celebrated method of control (of an unsavory insect responsible for livestock losses of $40 million a year [246]) and the chemical practices Carson castigates lies in the priorities and tastes of humans, rather than in the degrees of "naturalness" of these respective interventions. Had Carson chosen to cast the X-ray sterilization of males as unnatural, and rallied to this end the same rhetorical resources she uses in her discussion of pesticides, her description would have been quite different. The following is my imagined rendition of it:

> Rather than seeking to understand the intricate life cycle and ecology of this tiny insect, scientists invented a scheme that would allow them, by infiltrating the very heart of their natural reproductive cycle, to sever the link between generations. Day after day, in huge "fly factories," technicians bombarded male insects with mutagenic X-ray radiation and then, using twenty light planes working five to six hours daily, these insidious carriers of genetically altered material were dispersed over huge areas. Unsuspecting females mated with these seemingly normal products of the laboratory. While these unions produced eggs, these were, without exception, sterile. In less than two years, the species had vanished.

Though my version contains the same facts as Carson's celebratory account, its redirection of the rhetoric of "naturalness" turns a tale of creative triumph into a tragedy of technological hubris. This easy mutability of the "natural" into the "unnatural" highlights the extent to which it is human desires—rather than "nature" itself—that define which creatures belong in the balance of nature and what form that balance should take.

Murray Bookchin's analysis in *Our Synthetic Environment*, which did not rest on notions of a balance of nature as a guide to human interaction with the environment, is spared these particular paradoxes. Bookchin claims that the veneration of "nature as the only source of human health and well-being," and the "quasi-mystical" and unreserved valorization of the "natural state" as superior, are misguided, "an impediment to a rational outlook" (26–27). We should not, he asserts, be averse in principle to altering the environment technologically to create a new kind of community, if this is done with an eye first and foremost toward "promoting human health and fitness" (244). Of course, he cautions, the remaking of nature in the service of our own needs should be done with due caution and testing, since "an environmental setting developed by natural selection over many millions of years must be considered to have some merit" (30). But this is purely a pragmatic reticence, not a sentimental one: emotions in the presence of nature are not, for Bookchin, any indication of nature's special metaphysical status (as they were for the Transcendentalists with whom Carson sympathized), only a reflection of rational human needs. "Our nostalgia," he claimed, "springs neither from a greater sensitivity nor from the wilder depths of human instinct. It springs from a growing need to restore the normal, balanced, and manageable rhythms of *human* life—that is, an environment that meets *our* requirements as individuals and biological beings" (240, emphasis added).

CARSON'S CALL FOR REGAINING BALANCE

Just as Carson's use of "natural balance" masks the social (in this case, the criteria informing the acceptability of certain forms of pest management), her framing of a solution to pesticide problems largely in terms of achieving an appropriate relationship with "nature" shifts attention away from the social roots of the destruction she describes. Later in this essay I describe Carson's reticence to discuss the political and economic forces that encourage heedless pesticide use. This disabled her capacity to talk about fundamental social interventions in the pesticide problem, leaving a respect for the balance of nature and ecological interconnectedness as her primary resources for its solution. The result is a reasonable, even inspiring repudiation of human arrogance, whose mildness and abstraction are proportional to the book's missing politics.

Thus, Carson's call for a science of biological control emphasizes a philosophical and epistemic problematic: the need for replacing an arrogant domination of nature with new attitudes. "Nature herself," she says, "has met many of the problems that now beset us, and she has usually solved them in her own successful way. Where man has been intelligent enough to observe and to emulate Nature he, too is often rewarded with success" (80).

But are insufficient humility (64) and lack of knowledge of natural process the barriers to solutions? Are they the reason for the recurring situation Carson describes, where cheap, effective, and harmless forms of biological control are overlooked in favor of harmful expensive chemicals? After all, as Carson knew, it was in the late nineteenth century, long before the synthesis of artificial pesticides, that a form of biological control saved California's citrus groves from the cottony-cushion scale in less than a year. The solution—a predator species, the ladybeetle (Vedalia)—was an acclaimed success not because it offered a more "natural" or ethically superior solution, but because it was simple, cheap, and outstandingly effective.[22] Biological control was not a newly discovered technology whose promise lay in the future, as one might surmise from *Silent Spring*, but one that had been investigated by the USDA for seventy-five years prior to the book's appearance, underfunded and mismanaged into decline, then given the coup de grace by the rise of faster-acting, profit-producing insecticides after World War II.[23] Nor were the many problems that plagued chemical pesticides (resistance, resurgence, toxicity, bioaccumulation) a surprise that surfaced only with widespread agricultural use in the postwar years; most were recognized decades before *Silent Spring* was published.[24]

Silent Spring opened a space that might have been occupied by an attempt to answer the difficult and messy political and economic questions of how pest control might be guided by biological knowledge and democratically determined priorities, rather than the logic of capital accumulation. Instead, this space was more palatably filled with the hopeful ideal of biological control as Yankee ingenuity in service of a pastoral ideal. The pastoral mode, the most long-lived Western model for an appropriate relation to nature, proposes a middle ground between the wild and the overcivilized.[25] *Silent Spring*'s middle ground lies in the *status quo* it portrays as being disrupted by pesticides (as in the rustic idyll of the book's opening paragraphs), and in the science of biological control. The latter navigates between excessive technological hubris on the one hand, and a vulnerability to nature's wildness in the form of pests on the other, incorporating the best of human artifice and inventiveness while preserving a closeness to natural cycles and creatures. The relation to nature that Carson proposes is one of cautious "guidance" (243, 261), reasonable "accommodation" (261), sensitive "management" (80), and an ethic of "sharing" (261) rather than "brute force" (80). These are valuable orientations in themselves, but less so when they function as vague substitutes for attention to the relations of human groups to one another.

Science and Technology: Challenged and Affirmed

Part of *Silent Spring*'s success followed from its appearance at a moment ripe for a lucid, accessible, and nonthreatening statement around which anxieties about technology, and especially toxic hazards, could coalesce. In the larger background loomed the technological excesses of the two world wars, but more proximally, Americans had been anxious about fallout for well over a decade and had been shaken by the cranberry scandal of Thanksgiving 1959[26] and the thalidomide tragedy publicized a short while before the *New Yorker* version of *Silent Spring* appeared.[27] Carson makes masterly use of these preexisting anxieties. She loses no opportunity, for example, to underline the similarities between the chemicals she is introducing to public consciousness and the radiation people had grown to fear over the years,[28] or to heighten the aura of evil surrounding dangerous synthetic chemicals by muted but consistent reminders that these were fabricated by German chemists (32, 35).

More generally, Carson subtly but firmly challenges science's authority along almost every dimension. *Silent Spring* undermines scientific pretensions of omniscience, omnipotence, infallibility, and benignity. She reveals instead an enterprise that can at times be arrogant, gratuitously bellicose, childishly irresponsible, insensitive to nuance and complexity, unaccountable, and for all these reasons, in need of moderation by nonprofessional forms of knowledge and sectors of society.

Carson's discussion of pesticides and their impacts repeatedly highlights the limitations of scientific knowledge. These chemicals, which can defy detection or identification by science (45), give rise to "little-understood interactions, transformations, and summations of effect" (45) whose impacts are not well understood. To a public accustomed to confident scientific proclamations, Carson presented a stream of admissions of limitation and helplessness. Of a composite of river pollutants, says a Professor Eliass, "We don't begin to know what that is. . . . What is the effect on the people? We don't know" (45). Of aquatic pollution Carson informs us, "The chemist who guards water purity has no routine tests for these organic pollutants and no way to remove them. . . . Investigators knew of no way to contain the contamination or halt its advance" (46–48). On fields contaminated by heptachlor she reports, "Scientists were unable to predict how long [they] would remain poisonous or to recommend any procedure for correcting the condition" (63). Her chapter on aquatic pollution ends with what seems like more holes than knowledge about the "unseen . . . unknown and unmeasurable effects of pesticides." "The whole situation is beset with questions for which there are at present no satisfactory answers . . . we do not know the identity [or] . . . total quality . . . and we do not presently have any dependable tests for identifying them . . . we do not know whether the altered chemical is more toxic than the original or less. Another unexplored area is the question of interactions between chemicals" (139).

Carson challenged not only the completeness of scientific knowledge, but also its unproblematic translation into the useful products and desirable mastery of nature that are often posed as the basis for scientific authority. In writing so extensively about the hazardous products of past scientific laboratories, and by showing that what scientists regard as safe today might turn out to be dangerous tomorrow (199), she contributed still further to the post-Hiroshima erosion of science's claim to benefit society. She suggested problems in mapping theoretical knowledge onto the real world, as when an engineering model is applied to agriculture (20) or an attempt is made to transfer laboratory-derived knowledge to the field (116). Most basically, she turned the story of pesticides toward a critique of the very project of the "control of nature," which is apt to "boomerang" (78), marking her skepticism by writing of pest "control" and "eradication" in quotation marks (117). "The control of nature," she says dramatically at the book's conclusion, "is a phrase conceived in arrogance" (261), and a science which is predicated on control as its sole criterion, and produces and uses pesticides accordingly, belongs in the "stone age."

In this questioning of the project of control, *Silent Spring* contains premonitions of the critique of instrumental rationality later made popular by Marcuse, Roszak, and others, and of subsequent claims by some feminist critics of science. Carson also anticipates the sensitivity of some of these later works to the metaphors and desires informing technological prowess. She rejects, for example, a scientific relation to nature modeled on warfare, and at times seems to mock the pesticide industry ("a child of the Second World War") for its use of recycled military tools and imagery. Thus, she criticizes the spray planes sent on "a mission of death" to "wage" a "needless war" on blackbirds (118), and scorns those intoxicated with the "bright new toy" of chemical control (68).

As Carson undermines scientific authority, she also valorizes nonspecialist and local knowledge of nature. She does this implicitly, by drawing extensively on lay accounts of the destructiveness of pesticides (in the many letters and calls of concerned citizens she quotes), and by virtue of her own status as an expert despite her lack of a doctorate. She also does so explicitly by ranking the insight of "men of long experience with the ways of the land" favorably alongside that of scientists (67; see also 23), and by valuing the knowledge of those directly affected above the institutional prestige of professionals. "A glance at the Letters-from-Readers column of newspapers almost anywhere that spraying is being done makes clear the fact that citizens are not only becoming aroused and indignant but that often they show a keener understanding of the dangers and inconsistencies of spraying than do the officials who order it done" (106). It is for this reason that she trusts the suggestions of people at regional levels over the blanket recommendations of federal agencies (108).

Carson's call for democratic control and public accountability of scientists and the chemical industry is, for reasons I discuss below, partial and often indirectly phrased, but it is not entirely absent from the book. She highlights a lack of public oversight when she quotes an entomologist's complaint that regulatory entomologists function as "prosecutor, judge and jury, tax assessor and collector and sheriff to enforce their own orders" (22). Her typification of our era as one in which the right of industry to "make a dollar at whatever cost is seldom challenged" (23), and her noting of the impulse to shut out thoughts of "the sterile and hideous world we are letting our technicians make" (71) are (typically roundabout) appeals for greater public assertiveness. However, the book rests its confidence not so much on a more democratic process for setting priorities, but on the capacity of science to extricate society from the troubles it has created through new technologies of biological control. *Silent Spring* relied, finally, on scientific evaluation to establish the occurrence of environmental damage; it helped to make this a central component of subsequent environmental activism.

Politics, and Its Avoidance

In a brief review of Patricia Hynes's *Recurring Silent Spring*, a feminist analysis of Carson's life and of current pesticide issues, Peter Taylor remarks that a richly contextual account of the development of Carson's work and its reception "remains unwritten."[29] In particular, he would like to understand better how Carson was constrained by the audience she hoped to affect, contributing to the "political softness" of her account of the pesticide problem.

I agree with the need for such an account, but would want to reflect carefully on the question of political softness. Despite all its blind spots, *Silent Spring* is not entirely lacking in political bite, but more important, there is evidence that the book's moderation was in part a calculated move—Carson had a sharper critique of some of the political underpinnings of pesticide use than she let on. Was this softness, then, or tactical restraint, and can we condemn it if it resulted in the book's mass acceptance?

Before examining the politics that got left out of *Silent Spring*, let's look at what got in. Taylor overstates the claim that Carson "made no attempt to identify who profited from the situation she described and whose interests would be threatened by substituting biological for chemical control of pests."[30] Carson was explicitly critical, for instance, of the profit incentives of chemical manufacturers and resellers that skewed the definitions of certain plants as "weeds" (71), biased the gardening advice given to suburbanites (79), and encouraged irrational scientific research priorities (92) and spraying programs (69). In one exceptional paragraph, she offers an unusually direct explanation of why chemical control had outstripped biological control: chemical companies poured money into pesticide research, which promised them fortunes that biological control did not. This funding, she says, is the reason that "certain outstanding entomologists are among leading advocates of chemical control. Inquiry into the background of some of these men reveals that their entire research program is supported by the chemical industry. . . . Their professional prestige, sometimes their very jobs depend on the perpetuation of chemical methods. Can we expect them to bite the hand that literally feeds them? (. . .) Knowing their bias, how much credence can we give to their protests that insecticides are harmless?" (229).

Carson was even more outspoken where she did not have to weigh her statements so carefully as she felt she needed to in the book. In a talk at the Women's National Press Club on December 5, 1962, Carson told her audience of a reviewer who had been offended by the above charge. "I can scarcely believe the reviewer is unaware of [the support of university research by pesticide manufacturers]," she continued, "because his own university is among those receiving such grants." She then went on to call attention to research in the *Journal of Economic Entomology* whose academic writers acknowledged support from Shell, Velsicol, and Monsanto.[31] On another occasion, she was asked to respond to the American Medical Association's suggestion that doctors direct patients concerned about pesticides to the chemical trade associations for information. "I can't believe," she told her interviewer, "that the AMA seriously believes that an industry with $300 million a year in pesticide sales at stake is an objective source of data on health hazards."[32]

Carson's collaboration with Clarence Cottam, one of her closest confidants in compiling *Silent Spring*,[33] further indicates her considerable awareness of the role of profit incentives in creating the problem she documented.

Cottam, who resigned from his post as assistant director of the USDA Biological Survey, was probably Carson's key source of information on the fire ant eradication program Carson describes (98, 129–131, 146–156). He was sharply critical of the effort and not afraid to complain of "the hustling salesmanship of the free enterprise system,"[34] or to ask the Agricultural Research Service responsible for the program in February 1960, "who was profiting from the excess use of poison?"[35] Before its publication, Cottam reviewed Carson's manuscript and warned her what she might expect: "I am convinced you are going to be subjected to ridicule and condemnation by a few. Facts will not stand in the way of some confirmed pest control workers and those who are receiving substantial subsidies from pesticide manufacturers."[36] Writing in the *Sierra Club Bulletin* after the book's appearance, Cottam dismissed the suggestion that *Silent Spring* should have presented the "other side" of the pesticide issue. "Hasn't [the benefit of pesticide use] already been overemphasized by a multi-billion dollar industry employing the most experienced salesmen and lobbyists available?"[37]

There is clear evidence then that Carson knew whose interests were served by the situation she described, though she kept much of this critique out of the book. I suspect it would be fascinating to look through early drafts of *Silent Spring* and her correspondence to understand at a finer level the process of and pressures for exclusion.[38] Remember that she wrote in what some have called the "McCarthy era of the environmental movement," in which those who questioned the use of pesticides were specifically branded as being against the spirit of free enterprise.[39] After the appearance of the *New Yorker* articles, for example, Louis A. McLean, secretary and general counsel of Velsicol, the sole manufacturer of chlordane and heptachlor, sent a five-page registered letter to Houghton Mifflin suggesting they might want to reconsider publishing *Silent Spring*. His letter built up to the following statement.

> Unfortunately, in addition to the sincere opinions by natural food faddists, Audubon groups and others, members of the chemical industry in this country and in Western Europe must deal with sinister influences whose attacks on the chemical industry have a dual purpose: (1) to create the false impression that all business is grasping and immoral, and (2) to reduce the use of agricultural chemicals in this country and in the countries of Western Europe, so that our supply of food will be reduced to east-curtain parity. Many innocent groups are financed and led into attacks on the chemical industry by these sinister parties.[40]

While this was an extreme reaction, it indicates the climate in which Carson wrote and chose to moderate her claims.

Not all this moderation was calculated ploy, of course. Nothing in Carson's background had provided her with a framework for analyzing and talking about the structural determinants of environmental destruction. In this respect, it is useful to compare her work with that of Murray Bookchin, who in the same period drew from a Marxist and anarchist tradition[41] to present a more fully social and political critique of some of the same problems. Published half a year before *Silent Spring*, *Our Synthetic Environment*—while lacking Carson's lyricism and evocative images of poisoned wildlife—was a much more comprehensive, multidimensional, and above all politically far-reaching work.

Bookchin documents a range of challenges to human well-being that result from life in an industrialized society. His treatment of pesticides in agriculture (97–104) is thorough and brisk, touching on many of the same significant concepts as Carson, such as their physiological impacts, their tendency to concentrate in fat, the growing insect resistance, and in much bolder terms, the economic incentives that perpetuate the use of dangerous pesticides. This discussion forms only part of his chapter on chemicals in food, which also treats the dangers associated with the use of synthetic hormones and antibiotics, and it is one chapter out of eight devoted to other problems in agriculture, urban life, chemicals in food, the environmental causes of cancer, radiation, and human health.

Bookchin's work is also deeply social, never losing sight of how these various challenges to human well-being result from particular social arrangements. Bookchin's emphasis in his work almost inverts the dictum of Albert Schweitzer that Carson chose as her motto: "that we are not being truly civilized if we concern ourselves only with the relations of man to man. What is important is the relation of man to all life."[42] *Our Synthetic Environment*, Bookchin declared in his 1974 preface, "is concerned not only with the relationship of humanity to nature and the balance of nature; it is even more fundamentally concerned with the relationship between human and human. The book advances the notion that there can be no sound natural environment without a sound, ecologically oriented social environment."[43]

Environmental problems cannot be solved through "remedial legislation" (226–237) alone, he claimed, but demand an alternative to a capitalist economy[44] in which "the most pernicious laws of the market place are given precedence over the most compelling laws of biology" (26). The book contains in embryonic form the critique of domination that Bookchin developed extensively in his later works (52–53), and its vision of agricultural and urban regionalism and decentralized "communities of human scale" (244–245) anticipates bioregional thought. An America ready to take this book to heart to the same extent it did Carson's would have been and become quite a different place.

Bookchin's unabashedly political orientation highlights what *Silent Spring* did not do, though *Our Synthetic Environment*'s failure as a work of mass appeal might make us more forgiving of those "limitations." In the remainder of this section, I look more closely at how the absence or evasion of a political analysis is felt in *Silent Spring*: the rhetorical means by which Carson avoided the political implications of the problem she had exposed, and how these strained the logic of her text. My examples concern the book's lack of specificity about the identity of responsible agents; its consistent refusal to follow case studies through to their logical (political) implications; its gentle tread in impugning motives of greed in that minority of cases where a responsible agent is mentioned; and its penchant for passive sentence constructions in critical places, which helped to achieve the diffusion of identity and agency. It is as if Carson believed it was enough to voice the sheer facts of pesticides' destructiveness, carefully and fully, and wait for public opinion to do the rest.

With the exception of the Rocky Mountain Arsenal of the Army Chemical Corps (47), Carson does not name a single manufacturer of chemicals or other delinquent party. In some places in the book she must have consciously restrained herself, so appropriate would specific names have been. She mentions, for example, "a chemical plant" that had been dumping pesticide wastes for ten years, but does not name it (134). Why? Carson's avoidance of brand names in discussing a new carcinogenic chemical used against mites and ticks (199–200) requires a stream of nonspecific circumlocutions: "a chemical," "this chemical," "the chemical," "the product," "the suspected carcinogen," and so on. How much simpler, though more politically challenging, to have said Aramite.[45] More frustrating is the avoidance of brand names in her complaint about the innocuously named weed-killers sold for suburban lawns that do not list their ingredients (which include chlordane and dieldrin), nor mention their dangers (161). Wouldn't she have served her audience well by noting some of these brand names at this tantalizingly apt point, thus working directly to end, not just describe, the facade of benignity of which she is critical?

Another pattern of evasion recurs at a larger scale in Carson's consistently elliptical capping of her descriptions of irrational pesticide use. Repeatedly she builds a careful case to show that the instances of spraying she describes were not only harmful to humans and wildlife, but unjustified even in terms of biological effectiveness or economic payoff to farmers. Why did spraying take place nonetheless? Carson's scenarios demand an answer, but hers is vague or often lacking altogether. The reader is left to make their own inferences or, more likely, to ignore the troubling questions these narrative lapses signal. This kind of hanging question seems to be most comfortably accommodated at the end of sections. "The science of range-management," she says in the last sentence of chapter 6, "has largely ignored [the] possibility [of biological control of weeds by plant-eating insects] although these insects . . . could easily be turned to man's advantage" (81–82). "There is no dearth of men who understand these things," she says in the last sentence of another section, "but they are not the men who order the wholesale drenching of the landscape with chemicals" (73). "Funds for chemical control came in never-ending streams," she says elsewhere, "while the biologists . . . who attempted to measure the damage to wildlife had to operate on a financial shoestring" (89). Why the marginalization of effective biological control? the distance between those who know and those who order? the discrepancy between budgets for inventing chemicals and for studying their damage? Carson's silence on these questions buries the problem of the democratic control of science, technology, and production.

To the extent that Carson does trace the origins of the destruction whose "irrationality" she has exposed, her account of agency is feeble and diffuse and her blame is mild. People's destruction of the environment stems from their failure to "read" the "open book" of the landscape (65); facts about pesticides' destructiveness are denied out of "shortsightedness"[46]; spraying continues because of "entrenched custom" (74) or "surely, only because the facts are not known" (75).[47] "We are walking in nature like an elephant in the china cabinet," she quotes a scientist of "rare understanding" (77) as saying, implying "our" problem to be one of clumsiness.[48] And in choosing the "man with the spray gun" (83) as her icon of destructive pesticide use (rather than a symbol pointing back to the government agency, the chemical manufacturing tycoon, or the market), she locates agency in and directs expectations of responsibility toward only this most visible and least structural of entities.

Carson's frequent use of passive or negative sentence constructions further supports this masking of agency and blame. She uses a passive construction, for example, in explaining why a simple change in agricultural practices (a shift to a different variety of corn) was not the chosen solution to a problem with crows. Though this measure would have obviated the need for spraying with harmful pesticides, she explains, "the farmers *had been persuaded* of the merits of killing by poison" (118, emphasis added). Her excision of the subject closes down a crucial line of investigation. The following typical example of a negative formulation lessens blame even as it assigns it. "Because the spray planes were paid by the gallon rather than by the acre," Carson says, "there was no effort to be conservative" (145–146). How much more powerful would this sentence have been had its latter part been directly and positively phrased: "they tried to use as much as possible." (It also would have helped had she worked to unreify "spray planes" and see exactly which *people* were paid.)

Similar passive constructions pervade her description of the development of the fire ant eradication program mentioned earlier in connection with Carson's outspoken informant, Clarence Cottam. Initially the fire ant was not felt to be a particularly serious pest, but:

> with the development of chemicals of broad lethal powers, *there came* a sudden change in the official attitude toward the fire ant. . . . The fire ant suddenly *became the target* of a barrage of government releases, motion pictures, and government-inspired stories portraying it as a despoiler of southern agriculture and a killer of birds, livestock, and man. A mighty campaign *was announced*. . . . The fire ant *was pictured* as a serious threat to

> southern agriculture. . . . Its sting *was said* to make it a serious menace to human health, etc. (146–147, emphasis added)

Carson assigns agency for this development, but weakly and indirectly, relying on the following quote from a trade journal to hint at why a widely criticized program she calls "ill conceived and badly executed" received continuing support: "United States pesticide manufacturers appear to have tapped a sales bonanza in the increasing numbers of broad-scale pest elimination programs conducted by the U.S. Department of Agriculture." Imagine the rather different effect that might have been achieved by a description in which the italicized phrases were in active form, and in which Carson used her own voice to speak of the "creation" rather than the "tapping" of a sales bonanza.

Carson's account of the Japanese beetle spraying program in southeastern Michigan (85–95) further illustrates the mechanisms she used to depoliticize her writing, and how these undermine the logic of her account. She tells of a method of biological control for the beetle that was found to be efficient in the eastern states. Why then was this not employed in Michigan, rather than spraying, she asks? She eliminates two possible explanations: biological control was actually less expensive than spraying, and it would have been effective even at the limits of the beetle's range in Michigan. Her explanation of why the spraying she has now posed as irrational occurred nonetheless points to two agents: (1) those who "want immediate results at whatever cost"—a valid point about a (broad cultural?) bias toward immediate gratification, though one that begs the question of the attitude's origins and bearers; and (2) those "who favor the modern trend to built-in obsolescence, for chemical control is self-perpetuating, needing frequent and costly repetition" (95). The latter is an extremely nonspecific and indirect way of speaking about profits and who makes them, and neither assertion helps to understand how these interests have come to dominate the science, policy, and decision making surrounding beetle treatment.

Conclusions

In *Silent Spring*, Carson drew on familiar and widely shared conceptions of nature, and especially the balance of nature, to frame the pesticide poisoning she wanted to stop. She artfully harnessed and imaginatively inflected these to portray the damage the new chemicals caused, and to frame what it was they disturbed that should be restored. Her portrait of an increasingly and needlessly toxic landscape had the potential to dismay and anger; but its galvanizing capacity was directed largely away from questioning the ground rules of capitalist economy that shape destruction-inducing incentive structures in agriculture and pesticide production, and toward a diffuse critique of institutional oversight and scientific and technological arrogance. Thus, she both challenged and placated, extended and maintained, existing world-views.

Several factors conditioned the book in this way. First, Carson's perspective was partial; her location curtailed her analysis even as it provided its strengths. Neither her background, nor her identities as a scientist within a government bureaucracy and a nature writer, had provided her with the experiences or theoretical resources to push through the kind of political analysis that Bookchin did. But they gave her tools to stir readers and make a convincing appeal, of limited scope, for rationalizing scientific and bureaucratic relations to pesticides. Bookchin's viewpoint was partial in its own way; his radical proposal drew on, and was continuous with and limited by, a different set of conventions in Marxist thought. Both authors shared the blind spots of their times: neither considered, for instance, the impacts of the increasing portion of U.S. pesticide production exported to the Third World.

Second, I have already mentioned Carson's desire to make herself and her message less vulnerable to the antagonism she knew the book would evoke. A third factor in the book's tameness was its crafting by Carson to ensure the largest possible constituency for the changes she hoped to catalyze. She knew from her own experience how difficult it was to change one's view of the human relation to the natural world in response to the challenge of new information. In 1958, as she began gathering her thoughts toward writing about the impact of DDT, she described her own struggle in this respect.

> I suppose my thinking began to be affected soon after atomic science was firmly established. Some of the thoughts that came were so unattractive to me that I rejected them completely, for the old ideas die hard, especially when they are emotionally as well as intellectually dear to one. It was pleasant to believe, for example, that much of Nature was forever beyond the tampering reach of man: he might level the forests and dam the streams, but the clouds and the rain and the wind were God's. It was comforting to suppose that the stream of life would flow on through time in whatever course God had appointed for it—without interference by one of the drops of that stream, Man. And to suppose that, however the physical environment might mold Life, that life could never assume the power to change drastically—or even destroy—the physical world.
>
> These beliefs have been part of me for as long as I have thought about such things. To have them even vaguely threatened was so shocking that I shut my mind—refused to acknowledge what I couldn't help seeing. But that does no good, and I have now opened my eyes and my mind. I may not like what I see, but it does no good to ignore it, and it's worse than useless to go on repeating the old "eternal verities" that are no more eternal than the hills of the poets. So it seems time someone wrote of life in the light of the truth as it now appears to us. And I think that may be the book I am to write.[49]

Aware of how slow and painful it was to rework her own world-view in light of new facts, Carson must have realized how difficult it would be for her account of pesticides to alter the sensibilities of others appreciably, especially those who did not share her training or deep love of nature. Thus, she did not aim for a full-scale conversion of her audience's world-view, only "to persuade ordinary prudent men to be more cautious at the main points of stress upon the ecological system."[50]

Carson's achievement of what may seem to us a limited goal raises at least three kinds of fascinating and important questions. First is an evaluation of the ultimate efficacy of *Silent Spring* as a political intervention. My essay has repeatedly touched on the tension between the book's achievements and limitations. On the appreciative side, one could argue that changes brought about by *Silent Spring* were an important partial step toward a radically different world, and a necessary platform on which later, more far-reaching proposals could stand. Carson was the best messenger one could hope for—a skilled writer with the best available facts and a solidly innocuous reputation that gave her both the public's trust and access to an optimal platform of three consecutive issues of the *New Yorker* during the summer months. The angry response to her work in many quarters indicates that even Carson, who artfully rallied every conceivable appealing trope in the service of her cause, could not have pushed harder without losing her broad audience. And this audience was essential to what she achieved: a book on top of the bestseller list could not be ignored by the media, the scientific community, or the Kennedy administration. That the comprehensive, uncompromising *Our Synthetic Environment*, which was the best political *analysis* one could hope for at the time, had no political *impact*, is further indication that *Silent Spring* stood at the outer limits of acceptability. This evaluation should give us pause before speaking of *Silent Spring*'s political "tameness" or "softness," for isn't the capacity to reach out to people where they are and persuade them, a strength?

At the same time, we must ask difficult questions about *Silent Spring*'s limitations. To what extent, for example, is a partial analysis a wrong analysis? Specifically, did the book mystify the pesticide issue by largely ignoring the ways in which the problem is rooted in a capitalist economy? Must *Silent Spring*'s popularity be weighed against the degree to which it eclipsed Bookchin's more "politically pointed" work?[51] Did it make subsequent radical analyses more difficult? Was its reformist framework for thinking about pesticides, which sent future efforts down the track of remedial legislation rather than the fundamental democratization of research, technology and production, a distraction rather than a first step? After all, agriculture and the chemical industry could respond relatively easily to such legislation. Restrictions placed several years later on organochlorines (the earliest generation of synthetic pesticides such as DDT) did not halt their continued manufacture for export, or the development and profitable production of other pesticides, or recent attempts to genetically engineer profitable and hazardous pest- and pesticide-resistant crops, or most generally, the trend toward increasingly mechanized and chemicalized large-scale agriculture. Thirty years after *Silent Spring*'s publication, the executive director of the National Coalition Against the Misuse of Pesticides could still describe the crossroad facing America as the choice between "promoting safer alternative pest management techniques or simply substituting less toxic inputs into conventional pesticide-intensive practices," and complain about "the extremely limited support for alternative practices from most mainstream agricultural institutions."[52] Our celebration of the book's achievements must be tempered by a consideration of how little has changed fundamentally.

These two sides to an evaluation of *Silent Spring*'s impact are intended not to judge, dispirit, or confuse, but to clarify important topics: the limits of what could be said and widely heard in that particular historical moment; how Carson's critique embodied and perpetuated the limited viewpoint of its author and assumptions of its times; the mechanisms that favored the emergence and circulation of her more moderate criticism; and how such criticism could, despite these constraints, initiate more fundamental change. To investigate these more fully would require integrating the mostly internal reading of *Silent Spring* I have attempted here with an analysis of the following: (1) the network of factors facilitating the use of pesticides in America up until 1962[53]; (2) the processes of production and reception of both Carson's and Bookchin's work, with special attention to their respective blind spots and the pressures to blunt the strength of their critiques; and (3) *Silent Spring*'s impact on this network through its unfolding effects on subsequent legislation, scientific research, and the framing of environmentalist demands.[54] Some of these effects (and thus, the successes of *Silent Spring*) are more subtle and circuitous than appears at first glance: not simply the subsequent banning of DDT, for example, but the initiation of a slow process that raised the cost of approving a new pesticide so high that alternative forms of pest management have begun to be economically feasible.

The foregoing attempt at historical evaluation is closely allied to a second set of questions raised by *Silent Spring*. What are the implications of Carson's successful amalgam of change and continuity for political strategy on the Left today? What continuities with conventional understandings of nature are contained in protests that ostensibly break with them? When do we attempt to weed out these conventional elements, and when do we use them as persuasive resources? (Essentialisms of various sorts are, after all, powerfully galvanizing to public opinion.)[55] In what situations is a certain calculated political blandness strategically preferable to presenting radical utopias in attempts to mobilize a broad public? To what extent should we adjust the scope of calls for change in order to remain within the bounds of the popular imagination? When facing a specific political challenge, how are we to find answers to these questions, which cannot be decided on the basis of first principles? And is this too simple and arrogant a model of critical cultural production: does the confident "we" in these questions usually have this much choice and opportunity for premeditation as it considers which aspects of a preexisting perspective it should pour into less enlightened heads?

The third set of questions that emerges from this study concerns more general consideration of what has changed and what remained the same through past shifts in environmental world-view. For example, even in a transition with such drastic and thoroughgoing impacts on the human relationship to nature as the seventeenth-century rise of the mechanical world-view, one finds continuities and precursors. Cultural elements consonant with the emerging scientific paradigm had existed sotto voce in earlier periods, and the organic world-view marginalized by the newly dominant scientific world-view did not disappear entirely, but went underground or mutated into other forms; these persist as potential counterhegemonic cultural elements up to the present.[56] The complex admixture of the old and new in ideas, books, and programs for change is a truism of intellectual history. *Silent Spring*'s impact suggests it might be worth paying attention to how events, artifacts, or movements that balance dislocation and continuity in felicitous ratio facilitate these shifts, easing the emergence of the novel by couching it in familiar forms.

Acknowledgments

I appreciate Iain Boal's suggestion that Murray Bookchin's *Our Synthetic Environment* might be usefully compared with *Silent Spring*, and the comments of Leo Marx, Peter Taylor, Charles Weiner, Rosalind Williams, Anna Tsing, Barbara Goldoftas, and Danny Faber on drafts of this essay.

Notes

1. Rachel Carson, *Silent Spring* (New York: Ballantine, 1962). On the book's impact, see Frank J. Graham, *Since Silent Spring* (Boston: Houghton Mifflin, 1970); and Linda Lear, "Rachel Carson's *Silent Spring*," *Environmental History Review*, 17, no. 2 (1993): 23–48. Most of the biographical detail in the following several paragraphs is drawn from Lear's excellent article.

2. This can be read as a necessary but partial contribution to an as yet unwritten, broader contextual analysis of the construction and reception of *Silent Spring*. See Peter Taylor's call for such an analysis in his review of books by Rosaleen Love and Patricia Hynes, "Feminist Tales," *Science, Technology, and Human Values* 16, no. 4 (1991): 540–543.

3. Lear, "Rachel Carson's *Silent Spring*," p. 36.

4. Murray Bookchin, *Our Synthetic Environment* (New York: Knopf, 1974). The original edition appeared in 1962 under the pseudonym Lewis Herber. The 1974 edition appeared under Bookchin's name with an extensive introduction.

5. Described in Graham, *Since Silent Spring*, p. 56; see also Lear, "Rachel Carson's *Silent Spring*," n.49.

6. Page numbers in text hereafter refer to *Silent Spring*.

7. Frank N. Egerton, "Changing Concepts of the Balance of Nature," *Quarterly Review of Biology* 48 (1973): 322–350; W. F. Bynum, E. J. Browne, and Roy Porter, *Dictionary of the History of Science* (Princeton, N.J.: Princeton University Press, 1981), s.v. "economy of nature."

8. See Carson, *Silent Spring*, pp. 179–80.

9. Similarly, aspects of the inner ecology are discovered through their vulnerability. We become aware of the many ongoing miraculous processes that maintain cellular and physiological health by observing the consequences of their disruption.

10. For the kinds of attitudes and beliefs that *Silent Spring* helped to alter, see Nancy Kraus, Torbjörn Malmfors, and Paul Slovic, "Intuitive Toxicology: Expert and Lay Judgments of Chemical Risks," *Risk Analysis* 12, no. 2 (1992): 215–232, especially sect. 3.3 and n.11. Funded by Dow and Monsanto, this article tends toward a stance of "correcting" intuitive understandings so as to bring them into line with the more prochemical perceptions of professional toxicologists. As the authors point out, intuitive understandings of toxic exposure may be influenced by magical thinking about "contagion" and "contamination"; a similar line of thought has been developed by Mary Douglas in *Purity and Danger: An Analysis of Concepts of Pollution and Taboo* (New York: Praeger, 1966).

11. Carson quotes René Dubos: "Men are naturally most impressed by diseases which have obvious manifestations, yet some of their worst enemies creep on them unobtrusively" (169).

12. On aesthetic considerations, see Carson, *Silent Spring*, pp. 69–71. On nature as, quite literally, a resource akin to veins of copper and gold, see Carson's quotation of Justice Douglas on p. 72.

13. Donald Fleming, "Roots of the New Conservation Movement," *Perspectives in American History* 6 (1972): 7–91; see pp. 11–14.

14. Carson, *Under the Sea Wind* (New York: Dutton, 1941); *The Sea Around Us* (New York: Oxford University Press, 1951); *The Edge of the Sea* (Boston: Houghton Mifflin, 1955).

15. See Paul Brooks, *The House of Life: Rachel Carson at Work, With Selections from Her Writings Published and Unpublished* (Boston: Houghton Mifflin, 1955).

16. Roderick Nash, *The Rights of Nature: A History of Environmental Ethics* (Madison: University of Wisconsin Press, 1989) p. 79.

17. Fleming, "Roots of the New Conservation Movement," pp. 29–30.

18. Ibid., p. 32.

19. Carson, *Silent Spring*. See her discussions of the Clear Lake Grebes (51) and of pesticides in baby food (62).

20. An extension explicitly and extensively attempted in subsequent environmental ethics.

21. In other places, too, Carson celebrates the ability of biocontrol to "forge weapons from the insect's own life processes" (251), or looks to helpful insects as a means to "keep at bay a dark tide of enemies" (222).

22. Paul deBach and David Rosen, *Biological Control by Natural Enemies*, 2nd ed., (New York: Cambridge University Press, 1991), pp. 140–148.

23. Richard C. Sawyer, "Monopolizing the Insect Trade: Biological Control in the USDA, 1888–1951," *Agricultural History* 64, no. 2 (1990): 271–285.

24. Thomas R. Dunlap, *DDT: Scientists, Citizens, and Public Policy* (Princeton, N.J.: Princeton University Press, 1981); John H. Perkins, *Insects, Experts, and the Insecticide Crisis: The Quest for New Pest Management Strategies* (New York: Plenum, 1982); James Whorton, *Before Silent Spring: Pesticides and Public Health in pre-DDT America* (Princeton, N.J.: Princeton University Press, 1974); John H. Perkins, "Reshaping Technology in Wartime: The Effect of Military Goals on Entomological Research and Insect-Control Practices," *Technology and Culture*, 19, no. 2 (1978): 169–186; Robert van Den Bosch, *The Pesticide Conspiracy* (Berkeley: University of California Press, 1978); Edmund P. Russell, "Safe For Whom? Safe for What? Testing Insecticides and Repellents in World War II," presented at the meeting of the American Society for Environmental History, Pittsburgh, 1993; Angus A. MacIntyre, "Why Pesticides Received Extensive Use in America: A Political Economy of Agricultural Pest Management to 1970," *Natural Resources Journal* 27, no. 3 (1987): 533–578.

25. On the pastoral, and especially its adaptation to particularly American circumstances, see Leo Marx, *The Machine in the Garden: Technology and the Pastoral Ideal in America* (Oxford: Oxford University Press, 1964); Marx, "Does Pastoralism Have a Future?" in *The Pastoral Landscape*, ed. John Dixon Hunt (Washington, D.C.: National Gallery of Art, distributed by University Press of New England, Hanover, N.H., 1992), pp. 209–225.

26. The FDA banned cranberries sprayed with aminotriazol. See Brooks *The House of Life*, pp. 261–262.

27. Graham, *Since Silent Spring*, pp. 50–51.

28. Carson highlights their relatedness at every opportunity. She calls them "partners of radiation" (184); introduces chemical-induced mutation as "radiomimetic" (189); and equates a Japanese fisherman caught in fallout on the boat *Lucky Dragon* with the Swedish farmer exposed to indiscriminately sprayed pesticides ("For each man a poison drifting out of the sky carried a death sentence. For one, it was radiation-poisoned ash; for the other, chemical dust" (204). See also pp. 43, 168, 209. On the book's background in A-bomb anxieties see Graham, *Since Silent Spring*, p. 40. On the connection with fallout, see R. H. Lutts, "Chemical Fallout: Rachel Carson's Silent Spring, Radioactive Fallout and the Environmental Movement," *Environmental Review* 9 (Fall 1985): 211–225.

29. Taylor, "Feminist Tales."

30. Ibid., p. 542.

31. Graham, *Since Silent Spring*, p. 71.

32. Ibid., p. 58.

33. Ibid., p. 26.

34. Ibid., p. 27.

35. P. Daniel, "A Rogue Bureaucracy: The USDA Fire Ant Campaign of the Late 1950s," *Agricultural History* 64, no. 2 (1990): 99–114; see p. 110.

36. Graham, *Since Silent Spring*, p. 36.

37. Ibid., p. 63.

38. Carson had reason to fear both for her own well-being as well as for the survival of her message. Mundanely but critically, she was terrified of being bankrupted by libel suits; her publisher, who took out extra insurance with Lloyd's before the book's publication, also anticipated trouble. Lear, "Rachel Carson's *Silent Spring*," n.42.

39. Graham, *Since Silent Spring*, pp. 29–30.

40. Ibid., p. 49.

41. In his 1974 introduction, Bookchin traces his social analysis to Kropotkin, Mumford, and Paul Goodman. *Our Synthetic Environment*, p. lxxii.

42. Carson, address on receipt of the Schweitzer Medal of the Animal Welfare Institute, 7 January 1963. Quoted in Brooks, *The House of Life*, p. 316.

43. Bookchin, *Our Synthetic Environment*, pp. xviii–xix.

44. In his 1974 introduction, Bookchin has a clearer statement of what was only implicit in his original work. Environmental destruction is a "tendency inherent in the social system" of capitalism. It stems not from "moral delinquency" or even "greed," but "from a market-oriented system in which everything is reduced to a commodity, in which everyone is reduced to a mere buyer or seller, and in which every economic dynamic centers on capital accumulation." *Our Synthetic Environment*, p. xxxiii.

45. From Carson's notes on her sources (284), I infer this to be the name of the chemical.

46. In its full context, this phrase might be read as ironic.

47. Here, too, I detect a possible irony in her use of "surely, only . . ."

48. See also her use of a shotgun metaphor (67), implying that imprecision is the problem.

49. Quoted in Graham, *Since Silent Spring*, pp. 13–14.

50. Fleming, "Roots of the New Conservation Movement," p. 30.

51. Taylor, "Feminist Tales," p. 542.

52. Jay Feldman, "Thirty Years after *Silent Spring*, the Choice is Clear," *Global Pesticide Campaigner* 2, no. 4 (1992): 11–12.

53. A remarkably comprehensive analysis of this network is given in MacIntyre's "Why Pesticides Received Extensive Use in America," but he more or less brackets the "relatively fixed conditions of American political culture and market economy" (p. 575) that underlie the more proximal causes he discusses. It is precisely *Silent Spring*'s capacity to make inroads into these that is at issue.

54. For analyses of Carson's impact, see the sources cited in MacIntyre, "Why Pesticides Received Extensive Use," p. 551, n.75, and anthologized in David Wade Chambers, ed., *Worm in the Bud: Case Study of the Pesticide Controversy* (Victoria, Australia: Deakin University Press, 1984).

55. More fundamentally, the very notion of "essentialism," or at least its current functions as taboo and conversation stopper and its unproblematic opposition to constructionism, needs to be examined. See Diana Fuss, *Essentially Speaking: Feminism, Nature, and Difference* (New York: Routledge, 1989).

56. See Carolyn Merchant, *The Death of Nature: Women, Ecology, and the Scientific Revolution* (San Francisco: Harper and Row, 1980).

Chapter Eleven

The Commoner–Ehrlich Debate: Environmentalism and the Politics of Survival

Andrew Feenberg

Introduction

Early environmentalists attempted to awaken concern about a wide range of problems, from pesticides to population control, without always discriminating priorities among them. The writers may have ranked the issues differently, but as members of a beleaguered minority dismissed as cranks by majority opinion, they rarely had the time or inclination for quarrels amongst themselves. As is often the case with stigmatized out-groups, harmony prevailed precisely in proportion to the burden of exclusion carried by those brave enough to join.

That happy state of affairs did not survive the success of the environmental movement in the early 1970s. Significant disagreements emerged which are reflected in the movement to this day. The first visible signs of the depth of the split appeared in 1971, as Paul Ehrlich and Barry Commoner debated the relative importance of population and pollution control.

Paul Ehrlich was not the first to discover the population explosion (that honor is usually granted Malthus) but he has done more than anyone else in the United States to spread the notion. A professor of population studies at Stanford University, Ehrlich has been a tireless Cassandra of demographic disaster. Books such as *The Population Bomb* and *How To Be a Survivor*, speeches on dozens of college campuses, and his campaign for zero population growth reached a wide audience and helped to make ecology a legitimate public issue. On the jacket of his latest book, no less than Albert Gore is among those endorsing Ehrlich's work. Yet, Ehrlich's politics have always been ambiguous. By emphasizing population control as the key environmental issue, Ehrlich has identified himself with causes as diverse as no-growth ideology, Chinese population policy, New Left opposition to consumerism, and conservative attacks on Mexican immigration and high natality among minorities.

Barry Commoner is director of the Center for the Biology of Natural Systems at Queens College. He, too, began an intense campaign for the environment in the 1970s, culminating in his run for president on the Citizen's Party ticket in 1980. His 1971 bestseller, *The Closing Circle*, began a long polemic with the advocates of population control by arguing for a class politics of the environment. He soon became the chief public advocate of environmental socialism. Today he plays a leading role in the National Toxics Campaign. One aspect of his program, the emphasis on technical change, has become standard fare in the environmental movement.

The Commoner-Ehrlich debate quickly moved beyond scientific disagreement to embrace two radically different rhetorics and strategies. Their argument, which took place at the very beginning of widespread public concern over the environment, adumbrated the main themes of later controversies over humanism and antihumanism, democracy and dictatorship, and North/South disputes. These themes are reflected today in the very different emphases of deep ecologists (such as Earth First!) and environmentally conscious trade unions (such as the Oil, Chemical, and Atomic Workers). In Germany, some of the same disagreements were reflected in the split within the Green Party between the "Fundis" and the "Realos," the former demanding an end to industrial and population growth while the latter pursued red-green alliances with labor to reform industrialism.

Thus Ehrlich and Commoner did indeed prove to be prophets, but not so much of the environment as of controversies in the movement to save it. In this paper, I review their early debate and some of their more recent positions with a view to gaining historical perspective.[1]

An End to History

There is something surprising about these disagreements. After all, scientists are supposed to be better at building consensus than the rest of us. However, there is a significant precedent for the conflict over environmentalism: the scientists' movement for nuclear disarmament that followed World War II. That experience is especially relevant to environmental debates because leading environmentalists participated in it while others imitated it, consciously or unconsciously. Ehrlich, for example, attempted to give environmentalism some of the trappings of the postwar scientists' movement, as is apparent from the title of his 1968 bestseller, *The Population Bomb*. (His most recent book is called *The Population Explosion*.) In fact, he won the sympathy of the *Bulletin of the Atomic Scientists*, founded in 1945 to work for public understanding of the nuclear threat. Commoner had been an activist in the original scientists' movement as a senatorial assistant after the war who went on to fight for the Nuclear Test Ban Treaty and founded a journal called *Nuclear Information*. While Ehrlich attempted to revive something resembling the old scientists' movement on the basis of an apocalyptic rhetoric designed to strengthen the political authority of science, Commoner moved to a very different position.

The original scientists' movement arose from the anguished realization that the creation of the atomic bomb contradicted the supposedly humanitarian mission of research. Yet, the very fact that science had proved itself capable of such a feat promised scientists a larger voice in the disposition of the forces they had unleashed than they had ever enjoyed as benefactors of humankind. The opportunity to speak with new authority was immediately seized by scientists involved in the Manhattan Project.[2]

Physicists, many of whom had been sympathetic to socialism during the depression, quickly dropped public concern for class issues and set themselves up as spokesmen for science, a new force in human affairs with as yet unsuspected promise and power. The new scientific statesmanship hoped to gain a hearing by emphasizing the apocalyptic nature of the forces science had unleashed; calling on the human species to address the issue of survival; subordinating all particular individual, social, and national interests to this larger issue; and organizing a unanimous front of scientists to put the new authority of research to good use.

The scientists' movement brought a fear and a hope: the fear of the mortality of the human species, and the hope of a world government and end to the use of force in the affairs of nations as the only adequate response to the dangers of the nuclear age. Some suggested that all nuclear secrets be immediately shared with the Soviet Union as a quid pro quo for Soviet renunciation of the bomb. Others wanted the United States to surrender its nuclear arsenal to the United Nations, and to renounce further research on and production of such weapons. (A generation earlier, Nobel had similarly imagined that the discovery of the awful weapon dynamite would finally put an end to war.)

Of course, none of this occurred. Instead, cold war competition between the Soviet Union and the United States began surreptitiously in Hiroshima, and since then we have all grown accustomed to living under a nuclear sword of Damocles. It soon became clear that far from resolving world problems, the fear of nuclear destruction simply changed the stakes of the contest.

With ecology, the biological sciences are now supposed to open an escape hatch from the divisions of nation and class which drive human history. Like a natural disaster of planetary scope, the environmental crisis could unify humankind beyond historic rivalries in a more fundamental confrontation with nature itself. Accordingly, the environmental movement began as a politics of species survival, frightening people onto the common ground of a "no deposit, no return" world, but not surprisingly, the same old shit of history reappeared as a division within the environmental consensus. The millenial conflict of rich and poor invaded the common ground of the environment as it has every similar locale on which humanity has attempted to set up camp.

The Commoner–Ehrlich Debate

The scientific substance of the ecological debate concerns the causes of and solution to the environmental crisis. On the cause, experts are divided: some (like Ehrlich) asserting that the principal source of the crisis is overpopulation, others (like Commoner) blaming it on polluting technologies. The first argue that "the causal chain of deterioration is easily followed to its source. Too many cars, too many factories, too much pesticide, . . . too little water, too much carbon dioxide—all can be traced easily to *too many people*."[3] The second protest that "environmental degradation is not simply the outcome of some general expansive process, growth of population, or demand for goods, but of certain very specific changes in the ways goods are produced which are themselves governed by powerful economic and political considerations."[4]

On the solution, the same division appears, reflecting radically opposed policies corresponding to different class and national interests. Not surprisingly, those most concerned about the exhaustion of resources are the prosperous nations and social strata that consume such disproportionate quantities of them. Accordingly, they advocate controls over population and economic growth. On the other side, those principally concerned about polluting technologies worry most about the exhaustion of "garbage dumps" which they, too, claim is upon us. It is to be expected that the poor, who hope to gain from economic growth but who in the meantime cannot easily escape the health hazards and pollution with which it is associated, should be most attracted to theories that criticize not growth per se but its unintended consequences.

At first sight, this disagreement may seem artificial. Surely exponential growth in population must slow and stop before it destroys both the ecosystem and society, and it is equally indisputable that current industrial technology is dangerously unsound. Whatever the *main* source of the problem, certainly both population pressure and polluting technologies are contributing causes. Why can't slowing birth rates and changing technologies complement each other?

Ehrlich, whose whole strategy rested on augmenting the authority of the scientific community to the measure of the crisis it identified, was particularly anxious to preserve a united front of scientific opinion. He therefore proposed a public compromise with Commoner while, behind the scenes, the scientists could sort out their technical disagreements in their own good time. Ehrlich wrote, "We have made several personal attempts to persuade Commoner to avoid a debate on which factor in the environmental crisis is 'most important.' We felt that such a debate would be counterproductive for the goals which we all share. Unhappily, however, he has persisted in carrying out a campaign, both in speeches and the popular media, to dismiss the roles of population growth and affluence, and place the blame entirely on 'faulty technology.' "[5]

Ehrlich's complaint was overstated, but it is true that Commoner chose to polarize the issue. Why did he persist, in Ehrlich's terms, in "splitting the environmental movement?" In Commoner's words, there is "more than logic and ecology at work here." If the debate heats up, it is because human survival is a political issue; we are all concerned, but not all in the same way. As Commoner put it:

> There is a tendency, in some quarters, to treat the environmental crisis as though it were an issue in which everybody wins—whether rich or poor, worker or entrepreneur. Do the data on the United States situation support this view? Is it in fact true that environmental improvement is a good so universal in its value that it can override vested interests that contend so bitterly over other issues—such as jobs? The answer, I am convinced, is no. There is usually no way to work out an even-handed distribution of the cost of environmental improvement; something has to give.[6]

In sum, Commoner rejected the idea of a universal interest in survival around which the human race could unite. Ultimately, the policy issues that divide environmentalists correspond to class lines in what is essentially a new terrain for an old struggle.

THE BIOLOGICAL OR THE SOCIAL

Underlying the debate over the relative impact of overpopulation and polluting technology is the question of the relative significance of biological and social factors in causing environmental problems. For Ehrlich, the "population bomb" involves a biological process, human reproduction, gone completely out of control. For Commoner, environmental problems of all sorts, including overpopulation, are effects of social causes inherent in capitalism and colonialism.

Ehrlich's views have the virtue of being simple, clear, and easy to dramatize. Like any other exponential curve in a finite environment, that which traces the population explosion must finally level off. "Basically, then, there are only two kinds of solutions to the population problem. One is a 'birth rate solution,' in which we find ways to lower the birth rate. The other is a 'death rate solution,' in which ways to raise the death rate—war, famine, pestilence—*find us*."[7]

Writing in 1968, Ehrlich argued that the latter process had already begun.

He suggested three likely futures for the human race over the coming decades. His most optimistic projection included the death of "only" 500 million people in a ten-year "die-back" to a new balance of population and resources. This conclusion followed from Ehrlich's belief that in 1958, "the stork passed the plow" in the developing countries; the biological limits of agricultural production having been reached, the future looked grim. With disaster on the way, Ehrlich wondered whether after the "time of famines," the human race would be able finally to achieve a "birth rate solution" to its problems. He argued that human society could be saved only by a combination of moral, financial, and especially coercive legal incentives, applied on an international scale by the United States or a world government.

In sum, as a Malthusian, Ehrlich emphasized the objective, natural limits of the biosphere, or the absolute scarcities which confront the human race. His work popularized this approach, which quickly found echoes in a whole line of proclamations and essays announcing a new age of limits.

In 1972, the Club of Rome released a frightening doomsday study, concluding that "the basic mode of the world system is exponential growth of population and capital, followed by collapse."[8] *The Limits to Growth* predicted that rising population and industrial capacity would lead to increased demand for ever scarcer raw materials. Left to its own devices, within a century world industry would be spending so much money on these increasingly costly resources that it would be unable to renew depreciated capital. Finally, the industrial base would collapse along with services and agriculture, causing a drastic drop in population as the human race returned to barbarism. Was the modern industrial system destined to be a brief and tragically flawed experiment rather than the triumphant apotheosis of the species?

At about the same time in England, *A Blueprint for Survival*, prepared in 1972 by the editors of *The Ecologist* and endorsed by thirty-three leading scientists, called on Britons to slow their economic growth and halve their population to avoid "the breakdown of society and irreversible disruption of the life-support systems on this planet."[9]

Still more radical were arguments by Robert Heilbroner, who foresaw the end of liberal democracy as the turbulent era of environmental crisis brought out authoritarian tendencies in the human personality. "From the facts of population pressure," he writes in retrospect, "I inferred the rise of 'military-socialist' governments as the only kinds of regimes capable of establishing viable economic and social systems."[10] Retribalization appealed to him as a long-term alternative to a fatally flawed industrialism.[11]

The continuity with present positions is clear, although some recent debates carry us well beyond even the wildest speculations of the 1970s to the very borderline of madness. For example, the *Earth First! Journal* once published a discussion of the beneficial environmental effects of AIDS. One anonymous author wrote, "If radical environmentalists were to invent a disease to bring human population back to ecological sanity, it would probably be something like AIDS. As radical environmentalists, we can see AIDS not as a problem, but as a necessary solution (one you probably wouldn't want to try yourself)."[12]

Ehrlich's most recent book is far more moderate, perhaps in tacit recognition of the failure of his earlier alarmist predictions. (I say "tacit" because, rather surprisingly, despite the enormity of the errors in his earlier estimates of food and resource limits, he writes as though *The Population Bomb* has been confirmed on the whole by events.) He no longer emphasizes overpopulation exclusively or endorses coercive population control. Yet, population politics continues to be his central concern. After discussing the many sources of environmental catastrophe, he concludes, "Ending population growth and starting a slow decline is not a panacea; it would primarily provide humanity with the opportunity of solving its other problems."[13]

In contrast, Commoner holds that the environmental crisis, including the population problem, is due primarily to social causes rather than natural limits, for which a social solution is appropriate. The difference between his position and Ehrlich's can be clearly stated in terms of a formula Ehrlich himself devised for calculating environmental impact as a product of population size, affluence (that is, the amount of goods per capita), and the propensity of technology to pollute. Ehrlich, who assumed that the first was decisive, concluded that pollution derived ultimately from overpopulation. Commoner, however, argued from a close study of the relative impacts of the three factors that "most of the sharp increases in pollution levels [was] due not so much to population or affluence as to changes in productive technology."[14] "Population growth in the United States has only a minor influence on the intensification of environmental pollution."[15]

In support of his conclusion, Commoner noted that there was a manyfold increase in U.S. pollution in the twenty years between 1946 and 1966, while population went up only 42 percent. He argued that not population growth, but the massive transformation of industrial and agricultural technology after World War II was the main cause of rising pollution. "Productive technologies with intense impacts on the environment have displaced less destructive ones," for example, substitution of detergents for soap; synthetic fibers for cotton and wool; increased use of aluminum, air conditioners, more and more powerful automobiles; escalating use of fertilizers and pesticides.[16] The result of these changes was an immense increase in the environmental impact of modern societies, to the point where we now can foresee a breakdown of those biological processes which have continuously renewed the air, soil, and water for millions of years.[17]

What needed to be done? Commoner proposed transforming modern technology "to meet the inescapable demands of the ecosystem."[18] He placed a price tag on this program, and a high one at that, estimating that "most of the nation's resources for capital investment would need to be engaged in the task of ecological reconstruction for at least a generation."[19] Was this realistically possible? Commoner countered the sceptics by arguing that "no economic system can be regarded as stable if its operation strongly violates the principles of ecology."[20] In *The Closing Circle*, he showed that under capitalism, it is the search for maximum profits that motivates introduction of the new, dirtier production methods that have caused the environmental crisis.

For many years, in numerous articles and books, Commoner has argued that only a democratic socialist system can address environmental problems effectively. His latest book, like Ehrlich's, is somewhat more cautious. He still believes that the pursuit of short-term profits motivates bad technical decisions and, after the experience of the last twenty years, he is more convinced than ever that mere tinkering with pollution controls is insufficient. Instead, environmentally unsound technologies should be abolished outright and replaced with better ones.

However, Commoner has been chastened by the fall of communism in the Soviet Union and Eastern Europe and the breakdown of old assumptions about planning and markets that, like most leftists, he took for granted in earlier days. Now he writes that the market "is a useful means of facilitating the flow of goods from producer to consumer; but it becomes a social evil when it is allowed to govern the technology of production."[21] Like many socialists today, Commoner is looking for new solutions.

THE QUESTION OF DIMINISHING RETURNS

Now the political stakes in the debate are clear. Behind the contention over scientific issues, dispute over resource depletion and environmental degradation, and methodological disagreement over the biological or social character of the factors leading to crisis lies, quite simply, the old debate over capitalism and socialism. Was Commoner right to link the fresh new environmental movement with the tired old struggle for socialism? After all, environmental problems appear to be indifferent to socioeconomic systems. Commoner's critics saw little more than outmoded demography in his attempt to Marxify ecology. Ehrlich was particularly disturbed; he feared that Commoner's politics would shatter the unity (beyond class and ideology) of his movement for survival. As he put it, "There is no point in waving a red flag in front of the bulls."[22]

Ehrlich claimed that Commoner's politics was based on bad science.[23] Commoner was supposed to have underestimated the significance of population growth because he ignored the nonlinear relation between population size and pollution. Under the "law of diminishing returns," small increases in population might be responsible for disproportionately large increases in pollution. Commoner, he charged, crudely compared a small population increase with a large pollution increase and concluded that other factors must be decisive. In fact, the variables might interact in such a way that even a mere 42 percent increase in population could cause the manyfold increase in pollution cited by Commoner.

Take the case of food production. High levels of productivity may be obtained from good soil with only a little fertilizer, but there is only so much good soil. Once population grows beyond certain limits, farmers are compelled to plant mediocre soil in order to produce food for everyone. At this point, fertilizer use increases dramatically as attempts are made to prime bad land into giving reasonable yields. With increased fertilizer use comes increased water pollution. This, according to Ehrlich, is precisely what has happened in the United States.

Ehrlich's method was similar to that of *The Limits to Growth*. First, a natural limit on a presumably unsubstitutable resource is postulated, then a level of per capita demand is assumed and multiplied by actual and projected population size. As exhaustion of the resource approaches, efficiency declines, costs rise, and eventually the very survival of the population dependent on it is threatened.

Commoner replied that the problem of diminishing returns was simply irrelevant to major technological developments since World War II. In practice, returns "diminish" less for environmental than for economic reasons, reflecting not natural-resource limits but socially relative mechanisms of accounting and pricing. In Ehrlich's example, what *compels* the farmer to use excessive fertilizer? Certainly not absolute scarcity of land; at the time of the debate, there was still plenty of good land left in the United States, but the government actually paid farmers *not* to use it in order to maintain farm prices. This compelled farmers to push land in use to the limits of its capacity in order to make a profit. At those limits, something like a problem of diminishing returns occurs as ever greater increments of fertilizer are required to produce a constant increase in soil productivity, but this problem is strictly attributable to social causes which restrict planted acreage while requiring an increasing yield.

On balance, Commoner seems to get the better of this argument. It is, of course, possible to construct ideal models of the "world system" in which everything is held constant while population and resources are extrapolated to an inevitable clash, but this is not the real world, in which limits are relative to a multiplicity of factors. In the case of food, for example, in addition to limits on production there are losses of farm land due to urbanization; desertification due to bad agricultural practice; the inefficiency of culturally preferred high-protein diets; the voracious appetites of billions of pets whose owners can better afford to feed them than certain nations can afford to feed themselves; fertilizer costs which rise with artificial oil prices; and war and social disorganization in numerous Third World countries.

This crucial issue continues to divide Ehrlich and Commoner over twenty years after their original debate. Ehrlich still argues that overpopulation is at work in a wide variety of environmental problems and, for the most part, Commoner ignores the issue on the same grounds as before.

Ehrlich now defines overpopulation as an excess of inhabitants over the carrying capacity of the land they inhabit. No absolute limit exists to the number of people a given territory can support; environmental impact is relative to affluence and technology as well as human numbers. However, Ehrlich criticizes those such as Commoner who emphasize the possibility of accommodating larger numbers with improved technology; "overpopulation is defined by the animals that occupy the turf, behaving as they naturally behave, not by a hypothetical group that might be substituted for them."[24]

Fair enough, but Ehrlich frequently ignores the disproportionate impact of the sort of basic technological change Commoner advocates, while treating population control as the obvious solution to the problem of overpopulation. He relies once again on the supposed non-linear relation between population and pollution to explain why small reductions in population should have significant beneficial impacts. Once again we hear about diminishing returns without the evidence that would convince us that population control is really crucial to addressing environmental problems.[25]

Ehrlich's definition of overpopulation and the diminishing returns hypothesis work together to depoliticize environmental issues. He wants to argue for a politics of survival beyond considerations of class and national interest, but in fact he presupposes a specific constellation of class and national interests, that of modern capitalism and neoimperialism: "the animals that occupy the turf, behaving as they naturally behave." This is why he ends up seeking a biological solution. Although there are flaws in his approach that will be discussed later, Commoner achieves a more realistic assessment of the problems with a more socially conscious method.

THE PERSONAL OR THE POLITICAL

A method which treats society as a thing of nature, fixed and unalterable, ends by treating nature as a social object wherever it is subject to direct, conscious control. In the case of population politics, the biological locus of control is human reproduction, which individuals and governments can manipulate through voluntary contraception and involuntary sterilization.

By contrast, a method which emphasizes the social sources of the problems will prefer to act on the biological mediations indirectly through the forces governing institutional and mass behavior. Although the intended result may be the same—a better proportion between population and resources and a less polluting society, the means to the end will be quite different.

From a purely *technical* point of view, rapid, drastic and necessarily coercive reduction in the number of people is environmentally equivalent to changing the technology used by a much larger population. For example, Los Angeles's smog could be halved by halving its population (hence automobile use), but the same result also could be achieved at present population levels by halving emissions from the cars in use or by substituting mass transit for cars. Even though the environmental result is similar in these cases, there is no *moral* equivalence between two such very different policies as requiring smog-control devices on cars or legally limiting families to a single child.

This issue goes beyond the environmental question. Ehrlich claimed that growing population causes "not just garbage in our environment, but over-crowded highways, burgeoning slums, deteriorating school systems, rising crime rates, and other related problems."[26] Thus, zero population growth might help to stem the tide of urban blight, poverty, crime, and riots. Commoner did not deny that crowding can intensify social problems, but he argued that they demand a more radical solution than an end to crowding, namely, the elimination of the social conditions from which they arise. "To the degree that population size is reduced, to that degree may we be able to tolerate some of the technological, economic, and social faults that plague us; to the degree that we repair these faults, to that degree can the nation successfully support a growing population."[27]

Commoner concluded that the choice between these two routes is "a political one which reflects one's view of the relative importance of social control over personal acts and social processes."[28] Here is the nub of the disagreement. When the emphasis is placed on population, "social control over personal acts" appears as the solution. This approach is at once more individualistic and more repressive than emphasizing reform of the social processes which, Commoner claimed, ultimately determine both birth rates and technological choices.

Indeed, the dilemma of population politics is the absence of any significant realm of action other than appeals to individual conscience and coercion by the state. There is not much else to be done at the political level except attacking public opponents of birth control and lobbying for repressive legislation. One cannot very well demonstrate against babies or even against parents. Unless the state intervenes (as it has in China), the issue is private, each couple choosing how many children it wants as a function of its own values. This explains why Ehrlich's political program wavered between moralistic voluntarism and more or less harsh state action.[29] The resulting strategy offered a way in which without exercising social control over social processes, people could nevertheless mitigate the effects of an ecologically unsound technology by personally shouldering the burden and the costs.

The significance of the debate now becomes clear. In Commoner's words, people may choose a "new ecology-minded personal life-style . . . designed to minimize the two factors that intensify pollution that are under personal control: consumption and population size."[30] Or, "insofar as [they] are unwilling to undertake this personal action, they will need to seek relief by altering the economic, social and political priorities that govern the disposition of the nation's resources."[31] Commoner himself chose political rather than personal action, control over institutions rather than individuals.

Is a synthesis impossible? Can we not at the very least exercise voluntary control over personal behavior as well as political control over institutions? Ehrlich attempted just such a synthesis of the personal and the political in his 1971 book, *How To Be a Survivor.* There he broadened his approach to include not only population control, but egalitarian social reform, anti-imperialism, technological reform, and reduction through "de-development" of the excessively high living standard of the "over-developed" countries.

But synthesis is not so easy; divisive class and national issues cut directly across it, revealing it to be an eclectic combination of opposing strategies. For example, it must have been difficult to approach workers and the poor with a slogan such as, "Try to live below your means! It will be good for your family's economic situation, and it may also help to save the world."[32] In a society based on economic inequality, one cannot hope to organize a strong political movement around voluntary self-deprivation. The alternative, invoking the power of the state, usually has not served higher moral ends, but rather the interests of economic and political elites.

Meanwhile, business opposed much of Ehrlich's program. At first it felt threatened by the environmental issue, and its initial strategy consisted in distracting people from a crisis in which it did not believe or in ridiculing the proponents of environmental regulation. One mainstream commentator wrote sceptically on *The Limits to Growth*, "Conceivably, if you believe their predictions of extremely short time spans before the exhaustion of resources, there are many speculative killings to be made."[33]

Then, in 1971, the American Can Company extended its antilittering campaign to take in the whole environmental crisis. Keep America Beautiful, Inc. proclaimed: "People start pollution. People can stop it." Hundreds of millions of dollars of free advertising space were devoted to diverting environmental pressures away from business and toward individual action. This campaign was largely successful. Soon the public agreed with Pogo, the comic-strip character who said, "We have met the enemy and he is us."[34]

The businessmen who sponsored this campaign, and President Nixon who praised their civic consciousness, had no illusions about the implications of the environmental movement. They did not believe that it promised a universal good in which all could share equally. Rather, their hope was clearly that the political energy mobilized by the increasingly articulate critics of capitalist environmental practice could be focused on private options, leaving the basic economic institutions unchanged. Indeed, not only unchanged, but in a position to cash in on those great "speculative killings" in which, by this time, quite a few informed investors had come to believe.[35]

This anecdote suggests the divisive potential of the environmental crisis. Early in this discussion Commoner was quoted as saying, "Something has to give." The question is what, how, and why? This is the question I shall try to answer in the remainder of this essay through a discussion of the relevance of class, race, and national divisions to environmental policy.

Class Struggle Revisited

It should be clear by now that Commoner *wanted* the environmental movement to be "progressive" in the traditional sense of the term. He hoped it would become a factor in the struggle not only for nature but also for human beings, not only for survival but also for a more egalitarian society. Commoner had to prove that his radicalism was not gratuitous and that the allies he had chosen were indeed those most likely to work for environmental reform.

Commoner's argument is that pollution is a major short-term cost-cutter. During an initial "free period" in which the environment tolerates degradation:

> Pollutants accumulate in the ecosystem or in a victim's body, but not all the resultant costs are immediately felt. Part of the value represented by the free abuse of the environment is available to mitigate the economic conflict between capital and labor. The benefit *appears* to accrue to both parties and the conflict between them is reduced. But in fact pollution represents a debt to nature that must be repaid. Later, when the environmental bill is paid, it is met by labor more than by capital; the buffer is suddenly removed and conflict between these two economic sectors is revealed in full force.[36]

Thus, environmental politics is a zero-sum game in which the distribution of costs affects classes differently according to their position in the economic system. Starting from this premise, Commoner constructed what are, in effect, ideal-typical models of class-determined attitudes toward the environment.

The capitalist's relation to the environment is shaped by his short-term focus on profits and his ability to shift costs away from himself onto others. Pollution appears as an externality in his calculations, an externality suffered largely by others because he has the means to escape its worst effects privately (for example, by buying air-conditioning for his house and car, living in the suburbs or the country, or vacationing in unspoiled regions). Since environmental constraints often conflict with popular marketing strategies (such as increasing automotive horsepower) or threaten potentially profitable investment opportunities, capitalists will resist environmental controls until they become unavoidable, then attempt to get others to bear the burden. Commoner's theoretical prediction has been a fairly good description of business attitudes in the United States.

Workers' objective position with respect to the environment is quite different because for them, pollution is not an exogenous but an endogenous factor. Workers in the plant suffer the effects of pollution far more than executives in their air-conditioned administrative offices. Even during the "free period," workers and the poor "pay" for pollution through inconvenience and disease and as their "costs" rise, the issue is brought home in their daily lives. Here is the vital difference between the lower and upper classes in their relation to environmental degradation.

It was also the basis for Commoner's faith that workers, or at least their unions, eventually would lead militant opposition to environmental degradation. "The need for a new alliance is clear. Neither worker nor environmentalist can reach their separate goals without joining in a common one: to reconstruct the nation's productive system so that it conforms to the imperatives of the environment which supports it, meets the needs of the workers who operate it, and secures the future of the people who have built it."[37]

While Commoner offered good reasons for labor to become active in the environmental movement, today it is obvious that he overlooked the ambiguity of labor's situation. As he suggested, labor may fight to ensure that the burden of environmental restoration is shared more fairly by improving the conditions under which workers work and live. However, labor also can resist the unequal burden of capitalist environmentalism by resisting all environmental expenditures, shortsighted though that may be.

When he wrote *The Closing Circle*, Commoner was convinced that the intensified class conflict generated by the ecological crisis would be the greatest school in environmental policy that ever was. He believed that in this school, workers would learn to understand the economic mechanisms which cause the crisis and to reject equally the arguments of those who dismiss environmentalism and those who attempt to turn it into an issue of individual morality. In fact, labor environmentalism never played the central role he predicted. The failure of his strategy raises serious questions about his whole approach.

CULTURE AND CONSCIOUSNESS

The most obvious problem is his reliance on the traditional Marxist theory of class consciousness. One simply cannot predict the future beliefs of a class from their objective interests. The social theorist must explain the specific political and cultural factors that might, in any given case, distinguish the real consciousness of classes from the rational model constructed in theory, but Commoner omitted this second level of analysis, that of political and cultural mediations. The problem is especially serious because his theory of the environmental crisis hovered on the verge of a type of cultural criticism he himself did not wish to develop. He was so busy with his polemic against individualistic environmentalism that he rejected concern with culture which, he seemed to fear, would lead back to lifestyle politics.

However, the changes in production for which Commoner called amounted to far more than the American Way of Life equipped with emission controls; they presupposed radical cultural changes. The closest he came to facing this problem in the early seventies was in discussion of the distinction between the formal measures of "affluence," such as GNP, and the actual goods and services enjoyed by individuals. The good the consumer seeks, he argued, may be obtained in a variety of forms with very different environmental impacts.[38] What is at stake in such a distinction? Commoner noted that many older materials and technologies such as soap, wood, and bottles were displaced after World War II by more profitable substitutes such as detergents, plastics, and cans. Clearly, the return to capital would be affected by environmental restoration, but if consumers could obtain the same good in the old form, then they might actually benefit from technological "regression."

However, the matter is considerably more complex than this simple picture. First, the older technologies are more labor-intensive which suggests that labor, too, may have an interest in the application of new technologies to the extent that they increase productivity. Commoner discounted a large part of the increased welfare that workers are supposed to derive from such productivity increases, but in the final analysis the issue is not merely quantitative. Note, for example, his response to a question during testimony in House hearings on fuel and energy: "If you have followed the situation in some of the automobile plants in the last year or so, you probably realize that we may be reaching the human limit of automation. I think new ways of using human labor in a humane way ought to be looked into. In short, we should question the value of continued replacement of people by electronics."[39] Clearly, he was attempting to conjoin the environmental movement with a critical challenge to central values of capitalist culture, for both capitalists *and* workers.

Second, the resubstitutions for which Commoner called would change the form in which consumers obtained familiar goods. For example, he did not oppose individuals achieving a high level of geographical mobility in their daily lives; he just wanted them to do so through less polluting and wasteful means than the private automobile.

However, utilitarian considerations have little to do with the choice of the form in which goods are delivered in a modern consumer society. As is evident in advertising, forms are invested with meanings that often are wholly unrelated to the ostensible purposes of the goods they adorn, but that are nevertheless compelling for consumers. The sexual and status advantages of automobile ownership, while doubtless less important than the transportation it provides, are not trivial. They would be lost in the switchover to mass transit, a spare and utilitarian alternative by comparison. This fact in itself constitutes a significant cultural barrier to the changes Commoner advocated, but he had no way to discuss it because he worked exclusively at the level of imputed interests in abstract goods.[40]

This placed Commoner in a serious dilemma. As a Marxist, he rejected the sort of appeal to moral renunciation in which his adversaries specialized and attempted to attach environmentalism to the actual interests of the mass of humankind. However, the logic of Commoner's attack on capitalism led him to question its model of welfare and suggest an alternative that was not supported by the everyday consciousness of any group in the society. What was at issue, then, in Commoner's program was actually a radical change in the economic culture of capitalist societies, not just an environmentally sound version of it.

How much difference would there really be between Commoner's position, with its problematic connection to the real consciousness of workers, and a moral position that pretends to no such relation at all? This is a difficult question. To preserve the difference it would be necessary to theorize a process of cultural change that pointed the way toward a model of welfare more consonant with environmental survival than the current one. In these terms, it might still be possible to argue for a relationship between environmentalism and real (future) interests.

This is no merely verbal point. Where a clean and healthful environment is considered not as an exogenous dumping ground but as a component of individual well-being, different environmental practices would be followed spontaneously by the individuals in their pursuit of welfare and would not have to be imposed on them by market incentives or by political or moral coercion in opposition to their own perceived interests. Since technology is routinely adapted to changing social and economic conditions, there is no reason in principle that it should not be redesigned to conform to the requirements of such a culture.[41]

What is needed then is not a theory of individual lifestyle or simply of social control over production, but also of the processes of cultural change. Commoner was trapped in an overly rationalistic model that relied exclusively on scientific persuasion (at which, it should be said, he is very good). His adversaries meanwhile seized on all the symbolic machinery of environmental consciousness raising and turned it to account in pursuit of policies he deplored.

Of course, Commoner was right to reject the exaggerated environmental role they attributed to lifestyle politics, but personal involvement in environmentalism (through gestures such as participating in recycling or conserving water) is among the most effective means of cultural change available to the environmental movement, no matter what its policies or class content. Even if these gestures have a limited impact on the environment, they change people in ways that favor all forms of environmentalism and must not be rejected or ignored because they may be accommodated to reactionary policies. Significantly, as Commoner has become involved with movements against toxics and for recycling, he, too, has come to recognize the importance of voluntarism in the environmental movement (not, of course, for the sake of self-imposed poverty, but as a factor in technological change).

THE LIMITS OF MORALITY

Where did Ehrlich stand on these issues? The answer to this question is not at all clear. He claimed to seek a coalition based on those with "a stronger interest in survival than in perpetuating over-consumption."[42] Thus, the politics of ecology would appear to be sited in a world of overconsumers, some of whom are sufficiently enlightened to renounce their excesses for the sake of ultimate survival. But what of all the underconsumers for whom this dilemma does not exist? Do they have no place in a middle-class environmental movement?

It was by no means Ehrlich's intention to exclude them. His conception of the political base of the environmental movement was somewhat vague, but it was not exclusionary. "*All* Americans must be recruited into the de-development program; all must be assured of sharing the fruits of success. Those who have been effectively excluded from our society will surely block our attempts to save it—unless the doors are opened to them."[43] Ehrlich thus called for egalitarian reform of American society.

It is important to clarify the distinction between this approach and Commoner's. Commoner envisaged a resolution to the environmental crisis not through restrictions in the supply of material goods, but rather through changes in the definition and delivery of those goods. These changes would involve a shift between similar rewards offered in one or another form and, to some extent, a new perception of rewards where goods now go unnoticed and unappropriated (for example, in the case of clean air or water). Thus, Commoner remained within the traditional progressive framework. Ehrlich suggested a shift in the scene of fulfillment, from the material or economic domain to the spiritual or ideological domain. The difference is considerable since the primary cultural impulses of this society orient individuals toward the spontaneous pursuit of material ends, while spiritual ends must be imposed or self-imposed by law or morality in opposition to the individual's perceived interests.[44]

Two consequences follow from this distinction. First, Ehrlich's position was based on the universality of the current concept of affluence. He rejected Commoner's distinction between the actual economic good that individuals pursue and the form in which it is delivered to them by the dominant technology. Ehrlich called this "redefining affluence," and in rejecting it, he fell into uncritical acceptance of the dominant model of welfare.[45]

This model is presupposed by Ehrlich, the authors of *Limits to Growth*, Heilbroner, and others. For them, the environmental limits of what we currently understand as welfare are formulated as absolute limits on material progress as such. If contemporary ideas of wealth are in fact universal, then all adjustments to environmental constraints appear as economic regression. Far from identifying the actual, natural limits of the world system, this position only establishes the limits of a given type of capitalist economic culture, which it defends against environmental obsolescence through a larger provision of spiritual compensations.

This form of environmentalism leads inexorably from moral self-control to legal coercion. As Saint Paul wrote long ago, mankind is damned, not saved by the (moral) Law. The very need for self-control is a tribute to the power of temptation and the likelihood of sin. That which can be accomplished only against the material impulses of the species, surely will not be accomplished in the finite timespans of political action by the simple appeal to morality. Lurking in the wings is always the recourse to legal coercion or the power of the state (which, in Heilbroner's *The Human Prospect*, is explicitly charged with the salvation of the species). Here, the extreme consequence of the initial ideological choice is clearly drawn. The price of the perpetual maintenance of capitalist economic culture in a world where it has become environmentally absurd is a forced regression backward along the continuum of freedom and satisfaction.

Ehrlich's most recent book is not so ready to jettison democracy, but it is still confused on the question of affluence. At one point he indexes it to consumption while at another, he asserts that quality of life is independent of consumption and relative to the value of the actual goods delivered by it. This latter position is perilously close to the one he rejected as "redefining affluence" in his debate with Commoner. Now he advocates both reducing consumption and reducing the environmental impact of the technologies that provide consumer goods, both renouncing economic growth and changing production technology to accommodate environmentally sound development.[46] Ehrlich never succeeds in focusing the issue clearly or addressing the choices it implies.

We have arrived back at Commoner's original premise. The environmental movement must choose between a repressive policy of increasing control over individuals or a democratic policy of control over the social processes of production (and, I would add, culture.) The first choice suggests that the existing production system can be preserved, along with all its injustices, for a prolonged period in spite of the environmental crisis. The second choice suggests that the production system must be radically changed.

Race and Nation

Ehrlich's eclectic strategy was supposed to involve "everyone" in the work of environmental salvation, but his coalition got off to a bad start in at least one important respect. African Americans rejected zero population growth, which many saw as a racist attack on their survival.

Ehrlich denied that his movement was based on a specific class of prosperous, well-educated whites anxious to shift the ecological burden to poor blacks. He proposed, for example, that a "baby tax" he favored to discourage reproduction be accompanied by special exemptions for minorities. "The best way to avoid any hint of genocide is to control the population of the dominant group."[47] Why was the zero population growth movement still unable to attract African Americans? Perhaps part of the answer is to be sought in this newspaper advertisement for birth control: "Our city slums are packed with youngsters—thousands of them idle, victims of discontent and drug addiction. And millions more will pour into the streets in the next few years at the present rate of procreation. You go out after dark at your own peril. Last year one out of every four hundred Americans was murdered, raped, or robbed. *Birth control is an answer.*"[48]

This crime-in-the-streets, law-and-order approach selected the audience it wished to reach, and certainly it was not African Americans, who would appear to be the enemy—the human horde that must be stemmed. By whom? Evidently by those who "go out after dark at [their] own peril," a phrase that in the context of this ad seems to identify the respectable white population.

It is no wonder that African Americans rejected propaganda the ultimate implication of which was their forcible sterilization (a practice which, if rare, was by no means unknown in the United States). The attitude of some zero population growth advocates toward the Third World indicates, furthermore, that these perceptions may indeed have been justified. When the crowded slum was in a foreign country, there was no hesitation at all to invoke force in the name of population control for the poor.

In *The Population Bomb*, Ehrlich stated his agreement with Paul and William Paddock, authors of a book called *Famine—1975!*[49] The Paddocks proposed a triage approach to food aid based on traditional army medical policy. Those countries which had plenty of resources need not be helped because they could help themselves. Those on the borderline, which could survive but only with help, should be aided to the maximum. "Finally," Ehrlich noted in his summary of their proposal, "there is the last tragic category—those countries that are so far behind in the population-food game that there is no hope that our food aid will see them through to self-sufficiency. The Paddocks say that India is probably in this category. If it is, then under the triage system, she should receive no more food." He added, "In my opinion, there is no rational choice *except* to adopt some form of the Paddock's strategy as far as food distribution is concerned."[50]

While countries like India would be abandoned, others would be required to introduce strict, involuntary population control as a condition for receiving food. Ehrlich comments: "Coercion? Perhaps, but coercion in a good cause. I am sometimes astounded at the attitudes of Americans who are horrified at the prospect of our government insisting on population control as the price of food aid. All too often the very same people are fully in support of applying military force against those who disagree with our form of government or our foreign policy. We must be relentless in pushing for population control around the world."[51]

If this remark is "brutal and heartless" (the words are Ehrlich's own), what are we to say of Garrett Hardin's still more eloquent defense of the same position? "How can we help a foreign country to escape overpopulation? Clearly the worst thing we can do is send food. The child who is saved today becomes a breeder tomorrow. We send food out of compassion, but if we desired to increase the misery in an overpopulated nation, could we find a more effective way of doing so? Atomic bombs would be kinder." Hardin concludes, "Fortunate minorities must act as the trustees of a civilization that is threatened by uninformed good intentions."[52]

Ehrlich was more squeamish in *How to Be a Survivor.* There he proposed a massive aid program for the Third World, remarking that U.S. leadership of the world struggle to lower birth rates "means leadership by example."[53] He went on to say, "The population problem cannot be 'solved' by withholding medical services or food and letting people die of disease or starvation."[54] Although he still believed that force was the surest instrument of rational demographic policies, he no longer suggested that the United States should exercise this force alone. Instead, he called for "mutual coercion, mutually agreed upon."[55] A world government would be needed to wield power over human numbers.

At this point, we have come full circle. In 1946, J. Robert Oppenheimer wrote, "Many have said that without world government there could be no permanent peace, and that without peace there would be atomic warfare. I think one must agree with this."[56] In its conception of salvation through world government, the postwar scientists' movement achieved a kind of ideological apotheosis, echoing a humanistic opposition to war going back to Kant. The resurgence of such a conception in an environmental movement based on a similar ideological infrastructure is not surprising. It is implicit in the whole approach, which consists in identifying a common survival interest of the species superseding all particular interests.

However, in contact with vulgar realities, the universalistic scheme of world government suffered a peculiar degeneration in the late forties, a degeneration which may shed some light on the ambiguities of the new environmentalist formulations. Worried about the implications of a nonwhite majority in a world state, scientists proposed what today we would certainly condemn as racist measures, such as votes weighted in favor of rich white nations.[57]

Are these twisted proposals irrelevant relics of a bygone era? Or are they the typical consequences of the waves of impotent universalism breaking over the shoals of powerful particularisms? World government in the interests of population control is fraught with dangers anticipated in the earlier disappointing experience with the concept. Mutual coercion can be exercised only by approximately equal powers, but only the developed countries have the capacity to enforce their will. Furthermore, it is only in these countries that there is any significant popular support for coercing poor nations into population control programs. The kind of world government which would use force to impose demographic controls would be a government *of* the developing countries *by* the developed ones.

THE RIGHT TO DEMOGRAPHIC TRANSITION

Commoner rejected this policy which, in his words, "would condemn most of the people of the world to the material level of the barbarian, and the rest, the 'fortunate minorities,' to the moral level of barbarian."[58] His approach to the problems of the Third World was based on a key historical analogy, the rise and fall of population growth rates in the West since the seventeenth century. Here is a case of a veritable population explosion which peters out after several centuries of its own accord. This cycle occurred without a massive experiment in involuntary birth control such as many would like to impose on the developing countries.

The phenomenon of rapid expansion followed by stabilization observed in the West is known as "demographic transition." It seems to be the result of lowered rates of infant mortality. Wherever there is great insecurity of life, birth rates rise to ensure the survival of a few of the many children born, but where there is sufficient food and adequate hygiene, infant mortality rates fall and there is a tendency for birth rates to follow them. Thus, "although population growth is an inherent feature of the progressive development of productive activities, it tends to be limited by the same forces that stimulate it—the accumulation of social wealth and resources."[59]

On this basis, Commoner argued that the solution to the demographic problems of the Third World lay in increasing the standard of living there and making modern contraceptive methods available for voluntary use. This is in fact the solution preferred by most spokespersons for the countries concerned.

Commoner offered political and moral reasons for trying this solution before embarking on a program of "coercion in a good cause." He argued that the West is itself largely responsible for overpopulation in many developing countries, citing Nathan Keyfitz, who estimated that "the growth of industrial capitalism in western nations in the period 1800–1950 resulted in the development of a one-billion excess world population, largely in the tropics, as a result of the exploitation of these areas for raw materials (with the resultant need for labor) during the period of colonialism."[60] The capital accumulated in the colonized countries was exported back to the West, where it contributed to the industrialization process that eventually stabilized population there. Commoner argued that a "kind of demographic parasitism" ensued in which colonizers slowed their own population growth by processes that accelerated that of the colonized.[61] "Overpopulation" is thus really a euphemism for the proletarianization of the Third World.

Commoner claimed that we should recognize a "right of demographic transition" in the Third World countries devastated demographically by colonialism. This should involve some obligation on our part to help these countries achieve the levels of prosperity associated with a naturally falling birth rate. Most important, it should rule out the sort of drastic historical experiments in involuntary birth control suggested by the Paddocks and others.

Ehrlich wrote many sharp attacks on this position.[62] He was convinced that the demographic transition could not take place soon enough to save many developing countries from famine. Further, he argued that industrialization in the Third World would strain already diminishing supplies of natural resources.

It is difficult for a nonscientist to evaluate a scientific disagreement such as this one.[63] In fact, neither Ehrlich nor Commoner appears to have been entirely correct. Ehrlich's hope that the advanced countries could be induced to promote Third World birth control got off to a good start, but was soon overtaken by the Reagan administration's alliance with abortion foes. Meanwhile, food production continued to rise long after Ehrlich claimed it had stagnated. On the other hand, the classic demographic transition does not appear to explain slowing birth rates in the Third World as Commoner expected. Education of women, later age of marriage, and availability of birth control devices seem to play an unexpectedly large role, to some extent independent of economic growth. Actual events thus reflect neither the expected consequences of prosperity in the demographic transition nor the disaster scenarios of the early 1970s.

Unfortunately, Ehrlich's and Commoner's recent books do little to advance the argument. Both authors now have the benefit of twenty years of demographic debate, but while both profess to believe in voluntary birth control and agree on some basic facts (such as the likely point at which world population will level off, ten billion), they seem not to have noticed the developments most damaging to their own arguments. Ehrlich, for example, still expects "the stork to pass the plow," just a bit later than 1968 or 1975.[64] Commoner does not discuss the implications for women of the demographic transition argument on which he continues to rely.

Conclusion: Beyond the Politics of Survival

The Commoner-Ehrlich debate provides a window onto the deep and apparently unavoidable conflicts inherent in environmental politics, conflicts that were already implicit in the earlier scientists' movement for nuclear disarmament. The contemporary political sensibility must be informed by the nuclear—now also environmental—age, from which we learn the threat to survival contained in the very nature of our civilization. A society that can destroy life on earth by the careless application of fluorocarbon deodorant sprays is indeed beyond the pale of any rational calculation of its chances for survival. In principle, history is over in the sense that the old conflicts and ambitions must give way to a radically new type of human adventure or the species will surely die. Nevertheless, in practice the unfinished work of history continues, indeed intensifies the very horrors and upward struggles that threaten survival, yet also promise a precious spark of light to those hitherto excluded from the benefits of technical advance. Insensitivity to this ambiguity leads to a politics of despair that would freeze the current relations of force in the world—and with them the injustices they sustain—as a condition for solving the problem of survival. That this is an impossible route to salvation is abundantly clear from the whole experience of the nuclear and environmental movements.

What we most need to learn is that action to end history is still action in history for historical objectives. The human species is not yet the subject of the struggle to survive; thus, this struggle itself becomes a facet of the very class and national struggles whose ultimate obsolescence it demonstrates. From this dialectic there can be no escape.

The early seventies gave us a dress rehearsal of far deeper crises to come. If there was ever any doubt about the environmental crisis intensifying social and international conflict, that doubt should now be silenced. In short, the environmental crisis brings not peace but a sword, and precisely for that reason it is not a unifying messianic force through which the human race could join in an ennobling struggle beyond the petty conflicts of history. Rather, it is a new terrain on which the old issues will be fought, perhaps this time to a conclusion.

Notes

1. Recent books by Ehrlich and Commoner restate their positions in ways somewhat softened by time, bringing the debate up to date. See Paul and Anne Ehrlich, *The Population Explosion* (New York: Simon and Schuster, 1990); and Barry Commoner, *Making Peace with the Planet* (New York: Pantheon, 1990). I will comment on them where they are relevant, but I do not pretend to offer a full picture of these new books which, while sometimes more reasonable than the authors' original work, do not have the seminal importance of their initial contributions in defining the ideological polarities of the environmental movement.

2. For a detailed account of the scientists' movement, see Alice K. Smith, *A Peril and a Hope* (Cambridge, Mass.: The MIT Press, 1965). For a discussion of the concept

of the "end of history" in relation to science fiction, see Andrew Feenberg, *Alternative Modernity* (Berkeley: University of California Press, 1995), pp. 43–56.

3. Ehrlich, *The Population Bomb* (New York: Ballantine, 1968), pp. 66–67.

4. Barry Commoner, "Motherhood in Stockholm," *Harpers Magazine* (December 1973): 53.

5. Barry Commoner, Paul Ehrlich, and John Holdren, "Dispute: The Closing Circle," *Environment* 14, no. 3 (April 1972): 40.

6. Commoner, "Labor's Stake in the Environment/The Environment's Stake in Labor," presented at the Conference on Jobs and the Environment, San Francisco, 28 November 1972 (mimeo, p. 33).

7. Ehrlich, *The Population Bomb*, p. 34.

8. Donella H. Meadows et al., *The Limits to Growth* (New York: Universe Books, 1972), p. 142.

9. The Ecologist, *A Blueprint for Survival*, with an introduction by Paul Ehrlich (New York: Signet Books, 1974). Cf. Sicco Mansholt, *La lettre Mansholt* (Paris: Pauvert, 1972.)

10. Robert Heilbroner, *An Inquiry into the Human Prospect* (New York: Norton, 1974), p. 156.

11. Ibid., p. 141.

12. Miss Ann Thropy, "Population and AIDS," *The Earth First! Journal* (5 January 1987): 32.

13. Ehrlich and Ehrlich, *The Population Explosion*, p. 157.

14. Commoner, *The Closing Circle* (New York: Bantam, 1971), p. 175.

15. Ibid., p. 231

16. Ibid., p. 175

17. Ibid., p. 13

18. Ibid., p. 282

19. Ibid., p. 284.

20. Ibid., p. 272.

21. Commoner, *Making Peace with the Planet*, p. 223.

22. Paul R. Ehrlich and Richard L. Harriman, *How to Be a Survivor: A Plan to Save Planet Earth* (New York: Ballantine Books, 1971), p. 136.

23. Commoner, Ehrlich, and Holdren, "Dispute: The Closing Circle," pp. 23ff. The two articles published here, a review of *The Closing Circle* by Ehrlich and Holdren accompanied by Commoner's rejoinder, culminate the debate. A somewhat different version of the piece by Ehrlich and Holdren was published in *The Bulletin of the Atomic Scientists*, May 1972. The June issue of the *Bulletin* contains a re-rebuttal by Ehrlich and Holdren, entitled "One-Dimensional Ecology Revisited." See also Ehrlich's letter to *The New York Times*, 6 February 1972, sec. 7, p. 42. This section summarizes the central argument of these publications.

24. Ehrlich and Ehrlich, *The Population Explosion*, p. 40.

25. Ibid., p. 137. Ehrlich's position is more convincing in relation to African agriculture than Western industry. For example, he merely asserts that there is a nonlinear relation between population and acid rain without offering or citing any evidence (p. 123). It should be pointed out that Commoner does agree with Ehrlich now that a problem of diminishing returns exists with regard to fossil fuels; however, he is confident that solar energy can provide a relatively near-term alternative (Commoner, *Making Peace with the Planet*, pp. 197–198).

26. Ehrlich, *The Population Bomb*, pp. 24–25. There remain echoes of this law-and-order environmentalism in the Ehrlichs's current work, but he concedes reasonably that no causal relationship can be demonstrated between crime and crowding (Ehrlich and Ehrlich, *The Population Explosion*, pp. 11, 157).

27. Commoner, *The Closing Circle*, p. 212.
28. Ibid., p. 233.
29. Ehrlich, *The Population Bomb*, p. 175; Gladwin Hill, "Scientific and Welfare Groups Open a 4-Day Study of Population Growth," *The New York Times*, 9 June 1970, p. 32.
30. Commoner, *The Closing Circle*, p. 209.
31. Ibid., p. 212.
32. Ehrlich and Harriman, *How to Be a Survivor*, p. 149.
33. Leonard Silk, "Questions Must Be Raised about the Immanence of Disaster," *The New York Times*, 13 March 1972, p. 35.
34. "Group Seeks to Shift Protests on Pollution," *The Los Angeles Times*, 5 May 1971, sec. 1, p. 25.
35. As Henry Ford II put it, "The successful companies in the last third of the twentieth century will be the ones that look at changes in their environment as opportunities to get a jump on the competition." *The Human Environment and Business* (New York: Weybright and Talley, 1970), p. 62.
36. B. Commoner, *The Closing Circle* (New York: Bantam, 1971), p. 271.
37. Commoner, "Workplace Burden," *Environment* (July/August 1973): 20.
38. Commoner, Ehrlich, and Holdren, "Dispute: The Closing Circle," pp. 45–46.
39. Testimony of Barry Commoner, *Congressional Record*, 1972, 593.
40. For significant attempts to theorize such a challenge from an environmental perspective and in semiotic terms, respectively, see William Leiss, *The Limits to Satisfaction* (Toronto: University of Toronto, 1976); and Jean Baudrillard, *La société de consommation* (Paris: Gallimard, 1970).
41. For a fuller discussion of this point, see Andrew Feenberg, *Critical Theory of Technology* (New York: Oxford University Press, 1991), chap. 8.
42. Ehrlich and Harriman, *How to Be a Survivor*, p. 155.
43. Ibid., pp. 76–77.
44. For an illuminating discussion of the relation between material and spiritual values, see Alvin Gouldner, *The Coming Crisis in Western Sociology* (New York: Basic Books, 1970), pp. 326–331.
45. Ehrlich and Holdren, "One-Dimensional Ecology Revisited," *The Bulletin of the Atomic Scientists* (June 1972): 44.
46. Ehrlich and Ehrlich, *The Population Explosion*, pp. 58, 228–229, 219, 139.
47. Ehrlich and Harriman, *How to Be a Survivor*, p. 23.
48. Quoted in Commoner, *The Closing Circle*, p. 232.
49. William and Paul Paddock, *Famine—1975!* (Boston: Brown, 1967).
50. Ehrlich, *The Population Bomb*, pp. 160–161.
51. Ibid., p. 166.
52. Garrett Hardin, "The Survival of Nations and Civilization," *Science* 172 (1971): 1792. Note the metaphoric equivalence of the population bomb and the atomic bomb, characteristic of this trend.
53. Ehrlich and Harriman, *How to Be a Survivor*, p. 17.
54. Ibid., p. 52.
55. The phrase is Hardin's. Cf. "The Tragedy of the Commons," in *The Environmental Handbook*, ed. Garrett de Bell (New York: Ballantine, 1970), p. 45.
56. J. Robert Oppenheimer, *The Open Mind* (New York: Simon and Schuster, 1955), p. 12.
57. See Reinhold Niebuhr, *The Bulletin of the Atomic Scientists* (October 1949): 289; Edward Teller, *The Bulletin of the Atomic Scientists* (December 1947): 355, and *The Bulletin of the Atomic Scientists* (September 1948): 204; and E. M. Friedwald, *The Bulletin of the Atomic Scientists* (December 1948): 363.

58. Commoner, *The Closing Circle*, p. 296.

59. Commoner, *The Closing Circle*, p. 116.

60. Ibid., p. 243.

61. Ibid., p. 244.

62. Most recently in Ehrlich and Ehrlich, *The Population Explosion*, pp. 214–216.

63. For recent articles, see Hendry Peter, "Food and Population: Beyond Five Billion," *Population Bulletin* 43, no. 2 (1988); and Thomas Merrick, with PRB Staff, "World Population in Transition," *Population Bulletin* 41, no. 2 (1986).

64. Ehrlich, *The Population Explosion*, pp. 108–109.

Chapter Twelve

The Problem of Nature in Habermas

Joel Whitebook

1996 Introduction

Written almost twenty years ago, "The Problem of Nature in Habermas"—which is being published here more or less in its original form—belongs to an earlier phase in both my own thinking and in the development of critical theory and the ecology movement. Rereading it today, though I discovered a number of stylistic points that I would have like to have changed, I was also surprised to find how sound I still consider many of its formulations. Indeed, many of the questions posed in the article continue to engage my thinking today, and the main dilemma it addressed has by no means been resolved.

"The Problem of Nature in Habermas" was my first sustained confrontation with the debate, if it can be called that, between the first and second generations of the Frankfurt School. Having broken my theoretical teeth on Marcuse in the sixties, and having gone on to devour Horkheimer and Adorno's *Dialectic of Enlightenment* in the early seventies, I was grappling with the consequences of the Habermasian turn for critical theory. Although there may have been a shift of emphasis in the topics I have subsequently addressed (owing, in part, to the fact that I turned my attention to the problem of inner nature and became a practicing psychoanalyst along the way) the force field between Adorno's and Habermas's philosophy continues to animate my thinking.

This essay was, if not the first, at least one of the first articles to address systematically the ecological crisis from within critical theory. It must be recalled that in the early seventies, the ecology crisis was only beginning to emerge in popular consciousness. Three Mile Island, Chernobyl, and the *Exxon Valdez* had yet to happen, and an organized movement on the scale of the Greens had yet to appear on the political scene. The connection between the ecological crisis and the old Frankfurt School was, however, not difficult to establish. Indeed, in moving the domination of nature to the center of their analysis in the forties, the early critical theorists had almost predicted it. The difficulty was, rather, to envision a possible solution on the basis of their diagnosis. Like the solution to the crisis of modernity in general—of which it was a part—the solution to the ecological crisis appeared to necessitate an eschatological rupture in history *which would undo the domination of nature and establish a qualitatively new, noninstrumental relation with it.* Given Horkheimer and Adorno's analysis however, such a utopian solution seemed both virtually impossible and extremely hazardous; hence their impasse and political quietism.

I should point out that at the time I wrote "The Problem of Nature in Habermas," I was more willing to entertain the utopian gambit than I would be today. Twenty years of political history, as well as the sustained criticism of both the poststructuralists—who were also just making their appearance on the scene in the States—and the critical theorists, have changed my thinking on the subject of utopia.[1]

As is well known, Habermas attempted to break out of the theoretical and political impasse of the *Dialectic of Enlightenment* by introducing the distinction between instrumental and communicative reason. The advantages of this so-called linguistic turn were twofold: it allowed him to elucidate more successfully the theoretical and normative foundations of critical theory, and it also allowed him, as opposed to Horkheimer and Adorno, to identify certain normative advances in modernity which could provide a foothold for critique and action. Habermas was thus able to formulate a "radical liberal" program, which he came to understand as the completion of modernity's unfinished project of democratization, and which allowed him to circumvent the political paralysis of Horkheimer and Adorno.

Despite the decided advantages of this strategy, it resulted in a consequence that could be troubling to someone concerned with ecology. Habermas's basically Kantian, that is, anthropocentric, position appeared to relegate nature to the status of a meaningless object of instrumental control. Indeed, Habermas explicitly said as much and rejected the possibility, advocated by Marcuse, of a new, qualitative science in which nature would not be constituted as an object of instrumental domination. He argued that modern Galilean science provides the only valid (cognitive) relation to the natural world. Habermas thus seemed to have purchased his escape from Horkheimer's and Adorno's cul-de-sac at the price of surrendering nature to instrumental reason. At the same time, he was understandably hostile to any theoretical move that might threaten to undo the philosophical construction on which his theoretical and practical advances were based and hence trump his arguments for democracy—as the introduction of alternative, noninstrumental conceptions of nature certainly would. Any solution to the ecology crisis that could be conceived from within his anthropocentric framework would, therefore, have to retain nature as an object instrumental objectification.[2]

At this point, it is necessary to introduce two new characters into my narrative, Murray Bookchin, and Hans Jonas. I had discovered Bookchin's work in 1969—during the period when SDS was in the frenzied process of cannibalizing itself—and it seemed to provide a promising alternative to the self-destruction of the New Left.[3] Bookchin was probably the first figure on the Left to recognize the significance of ecology for the trajectory of capitalist development, hence for radical politics. It is remarkable indeed that he had already written about the importance of ecology in the fifties. Under his influence, it became increasingly central to my own theoretical and political perspective as well.

Similarly, Hans Jonas, with whom I studied at the New School for Social Research in the seventies, was one of the first philosophers in the academy to address not only the ecological crisis, but the whole array of novel issues introduced by the revolutions in technology, biology, and medicine. Both Bookchin and Jonas argued that the nature and enormity of the ecological crisis demanded a rethinking of the basic structure of modernity, containing as it does the domination of nature as one of its constitutive features. Philosophically, this would mean challenging the anthropocentrism of modern ethics, that is, the position which sees nature as a meaningless manifold and all value emanating from the side of the human subject. To conceive a solution to the ecology crisis, they argued, it would be necessary to conceptualize not just the good-for-humanity, from which the good-for-nature could then possibly be derived, but the good-for-nature in its own right.

"The Problem of Nature in Habermas," then, represented my attempt to think through the issues that the ecological crisis raised for critical theory by bringing the anthropocentric position, represented by Habermas, into confrontation with the antianthropocentric position, represented by Marcuse, Bookchin, and Jonas. As the reader will discover, however, the results were anything but conclusive. On the one hand, it seemed that nothing short of a comprehensive structural transformation of our relation to the natural world would be adequate to the character, severity, and depth of the ecological crisis. On the other hand, the possibility of conceiving such a transformation in a way that would not sacrifice the indisputable achievements of modernity—its productive capacity and advances in democratization—seemed slim indeed.

While that dilemma remains, in principle, no less acute now than it was twenty years ago, it has been ignored for the most part by recent critical theory.[4] Owing to the demise of Marxism, the renewed encounter with liberal theory, and the increased defense of democracy and human rights that has resulted from it, contemporary critical theorists have been occupied almost exclusively with fortifying their normative theory. They have had little inclination to take up an issue like the problem of nature, which might threaten the philosophical construction on which that normative theory is based. While no one can argue against the defense of democracy and human rights, this entire strategy has had a curious result: namely, that critical theory, which aspires to provide a comprehensive theory of the crisis of modernity, has little to say about one of its most decisive features, ecology. The real challenge would be to think both democracy *and* ecology.

* * *

Introduction

The way Habermas has elaborated the tradition of critical theory makes his contribution difficult to evaluate. While he has undoubtedly rectified some of the most glaring theoretical defects of his predecessors, he has also markedly altered the spirit of their project. He has gained the theoretical advantages of his own position at the price of breaking with Adorno, Horkheimer, and Marcuse on fundamental issues. This does not mean that these theoretical advances could have been achieved in a different manner, nor that the spirit of the early Frankfurt School ought to be preserved. It means that Habermas's theory so differs from that of his predecessors as to seriously raise the question of continuity.[5]

Two standard criticisms are usually raised against the early Frankfurt School: (1) their pessimism could only lead to resignation, and (2) the foundations of critical theory were never clarified so that the resulting critique of society was itself never grounded. With respect to both these problems, Habermas's theory represents an advance. As for the first point, Habermas argues that the pessimism of the early Frankfurt School was not simply the result of a sober examination of historical forces, but, insofar as it resulted from tacit (and incorrect) theoretical presuppositions, was presumed at the outset. Concerning the second point, Habermas develops a transcendental argument to provide the epistemological and normative foundations for critical theory. It is here that his theoretical advances often result in blunting the thrust of the early critical theorists' vision. Given their analysis, a solution to the historical impasse of our time—however unlikely—requires reconciliation with nature.

"The Problem of Nature in Habermas" was first published in *Telos* 40 (Summer 1979): 41–94; reprinted by permission, with minor corrections.

While providing a superior theoretical grounding for critical theory, Habermas's transcendentalism necessarily precludes any reconciliation with nature. Theoretical rigor is obtained at the expense of the original utopian ideal.

The Impasse of the Frankfurt School

During its exile, the Frankfurt School sought to comprehend the new and deeply disturbing world situation. They located the cause of disparate phenomena such as European fascism and the American culture industry in the same overriding historical tendency: that is, the Enlightenment. While the Enlightenment's goal of improving "man's estate" through the progressive replacement of myth by reason and the conquest of nature may be laudable, its strategy is flawed: the domination of outer nature necessitates the domination of inner nature, that is, the recasting of man's instinctual organization so that he becomes capable of exercising the type of renunciation necessary for the transformation of external nature: "The subjective spirit which cancels the animation of nature can master a despiritualized nature only by imitating its rigidity and despiritualizing itself ill turn."[6] This thesis explained the reemergence of barbarism and was the source of the early Frankfurt School's pessimism.

By the middle of this century, the goal for which the domination of nature had been undertaken, the creation of the material preconditions for a free society, so it was assumed, was being achieved. But, because of the concomitant reification of the "subjective spirit," those preconditions could not result in a more humane social order. Generalizing from this, it was argued that the attempt to create a free society was inevitably a self-vitiating enterprise. *To the extent that the material preconditions for a free society have been created, the subjective conditions necessary for its realization will have been distorted.*

Given this analysis, the pessimism of the early critical theorists can be challenged in several ways. One, and by far the most difficult option, consists in trying to make plausible a nonregressive reconciliation with nature. In fact, this seems to be the only way out as long as one accepts their main thesis. It entails developing a new mediation between society and the natural world, for example, a "new science" that does not approach nature as a purposeless object of domination. This process of reconciliation would have to be nonregressive in two senses. First, it would have to avoid slipping back into prerational, for example, mythical, forms of thought.[7] Second, it could not regress behind the present level of technical proficiency or else one of the conditions of a satisfactory solution, the material prerequisites for a new society, would no longer obtain. Adorno and Horkheimer accepted these conditions for a satisfactory solution to the "riddle of history," but were skeptical about the possibility of their fulfillment. Marcuse, on the other hand, has hinted at the possibility of such a utopian solution by means of a "qualitative physics" and a new sensibility. Unfortunately, his discussion has never gone beyond mere suggestions and he has not grappled with the ensuing difficulties.[8]

A second alternative would be to challenge the main thesis. This is the strategy followed by Habermas. Thus, the basic error of the earlier critical theorists was that their philosophy was (at least implicitly) *monistic*.[9] According to Habermas, while there is an intimate connection between the domination of external and internal nature, the two processes do not *follow the same logic*. Horkheimer's and Adorno's failure to differentiate satisfactorily between the two led to their fateful impasse. To correct this situation and to avoid those mistakes, Habermas introduces his dualistic framework: while the logic of instrumental rationality governs the domination of external nature, the logic of communicative rationality governs that of internal nature.

The most notable difference between the two logics is that, whereas the former *aims at reification*, for the second, reification is a *possible but pathological outcome*. The domination of external nature aims at the reification of the object in order to render it susceptible to instrumental manipulation. Taking modern science as its most refined example, instrumental rationality can in principle be reconstructed as a formal, deductive system. This degree of formal rigor can be achieved because science abstracts from the existing intersubjective communication in scientific practice, and can thereby methodologically bracket the ambiguities inherent in ordinary language communication. Karl-Otto Apel attempts to criticize scientism precisely by returning to this suppressed dimension of "the *a priori* of communication" underlying science.[10]

Instead of the application of technical *rules* to heterogeneous objects, the appropriation of internal nature involves the transformation of drives through the internalization of intersubjective *norms*. The proper *telos* of this process is not reification, but autonomy, individuation, and socialization, which means the ability to maintain one's identity by simultaneously identifying with and differentiating oneself from other subjects in a social, that is, communicative, context. Reification—for example, the rigid compulsions of a neurotic—is a possible outcome of this process, but in this case it is a pathological rather than the desired one. As it moves in irreducibly plural, that is, "dialogical," milieu of ordinary communication, communicative rationality can never attain the same degree of formalization as instrumental rationality.

Because they fail to differentiate sufficiently between the domination of inner and outer nature, Horkheimer and Adorno do not have the conceptual resources necessary to formulate an adequate notion of an emancipated self. Or, to put it differently, were they confronted with the emancipated self presupposed by their argument, they would have to disavow it. Horkheimer and Adorno, of course, know that the self is formed through the domination of inner nature and that prior to its achieving a degree of instinctual mastery it does not make sense to speak of a self: "man's mastery of himself . . . grounds his selfhood." However, since they conceive of the domination of inner nature on the model of instrumental rationality, *the telos of which is reification*, all self-formation must be equivalent to reification. Given their argument, a nonreified self is a logical impossibility.

Not only do they fail to formulate a positive notion of the self, but insofar as it imposes its tyrannical unity on the manifold of instinctual contents, Horkheimer and Adorno tend to view the self as the "enemy": "Men had to do fearful things to themselves before the self, the identical, purposive, and virile nature of man was formed and something of that recurs in every childhood. The strain of holding the I together adheres to the I in all stages: and the temptation to lose it has always been there with the blind determination to maintain it."[11] As the self *per se* is taken as the agent of repression, it can no longer be what is to be emancipated. Instead, what is to be emancipated becomes the repressed inner nature, which is tyrannized by the self. It can be granted that the self will be impoverished and rigidified to the extent that it denies its inner nature, and that an emancipated self will have reconciled itself with its own natural substrate. Nevertheless, emancipation is something that is undertaken for the sake of the self, and not for the sake of our biological endowment, as Horkheimer and Adorno imply.

Because he loosens the connection between the domination of internal and external nature and grants a degree of relative autonomy to the communicative level, Habermas can conceptualize moral progress. Whereas for Horkheimer and Adorno, technical progress necessarily entails moral regression, Habermas can envision simultaneous progress on both levels. Unfortunately, his conception of progress leaves much to be desired from the viewpoint of those, who retain some identification with the old Frankfurt School. The problem revolves around the question of *disenchantment*. The New Left and the counterculture of the 1960s more or less self-consciously sought the basic reenchantment of the world in a variety of forms,[12] and, while he does not thematize this point, Habermas is a thoroughgoing disenchanter. In this respect, he differs considerably from his predecessors. He counsels an end to utopian excesses, a recognition of the progressive features of modernity and the attempt to achieve a just and rational society from within the modern *Weltbild*. Indeed, an emancipated society for Habermas consists in the completion, and not the transfiguration, of the modern project.

Habermas's Reformulation of Critical Theory

The theoretical advances of Habermas's position and his doctrine of nature stem from the same source, namely, his transcendentalism—or "quasi-transcendentalism," as he calls it. Through reflection on the evolution of the species, Habermas claims to have determined the categorial frameworks—instrumental and communicative—within which the basic modes of human knowledge and action develop. He then employs those categorial distinctions to elucidate the heretofore unclarified foundations and status of critique. While Habermas's account grounds critical theory with an increased theoretical rigor, it also condemns nature to being exclusively an object of domination. A consequence of his analysis is that nature can only be known as an object of possible technical control.

Habermas's transcendentalism can best be understood by comparing it with Kant's, the classical prototype of transcendental theorizing. While both share a similar *desideratum*,[13] the means by which they seek it differ significantly. The feature that most distinguishes Habermas's transcendentalism from Kant's is that, whereas Kant's transcendental subject is *singular*, Habermas's is *plural*. Kant sought to determine the conditions of the possibility of objects and knowledge of experience through an investigation of the synthetic acts of a transcendental consciousness. To achieve the same end, Habermas examines the *a priori* interest structures that inform the dimensions within which the human species evolves and which also govern the various forms of human knowing and acting. And, inasmuch as modes of scientific activity have been differentiated in the course of evolution which represent methodologically rigorous means of pursuing these interests, transcendental reflection can also proceed through an examination of the methodological *a prioris* of the various scientific domains. "I do not assume the synthetic achievements of an intellectual ego nor in general a productive subjectivity. But I do presuppose, as does Peirce, the real interrelationship of communicating investigators, where each of these subsystems is part of the surrounding social systems, which in turn are the result of the sociocultural evolution of the human race."[14]

The unity of Habermas's plural, transcendental subject can be clarified by considering the conditions of the emergence of the human species.[15] At some point in time, and strictly in accordance with the laws governing the evolution of prehuman nature, a unique event occurred, namely, the emergence of man as a *zoon logikon*. With that event, a qualitatively new type of law, the law governing *logos*, was introduced into the course of what had thereby become socionatural evolution. Man's lack of a specialized instinctual endowment and the resulting period of extended dependency and maturation made the use of language for communication both possible and necessary for the species. Is it not plausible, as Habermas maintains, that the laws which were introduced into the course of evolution with the emergence of man as the *zoon logikon*, and which constitute conditions of the possibility of human association, also constitute the fundamental norms of that association? "The human interest in autonomy and responsibility is not mere fancy, for it can be apprehended a priori. What raises us out of nature is the only thing whose nature we can know: language. Through its structure, autonomy, and responsibility are posited for us. Our first sentence expresses unequivocally the intention of universal and unconstrained consensus."[16] We can see in this passage the convergence of

Habermas's epistemological and normative concerns. For transcendental investigation reveals not only conditions of the possibility of the evolution of the species, but also the norms that Habermas employs as the basis of his communicational ethics.

While the norms of human communication and association are "posited" with the first sentence, they need not be realized in any given empirical instance of communication nor in any given language game. The point is that they are counterfactually presupposed in all communication in that tacit reference must be made to them for communication to take place.[17] Indeed, even the ability to violate them, as in the case of Iying, presupposes their existence. Regarding the unity of Habermas's plural subject of thought and action, it is transcendentally unified and empirically diverse. With respect to the conditions of the possibility and anticipated goal of evolution, the subject is unitary. But that presupposed and anticipated unity may be more or less of an actuality at any point in empirical history.

Habermas thus holds that Hegel is correct against Kant in maintaining that, with respect to the realm of history, and as opposed to the realm of nature, it is impossible to locate unambiguously a transcendental subject outside of history, who stands over the object it constitutes. Instead, the genesis of the transcendental subject occurs in the very realm that comprises its object domain, that is, the realm of history. An examination of human evolution from the perspective of language is the only type of theory, so it is argued, that is adequate to the "peculiar relation between subjectivity and intersubjectivity between the 'transcendental' and the 'empirical.' " Inasmuch as it can be simultaneously its own metalanguage, ordinary language is the only phenomenon available to us which contains at the same time an empirical and a transcendental moment.[18]

But if Hegel is correct in his move from a philosophy of consciousness to a philosophy of spirit (which Habermas interprets linguistically), Marx, in turn, is right, against Hegel's idealism, in insisting that "nature is the absolute ground of mind": "The seal placed on absolute knowledge by the philosophy of identity is broken if the externality of nature, both objective environmental and subjective bodily nature, not only seems external to a consciousness that finds itself within nature but refers instead to the immediacy of a substratum on which the mind contingently depends."[19] To be sure, Habermas must "reconstruct" Marx, by bringing out the latent transcendentalism in Marx's own position, in order that Marx's materialism does not degenerate into a bad objectivism. Yet, Habermas's modification of the traditional transcendental project follows from his basic acceptance of Marx's materialistic critique of philosophy.

For Habermas, after Hegel's last heroic effort, and Marx's critique of Hegel, it has become apparent that philosophy cannot fulfill its traditional ambition of, as it were, getting behind reason's back and securing its own foundations.

Philosophy, in short, can no longer be practiced as a "philosophy of origins" (*Ursprungsphilosophie*).[20] Habermas is quite willing to accept a degree of "unavoidable circularity"[21] with respect to foundational questions, and simply attributes this to the finitude, that is, the material groundedness, of the human mind. Despite the impossibility of ultimate foundations, Habermas is nevertheless satisfied that, in the course of evolution, mental structures developed which—owing to their having emerged through a process of natural selection—somehow "fit" the world. The task of theory is not to deduce the validity of those structures in a way that would require an ultimate standpoint, but to explicate that "fit."

Yet, if Habermas's transcendental claims are not philosophical *strictu sensu*, what is their exact status and how does he arrive at them? This is where we encounter the unique terrain of critical theory, as Habermas conceives it, which lies somewhere "between philosophy and science": "Representations and descriptions are never independent of standards. And the choice of those standards is based on attitudes that require critical consideration by means of arguments, *because they cannot be either logically deduced or empirically demonstrated*. Fundamental methodological decisions, for example such basic distinctions as those between categorial and noncategorial being, between analytic and synthetic statements, or between descriptive and emotive meaning, *have the singular character of being neither arbitrary nor compelling. They prove appropriate or inappropriate*. For their criterion is that the metalogical necessity of interests, *that we can neither prescribe nor represent, but with which we must come to terms*. Therefore my . . . thesis is this—the achievements of the transcendental subject have their basis in the natural history of the species."[22]

Habermas seems concerned to avoid two equally unacceptable alternatives. He does not want to assert that the validity of fundamental norms can be demonstrated with the rigor sought by "traditional theory." Yet, he is equally opposed to the skeptical contention that these concepts are arbitrary, in the sense of either being the mere products of convention or of biological adaptation *in any simple sense*. While the validity of these principles may not be demonstrable in a completely compelling fashion, there are nevertheless "good reasons" for their acceptance.

To the extent that these transcendental structures "have their basis in the natural history of the species," they possess a certain facticity and can be examined through empirical anthropology. If, however, these transcendental structures were *only* the products of evolution, then a naturalistic interpretation of reason—with its inescapable skepticism—would be unavoidable. Reason would simply be an "organ of adaptation for men just as claws and teeth are for animals," and no claims for its autonomy could be made. Habermas argues that, although reason has its genesis in natural evolution, at some point in that process, reason transcends the conditions of its genesis and achieves a degree of autonomy.

Methodologically, the theory that elucidates and justifies the transcendental interest structures within which the species develops, although it contains an empirical moment, cannot simply be an empirical theory. The employment of a strictly empirical theory for the purposes of an "anthropology of knowledge" would involve a logical dilemma. For empirical anthropology is a "founded science," which is itself constituted within "the framework of the objectifying sciences," and cannot therefore be used as a "founding science" to establish that framework.[23] This dilemma has its origins in the fact that the knowing subject, which constitutes nature as an object of knowledge, is itself a product of nature.[24] In other words, the difficulty arises from the fact that the constituted constitutes the constituter. An anthropology of knowledge, if it is to avoid a vicious circle, cannot simply be an empirical theory, but must contain a "reflective" moment as well.

To take account of this state of affairs, Habermas makes a threefold distinction between objective nature, subjective nature, and nature in itself (*natura naturans*).[25] Nature in itself can be further differentiated into (evolutionarily) prehuman nature and something similar to the Kantian *Ding-an-sich*. Thus, prehuman nature produces the human species in the course of natural evolution, and that created species possesses a subjective nature which constitutes objective nature as an object of possible experience and knowledge. The particular makeup of the species is such, according to Habermas, that objective nature is constituted as an object of possible technical control. Furthermore, knowledge of objective nature is, in a Kantian fashion, knowledge of a system of apperances, and something like a *Ding-an-sich* must be posited as lying behind our apprehension of objective nature. Unlike the Kantian *Ding-an-sich*, however, Habermas's is not a quasi-object affecting our receptive apparatus. Rather, it is simply a theoretical posit which must be made to indicate the externality, contingency, and facticity of nature which conspire to confound any arbitrary interpretations we seek to impose on it.

Habermas must posit the existence of a prehuman nature and a *Ding-an-sich* to take account of the fact that, as McCarthy puts it, "cognition appears to be bound on both sides by contingent conditions."[26] Leaving aside the question of the *Ding-an-sich*, a serious difficulty intrinsic to Habermas's attempt to synthesize materialism and transcendentalism arises concerning prehuman nature. Does the "materialist" claim that prehuman nature produces subjective nature, as McCarthy asks, not throw Habermas back into a precritical ontology that violates his transcendental posture? The question reemerges at the level of philosophy of biology, where Habermas needs to, but cannot, account for the transition from prehuman to human nature.[27] In both cases, he wants to say more than can legitimately be said from within the confines of his position.

The terrain that Habermas wants to stake out for his transcendental-foundational discourse thus lies somewhere between empirical science and first philosophy. *He wants to oppose the objectivistic misconception of the sciences through reflection on the conditions of the possibility of the various sciences. He does not, however, want to have to move to the topos of traditional* first philosophy, *with all of its attendant dilemmas. But can this middle ground be maintained?* This, in turn, rests on the successful differentiation between empirical and transcendental anthropology. The following question, however, can be asked of *any given empirical theory*, for example, Gehlen's, that Habermas might employ for transcendental-foundational purposes: *By what right does it gain its transcendental status?* The claim is, of course, that in the course of anthropological research certain "reflexively traceable rules" have been discovered that function as transcendental frameworks within which the different modes of thought and action are constituted.[28] With our sensitivity to historicist concerns, however, we want to inquire into the status of those theories that claim to have discovered the invariant features of human history. Are they not themselves historically specific, fallible theories that can be superseded at any time?[29]

To put the question in more explicitly Hegelian terms, we can ask Habermas the question Hegel posed to Kant: What are the conditions of the possibility of discovering the conditions of the possibility of knowledge? Habermas's criticism of the objectivist understanding of science is that objectivism dogmatically assumes the validity of first-order scientific knowledge without reflecting on the presuppositions of that knowledge. However, we can ask Habermas, in turn, why reflection should stop at the level of transcendental anthropology and not proceed further? More traditional transcendental philosophers at least claim to carry reflection to a point where it becomes theoretically self-evident that a terminus has been reached. Indeed, the presence of such a terminal point for reflection can be taken as a hallmark of transcendental philosophy in the strict sense. In light of this consideration, does Habermas's failure to carry reflection beyond the level of transcendental anthropology, which certainly does not contain a self-evident point of termination, not constitute an objectivism of the second order—a transcendental objectivism as it were? In short, is Habermas himself not guilty of "arbitrarily arresting reflection?"[30]

Habermas would doubtless answer—and here his acceptance of the Marxian critique of the limits of philosophy is fully visible—that the questions presuppose that the traditional ambitions of first philosophy can be fulfilled, a presupposition that we have seen he denies. If we are seeking the sort of theoretical satisfaction promised by "traditional theory," Habermas admits he cannot provide it. It must be understood that, if philosophy is taken as that mode of theorizing that adheres to the standards of rigor that Habermas says cannot be met, then his claims about the unrealizability of those standards must in some sense be "extraphilosophical"; one cannot offer a strict philosophical proof of the impossibility of philosophy. The status of Habermas's claims concerning the impossibility of *first philosophy* self-avowedly suffers from the same limitations as what they assert.

For Habermas, "coming to terms" with the transcendental standpoint—a phrase with almost therapeutic overtones—means that, while the epistemological and transcendental grounding of the sciences cannot be accomplished with the rigor traditional philosophy sought, they can nevertheless be accomplished in some sense. If Hegel is correct about the *aporetic* character of the *Erkenntnisproblem*, "then the critique of knowledge can no longer claim to fulfill the intention of First Philosophy. But it is not at all clear why abandoning this should entail abandoning the critique of knowledge itself."[31] With Wittgenstein, Habermas shares the "therapeutic" intention of disabusing us of the wish for a fully satisfying *first philosophy*. Against Wittgenstein, however, Habermas does not want to embrace the skeptical consequences of either remaining silent about the most important question or accepting a plurality of incommensurable language games. Habermas's point is, rather, that a self-consciously less ambitious, quasi-transcendental grounding provides us with plausible, if not completely compelling, foundations for our knowledge. And while this type of theorizing may not offer the satisfaction held out by *first philosophy*, it does nonetheless avoid the despair of skepticism.

To "come to terms" with the transcendental standpoint is to see the essential rightness of this position and the "impossibility of getting beyond these transcendental limits."[32] It is somewhat ironic, however, that Habermas, who is often accused of hyperrationalism by the hermeneuticists, requires so large an element of judgment at the very base of his scheme. The way in which one "comes to terms" with the transcendental standpoint ultimately bears a closer resemblance to aesthetic taste or Aristotelian phronesis than to emphatic philosophic proof. While Habermas's transcendental scheme is meant to serve a theoretical function of grounding our knowledge, which is analogous to traditional *Ursprungsphilosophie*, the scheme itself is not grounded in as emphatic a fashion as one generally finds in *first philosophy*.[33]

Some Difficulties with Habermas's Position

The practical implications of Habermas's philosophical scheme become clear in his colleague Apel's article on "The Conflicts of Our Time and the Problem of Political Ethics."[34] Apel explicitly locates his concerns in the context of the planetary ecology crisis and inquires into the prospect of a "philosophically grounded political ethics" that is adequate to that crisis. The major difficulty that such an ethics must confront results from the unprecedented power and global expansion of modern science and technology. A dilemma occurs in that at the same time as the global effects of modern science and technology make political ethics more necessary than ever, the hypostatization of scientific rationality as the only valid mode of rationality, that is, scientism, appears to preclude the formation of such an ethics. Consequently, the desideratum for communications theory is to dissolve the blockage to rational ethics caused by the hypostatization of scientific rationality and the formulation of a positive doctrine of ethics.

While their particular theories differ in some details, Apel and Habermas agree that a transcendental critique of the pragmatic dimension of language can accomplish *both these tasks simultaneously*. With respect to the dehypostatization of scientific rationality, such a critique demonstrates that the language of science, as a specialized form of language, presupposes the general structures of communication as such. The argument is that scientific rationality is not the ultimate form of rationality, but is a specialized language game abstracted out of ordinary language and, as such, presupposes a more fundamental form of rationality embedded in ordinary language. Unlike the romantic attack on science, the point is not to disparage the achievements of *science*, but to criticize *scientism*, that is, the grandiose claims made for the specialized form of rationality embodied in modern science.

The same critique which removes the obstacles to a rational ethics also indicates what direction a positive doctrine of ethics should take. Not only is scientific rationality derivative from and bound by a more fundamental form of communicative rationality, but communicative rationality also has a normative dimension within it which can become the basis for an ethical doctrine. It can be shown, so the argument goes, that the conditions of the possibility of communication are, in part, normative. The individual's ability to use language, that is, "communicative competence," presupposes the existence of certain norms that need not be realized in any given act of communication. In language, Apel and Habermas claim to have located that *sui generis* phenomenon which, while factually given, nonetheless has a normative dimension. Thus, it is the only realm where the problem of the "is" and "ought" can be overcome. Communicative ethics promises to eliminate the impasse of recent ethical theory by removing the scientistic obstruction to normative theory and by locating language as the field where the fact-value problem can be resolved.

Although communicative ethics may hold promise for overcoming scientistic impediments to normative theory and for formulating positive ethical doctrines in certain contexts, its adequacy to the unprecedented ethical problems raised by the ecology crisis remains questionable.[35] This is due to the fact that, to use the terminology of traditional ethical theory, communicative ethics is *thoroughly anthropocentric*. As opposed to all forms of *naturalistic*, ethics *anthropocentrism* holds that man is the only locus of value and the only being that commands respect in the universe. Communicative ethics represents a variation on the anthropocentric theme in that it maintains that man, by virtue of his communicative capacity, is the only value bearing being that can be identified. Thus, communicative ethics, as a form of anthropocentrism, rules out any conception of nature as an "end-in-itself." This is to be expected. Habermas's transcendental stance prevents the sort of direct access to nature that would make any claim for nature as an end-in-itself possible. As with Kant, theoretical transcendentalism and ethical anthropocentrism go hand in hand.

In addition to his acceptance of modern science and modern philosophy's focus on subjectivity, there are good reasons why Habermas and most contemporary philosophers want to exclude the idea of nature as an end-in-itself from ethical theory. The dignity and rights of the moral and legal subject have been secured by severing the subject from the realm of natural existence. Because they are characterized by self-consciousness or language, subjects are considered qualitatively different from the rest of natural existence. This is why they command respect and ought to be treated as ends-in-themselves. It is often feared that anything that threatens to disturb this distinction—*which the concept of nature as an end-in-itself certainly does*—also threatens the dignity of the subject.

However, the dignity of the subject—the discovery of which, as Hegel continually stresses, is one of the momentous achievements of modern philosophy—*is attained at the price of denying all worth to nature*. Habermas appears willing to accept this arrangement. Given the main thrust of his philosophy, Habermas would be unwilling to tamper with the constellation on which the dignity of the subject has traditionally rested, even for so grave an issue as the ecology crisis. Whatever solution he formulates must leave that constellation intact. For a variety of reasons—for example, his adherence to the basic posture of modern philosophy; his desire to "save the subject" from the multiple threats of vulgar Marxism, mainstream scientism, and the monism of the early Frankfurt School; his suspicion of *Naturphilosophie* owing to the pernicious role it has played in German intellectual history; and so forth—Habermas is unwilling to entertain the concept of nature as an end-in-itself, regardless of how tempting it might be for an ecological ethics.[36] The "good-for-nature" must somehow be derived from the "good-for-man."

While no one would want to violate the dignity of the subject, the following question must nevertheless be raised: *Can we continue to deny all worth to nature and treat it as a mere means without destroying the natural preconditions for the existence of subjects?* Likewise, can the worth of nature be secured without devaluing the dignity of the subject? Hans Jonas has argued that the gravity of the ecology crisis and the fact that the biosphere "has become a human trust" require a rethinking of the "givens" of modern philosophy so that nature will no longer be condemned to being a mere object.[37] And from within the critical theory camp itself, Apel has argued similarly that the "unchecked onslaught of technology upon nature [which] threatens to destroy the life space of all living creatures" should prompt a reconsideration of the "objective-teleological conceptions" that were denigrated with the "triumph of mechanistic thought." Unfortunately, Apel does not pursue this line of thought far enough. If he did, it would produce serious difficulties for his and Habermas's scheme of cognitive interests and their communicative (i.e., anthropocentric) approach to ethics.

Several of the consequences that followed from Kant's transcendentalism also appear to follow from Habermas's: a dialectics or philosophy of nature[38] becomes impossible and nature can only be validly apprehended as an object of the natural sciences.

Kant's "Copernican Revolution" is seen by the proponents of transcendental philosophy as having established inviolable standards for philosophizing. Among the central components of this "revolution" is the insight that, due to the fact that the *a priori* structures of knowing constitute the objects of knowledge, our access to those objects is *necessarily oblique*. Any theory that does not take this obliqueness sufficiently into consideration and does not deal adequately with the constitutional moment of knowing in its philosophical construction is *theoretically naive*. A "dialectics of nature" that purports to know the movement of nature in itself, independently of the *a priori* structures of knowing, makes precisely the sort of claim to direct access to the "othersidedness" of nature which the transcendental turn rules out. From the standpoint of transcendental philosophy, a "dialectics of nature" is theoretically naive.[39]

Moreover, for both Kant and Habermas, the knowledge that can be validly claimed of nature within the framework of transcendental philosophy is the type of knowledge contained in modern science, that is, knowledge of nature as a thoroughgoing mathematical manifold, devoid of meaning, value, and purpose. Although Kant himself does not lay great stress on the utilitarian possibilities of a mathematicized science, Habermas argues that nature thus constituted is essentially an object of possible technical control. This, however, concerns the *constitutional features* of modern science, and not the subjective motives of the practicing scientist. However gratuitous the intentions of any given scientist may be, Galilean science is—by virtue of its mathematical character, nomological form, experimental method, and so forth—*constitutionally instrumental* and at least *potentially technological*. While it would be mistaken to make too immediate an identification between Galilean science and modern technology, it is nevertheless the case that the technological revolution was one of the innermost possibilities of a mathematized science.

Habermas does not shy away from the consequences of his position: "The resurrection of nature cannot be logically conceived within materialism, no matter how much the early Marx and the speculative minds in the Marxist tradition (Walter Benjamin, Ernst Bloch, Herbert Marcuse, Theodor W. Adorno) find themselves attracted by this heritage of mysticism."[40] The "resurrection of nature" refers to the transformation of our relation to and knowledge of nature such that nature would once again be taken as purposeful, meaningful, or as possessing value. It would mean the undoing of the deteleologization of nature that occurred with the rise of Galilean science. Just as Habermas believes the ethical and legal principles of modernity are somehow "timeless," he also believes that there is something unsurpassable about modern mathematical science.

For Habermas our relation to nature, insofar as nature is taken as an object of cognition, can only be one of *instrumental control*. This is not a contingent fact over which we might exert any influence. On the contrary, it derives from the invariant features of our anthropological endowment, which constitutes nature as an object of possible experience. Modern science and technology represent humanity's most refined and sophisticated means of pursuing its interest in instrumental control—an interest rooted in the makeup of the species.[41] If this argument is correct, then a qualitatively new relation to the natural world, and a new science corresponding to it, such as Marcuse envisages, are indeed impossible.

Two strategies can be adopted to challenge Habermas on this claim: (1) I have tried to show where Habermas's transcendental scheme, on which this objectification of nature rests, is vulnerable. The point is to question the thesis that an insurpassable instrumental relation to nature is rooted in our species endowments and is therefore inevitable. (2) Habermas's claims can also be questioned with respect to the philosophy of science. While he does not seek to provide us with a detailed philosophy of science, his theory of research-guiding interests purports to delineate the various domains of scientific research and the constitutive principles operative therein. If it could be shown then that a particular branch of scientific practice could not be accounted for satisfactorily by Habermas's theory—that the principles presupposed by that science contradicted his transcendental scheme—then that branch would constitute an anomaly for Habermas's philosophy of science. Biology may very well constitute one such anomaly.[42]

Biology as a Possible Anomaly for the Habermasian Position

It should be recalled that Galilean science is as much a program for the "mathematical march of mind"[43] through the universe as it is a completed project. And this march has proceeded much further in the realm of inanimate matter than in the realm of vital phenomena, where some of the most elementary steps have yet to be taken. One may ask whether this is a contingent state of affairs that awaits a "Newton of a blade of grass," or whether there are good reasons to believe that the program cannot be executed. Or, one might make the stronger assertion that even if life had been exhaustively treated in mechanistic terms, it would not have thereby been explained, but explained away. Just as an explanation of mind in terms of the brain would be an explanation of mind in terms of nonmind, so an explanation of life in terms of mechanism would be an explanation of life in terms of nonlife.

Proceeding through an examination of biology is the most fruitful way to criticize Habermas on these matters and has decided advantages over other approaches. Unlike Marcuse's, for example, this approach is not forced to confront the imponderable task of creating an "alternative science" *ex nihilo*. Furthermore, it involves exploring an avenue that remains immanent to modern science. By beginning from a position immanent to science, this approach would at least be in no *immediate* danger of contradicting the current standards of intersubjective rationality.

The surrender of nature to the domain of instrumental reason is even more thoroughgoing in Habermas than in Kant, who at least evinced some uneasiness with the "mechanization of the world picture" and wrote the third critique as a result. Curiously enough, Kantian as Habermas is, there is no analogue to the *Critique of Judgment* in his *opus*; it is as though he perceives no pressing problem.

Kant, it will be recalled, was faced with the following dilemma: after having demonstrated that nature, as an object of possible experience, was a causal nexus through and through and therefore contained no room for purposefulness, he had to account for the seemingly teleological phenomena that confronted the scientist in his investigation of nature. Prominent among these was, of course, the apparent purposefulness of "organized beings," that is, living things. Kant's solution was to introduce the distinction between "determinant" and "reflective" judgments.[44] Owing to the nature of the human mind, purposefulness has to be assumed, as a *heuristic*, in the investigation of vital phenomena. The concept of purposefulness cannot, however, appear in satisfactory scientific theories in their final form, but has to be jettisoned along the way. The concept pertains, in other words, to the logic of discovery and not to the logic of validation. Whatever one makes of Kant's critique of teleological judgment—and it is a highly perplexing theory indeed[45]—the fact remains that he attempts to account for the apparent purposefulness that manifests itself at the subhuman level.

The questions examined by Kant concerning the nature of living phenomena were again taken up by the vitalists at the end of the last century and have come down to us in the contemporary controversies over the role of systems theory in biology. The vitalists maintained that there is an "extra something," a vital element, in living things which make them *in principle* not susceptible to an exhaustive treatment in mechanistic terms. Against this contention, the mechanists argued that, as this vital element has never been observed, it is a "metaphysical" posit which deserves no place in modern science. The assumption of a vital principle, they further argued, is unnecessary for biological research and, *given sufficient time*, an exhaustive treatment of life in purely mechanistic terms will be forthcoming.[46] The claims of contemporary reductionists are, for the most part, of a programmatic nature, and, to date, biology remains largely an autonomous science: "Indeed, no serious student, reductionist, or antireductionist, questions the truth of this [i.e., that major biological theories have yet to be reduced] and that biology remains at present an autonomous science is not a matter of dispute."[47] The interesting question for the philosophy of biology is whether there is good reason to believe the reduction can in fact be accomplished.

A major focus of the current debate concerning reductionism is the adoption of systems theory by biologists in recent years. There can be no question as to the impact of systems-theoretical approaches on contemporary biology.[48] However, while there can be no questions concerning the advantages gained by the adoption of systems theory for biological research, how to interpret the philosophical significance of this development remains controversial. The controversy revolves around the status of systems theory as a mode of rationality. Three basic positions are discernable in the debate:

(1) The first position, which is the most prevalent among neoempiricist philosophers of science, holds that systems theory at last provides the sophisticated methodology required to complete the reductionist program and the incorporation of biology into a unified physiochemical science.[49] In this case, no claims are made for systems theory as a unique mode of rationality, but it is simply viewed as a sophisticated form of mathematical-functional analysis. The sophistication of systems theory, it is argued, is equal to the complexity of those phenomena—for example, organismic, teleological, self-organizing, and so forth—which were formerly held to uniquely characterize life and to be resistant to exhaustive analysis on mechanistic assumptions alone.[50]

(2) The second position, which is espoused by the more enthusiastic systems theorists, is that systems theory represents a third viewpoint that can overcome the old opposition between vitalism and mechanism. It is maintained that systems theory can account for the same "vital" phenomena which vitalism sought to explain, but, in accordance with mechanism, it does not have to appeal to any "metaphysical" or "mystical" entities.[51] In contrast to the first position, it should be stressed that—while it may be a *physical* theory—systems theorists do not consider their theory as either *mechanistic* or *reductionistic*. On the contrary, as an emergentist theory of hierarchical wholes, systems theory is considered an alternative to both reductionism and mechanism.[52] To use the language of critical theory, it is considered an alternative to instrumental reason. Whereas the first position maintained that systems theory could be employed to reduce biology to physics and chemistry, here biology's employment of systems theory is adduced as evidence for the autonomy of the biological sciences.

(3) The third position, which harks back to the older *Naturphilosophie*, is the most speculative of the three and the least fashionable today. While it admits that vital phenomena may be *amenable* to a systems-theoretical treatment, it maintains that such treatment cannot do complete justice to living things; after systems theory has, as it were, cast its theoretical net, certain distinctive features of life, so the argument goes, will remain beyond its theoretical reach. It is argued that systems theory remains within the framework

of Galilean science insofar as it accepts the constitution of nature as a mathematical manifold. Indeed, instead of presenting some sort of transcendence of Galilean science as proponents of the second position often contend, systems theory, by virtue of its increased mathematical sophistication, simply represents its most sophisticated refinement to date. The argument is that essential attributes of life cannot be adequately apprehended in spatiotemporal terms alone, and that life already represents an anomaly for nature conceived of as only *res extensa*. Thus, Jonas argues that, although living things can be treated in purely external terms - as objects among objects—they must also be understood *from the inside*.[53] Essential attributes of vital phenomena elude a purely external approach: living form cannot be reduced to cybernetic information, inwardness cannot be understood as spatial interiority, and concern for existence cannot be explained in terms of self-regulating systems. Moreover, Jonas maintains that it is only because we are "peepholes into the inwardness of substance," and can know ourselves from the inside, that we can understand biological phenomena on analogy to our own experience as living beings. The proper understanding of biological phenomena, in other words, presupposes a certain communality of life.

Returning to Habermas, if the entire realm of human cognition can be *exhaustively* subdivided, as he claims, in terms of the three knowledge-constitutive interests—instrumental, communicative, and emancipatory—then our question becomes: where are the biological sciences to be located within his scheme? We know that Habermas emphatically denies the possibility of conceptualizing nature—presumably including living nature—under the categories of communicative rationality: "Nature does not conform to the categories under which the subject can conform to the understanding of another subject on the basis of reciprocal recognition under the categories that are binding on both of them."[54] If the only two possible alternatives available in this case are communicative and instrumental rationality, and if the domain of communicative rationality is exclusively reserved for speaking subjects, then it follows that the biological sciences must be assigned to the domain of instrumental rationality. Our only possible cognitive relation to other living beings in this scheme is one transcendentally oriented to technical domination. Moreover, it must be assumed that, in order to preserve his scheme of cognitive interests, Habermas would have to subscribe to the reductionist program for biology—a position which is problematic on both empirical and theoretical grounds. He would have to maintain that, through the application of systems theory, the apparently teleological features of living things could be accounted for, and biology thereby fully incorporated into a unified physical theory—that is, into instrumental reason. And, the claims of the more naive systems theorists notwithstanding, it is obvious from his discussions with Luhmann that Habermas has no illusions about the "noninstrumental" character of systems theory.

Habermas thus divides the *scala natura*—albeit from the side of epistemology and methodology—at the level of human intentionality or communicability. While everything on the subhuman level, including life, is assigned to the realm of instrumental reason, the domain of human communicability remains the last preserve in an otherwise mechanized universe. Habermas is, in short, an antireductionist for the human sciences and a reductionist for the life sciences. He is not basically in disagreement with the neoempiricist program for a unified science as long as that program is contained to the realm of nonspeaking nature and is interpreted in transcendental-instrumental rather than a realistic fashion. Today there is nothing particularly unusual about this stance. On the contrary, it is more or less the strategy adopted by most contemporary antipositivist philosophers, for example, phenomenologists, ordinary language philosophers, action theorists, and so forth, who are concerned to "save the subject."[55]

Habermas, however, crosses over the "ontological hiatus" between speaking nature and nonspeaking nature in a way which, on his own terms, is literally *illicit*. In a discussion of ethology, he maintains that we are to understand the seemingly teleological behavior of animals by reasoning "privately" from human intentionality.[56]

Similarly, in a response to the question whether Darwinian theory belongs to the empirical-analytic sciences and is for that reason constituted with an interest in technical control, Habermas answers: "Since the evolution theory has a methodological status which is quite different from a normal theory in, say, physics, I think that the categorical framework in which the evolution theory has been developed since Darwin presupposes some references to a preunderstanding of the *human* world and not only of nature. The whole concept of adaptation and selection presupposes some elements which are more characteristic for the human sciences than for the empirical-analytic sciences, strictly speaking. So in my opinion, the evolution theory is no example of empirical-analytic science at all. But as far as biochemical theories about mutations go into evolution theory, we have, of course, a usual empirical-analytical theory. However, this is not what is characteristic for the design of evolution theory. This is only a component of the evolution theory. Modern genetics is not dependent on the evolution theory framework. Modern genetics is, I propose, a strictly objectifying theory which makes no use of concepts inherently related to our preunderstanding of what social life or cultural life is."[57] How one should interpret this passage which is, unfortunately, unclear. Either Habermas is saying that evolutionary theory is an immature and inferior theory to the extent that it still contains anthropomorphic elements such as the concept of adaptation, in which case he must be committed to the controversial thesis that evolutionary theory can be reduced to molecular biology. Or, he is saying that those anthropomorphic elements are uneliminable and must be understood "privately," in which case one does not know how to locate neo-Darwinian theory, for example, in his scheme of knowledge-constitutive interests.

While the idea of privative reasoning may make sense in the context of Aristotelian metaphysics, *where a continuity of being is presupposed*, in Habermas's case, where *no such continuity is assumed*, it strikes me as peculiar. Earlier it has been shown that Habermas's attempt to combine materialism and transcendentalism leaves a radical hiatus between prehuman and human nature. This hiatus cannot be legitimately crossed until an adequate account of the production of human nature by prehuman nature has been provided. Here, however, we find Habermas traversing that hiatus, albeit in a backward direction. Two questions thus arise: How can he methodologically justify this kind of private reasoning which appears to violate the fundamental division of his philosophical construction? And, does this type of reasoning not presuppose that there is some continuity between nonspeaking and speaking nature—some "communality of life"—by virtue of which "private" reasoning can legitimately take place? And once we can reason privately in a backward direction, what is to prevent us from reasoning in a forward direction and conceiving of prehuman nature as *incipient spirit?*

This notion of incipient spirit might, in turn, form a basis for an ecological ethics from a naturalistic perspective.

Critical Theory and Ecological Ethics

Habermas has nowhere discussed "ecological ethics." It would therefore be fruitful to construct what his position might be given the main tenets and constraints of his theory. More specifically, how, given his thoroughgoing anthropocentrism, which rules out the possibility of nature as an end-in-itself, might he argue for the protection of the natural environment? How, in other words, would he think a resolution to the environmental crisis without appeal to anything like "the resurrection of nature"?

At the most general level, the point would have to be that an end to the disharmony between man and man would entail an end to the disharmony between man and nature. A solution to the ecology crisis would therefore follow from a solution to the "social question" and it would not be necessary to develop a qualitatively new relationship to the natural world. Were this in fact the case, a communicative ethics, inasmuch as it would provide us with the proper principles for adjudicating the conflicts between humans, would be sufficient for deriving an ecological ethics. The proper norms for regulating the relation between society and nature would somehow follow from the communicatively conceived idea of the human good life without reference to nature as an end-in-itself. It would have to be shown that the preservation of the natural environment ("the good-for-nature") was somehow entailed by the communicatively conceived good-for-man.

Put in somewhat more concrete sociohistorical terms, Habermas might argue as follows. *The ecological crisis is, at its roots, caused by the strain that the incessant expansion of the economy—which enlists science and technology for its purposes—places on the natural environment.* Habermas observes that "traditional societies," in which the economy is *embedded* in a larger ethicoinstitutional matrix, always placed intrinsic limits on economic and technological growth: "The expression 'traditional society' refers to the circumstance that the institutional framework is grounded in the unquestionable underpinning of legitimation constituted by mythical, religious, or metaphysical interpretation of reality—cosmic as well as social—as a whole. 'Traditional societies' exist as long as the development of subsystems of purposive-rational action keep within the limit of the legitimating efficacy of cultural traditions. This is the basis for the 'superiority' of the institutional framework which does not preclude structural changes adapted to a potential surplus generated in the economic system but does preclude critically challenging the traditional form of legitimation."[58] With the rise of modern, capitalist society, the economy becomes disembedded from the ethicoinstitutional framework—that is, the differentiation of the market—and economic activity becomes "denormatized" or "emancipated." The result of this process, as Aristotle had already apprehensively predicted, is that a new dynamism is unleashed in the economy in which growth, including the expansion of human needs, tends toward the unlimited: "It is only since the capitalist mode of production has equipped the economic system with a self-propelling mechanism that ensures long-term continuous growth (despite crises) in the productivity of labor that the introduction of new technologies and strategies, that is, innovation as such, has been institutionalized. . . . Capitalism is the first mode of production in world history to institutionalize self-sustaining economic growth."[59]

The disembedding of the economy and its subsequent expansion tends to dissolve the traditional frameworks in which it was formerly situated, and which served to legitimate the social order as a whole. And, interestingly enough, the very thing that served to undermine the previous forms of legitimation itself becomes—after bourgeois consciousness "grew cynical"—the new principle of legitimation: in advanced capitalist society, institutionalized economic and technological progress becomes the new principle of legitimation, which is a peculiar principle of legitimation indeed. Viewed from the perspective of not only the philosophical tradition, but of traditionalism in general, this amounts to something like a huge "category mistake." Economic and technological progress belong to the realm of the *technical*, not the *practical* or the *normative*, and therefore cannot properly serve as legitimating, that is, normative, principles. Thus, it is Habermas's contention that (late) capitalist society, through the hypostatization of science and technology into ideology, has for the first time in history attempted to suppress the normative dimension of sociation. Due to this state of affairs, questions bearing on the character of good life are not submitted to public political debates about the "good life," but are left to the administrative decisions of technical experts.

Contemporary modes of positivist and neopositivist thought contribute to this suppression of the normative dimension in that they maintain that questions of value lie beyond the purview of rational adjudication and can therefore only be decided by extrarational means. The theoretical level, where it is held that questions of value cannot be rationally discussed, and the practical level, where what were traditionally ethicopolitical questions are answered through technical means, thus serve to reinforce each other in advanced capitalist society. Communication theory, as I have tried to show, seeks, through its critique of scientism and its reinstatement of the legitimacy of practical reason, to defend the possibility of rational discussion of the good life and the goals of social development. And, if this renewed normative political discussion were translated into a program of action which eventuated in a radical transformation of society, one of the results of that transformation would presumably be the resubordination of the economy—as well as of scientific and technological activity—to the collectively conceived idea of the good life. The economy, in short, would become "re-embedded." However, in contrast to premodern societies, where the institutional framework in which the economy was embedded *ultimately rested on tradition*, here the institutional framework would be grounded in a *rationally formed consensus*.

It is important to stress that, as opposed to Marcuse, it is not incumbent on Habermas, given the strategy outlined, to formulate a new mode of mediation between society and external nature, that is, a "new Science," nor does he think such a science possible. Habermas maintains that modern science and the outlook toward the natural world sedimented in it are rooted in the anthropological endowment of the species and therefore cannot be transcended.[60] The impossibility of a New Science does not, however, present a difficulty for him. He contends that the pernicious effects of modern science and technology, presumably including those on the environment, do not derive from modern science and technology *per se*, but modern science and technology *in conjunction with the chrematistic forces of the market*. Modern science and technology are not themselves intrinsically expansionistic and rapacious, but are driven by the *chrematistic* forces of the capitalist economy. In a different socioeconomic context, where their use was detached from the requirements of capitalist accumulation and directed toward the realization of a clearly conceived vision of the good life, the destructive aspects of modern science and technology could be contained and their beneficial aspects could be selectively chosen.

I have tried to show how, given the perspective of a communicative ethics, a solution to the ecology crisis might possibly be conceived. Most importantly, if a solution to the crisis in the social sphere would entail an end to the lethal stress that that sphere places on the biosphere, there would be no need to go beyond the limits of an anthropocentric framework. One need not, that is, envision the "resurrection of nature." Nature will remain objectified from the constitutive standpoint of modern mathematical science—which means that it will remain constituted as an object of instrumental control—and no rational possibility for taking nature as an end-in-itself will exist. *The major difference will pertain to the socioeconomic context within which nature is objectified and not to the way in which nature is objectified.*

This solution is certainly plausible and internally coherent. Moreover, it is not in danger of violating some of the most fundamental principles of modern rationalism, thereby risking an irrationalist regression, as are some of the more radical proposals for the ecology crisis. However, despite these advantages, two questions must be examined. First, with respect to ethical theory, can imperatives with strength adequate to the unprecedented tasks posed by the ecology crisis be obtained on anthropocentric premises? Could, for example, binding imperatives which oblige us to protect species from extinction be formulated on purely anthropocentric grounds? Secondly, even if it could be shown *theoretically* that it is not necessary to move from the standpoint of anthropocentrism to formulate solutions to the environmental crisis, a question would still remain at the level of social psychology. For it is difficult to imagine how the conflict between society and nature is going to be resolved without a major transformation in our social consciousness of the natural world—for example, a renewed reverence for life.

Conclusion

In conclusion, I would like to step back and consider the larger issues at stake in this discussion. These issues concern the deep ambivalence toward modernity that can be traced throughout the history of the Western revolutionary movement. On the one hand, one often finds left radicals speaking as if they, and not the powers that be, were the true defenders of the values of the modern Enlightenment—for example, autonomy, conscience, individuality, tolerance, and so forth—and that the goal of a revolution would be the completion of the Enlightenment's program rather than its negation. In this case, the current social crisis is seen as resulting from an uncompleted modernity which is, for that reason, at war with itself. On the other hand, left radicals also speak at times as if modernity itself were corrupt through and through, and what is therefore required is a transvaluation of contemporary values and a radical break in the historical continuum. In this case, the values of the Enlightenment are viewed as in no sense valid, and are seen as nothing more than the ideological rationalizations for a thoroughly corrupt social order. Needless to say, these positions represent ideal types and one rarely, if ever, finds them in their form.

Habermas, who is far less ambivalent than his predecessors at Frankfurt on this score, is a primary example of the first problem. He seeks to criticize modern *bourgeois* society in terms of its own ideals in order to facilitate a process of political enlightenment which could, in turn, lead to the transformation of that society. In this approach, Habermas, like the younger Marx, assumes that valid norms of sociation have already been recognized, and what is required is a program of immanent critique.[61] The modern state, which attempts to legitimate itself through appeal to rational norms, is an inadequate instantiation of intrinsically valid principles, and the fact that it is in conflict with itself presents the point of entry for critique. Habermas goes so far as to argue—the thesis of one-dimensional society notwithstanding—that social criticism is *easier* in *bourgeois* society than in previously existing societies. This is due to the fact that *bourgeois* society is the first to attempt to legitimate itself through reference to rationally validated norms rather than to dogmatic tradition.[62] We can see here that one of the unsolved questions of the early Frankfurt School—that is, where do standards of critique come from?—is answered for Habermas: the standards are the "de-ideologized" norms of bourgeois society itself.

The norms of *bourgeois* society do, however, have to be "de-ideologized," and this "de-ideologization" is one of the many theoretical tasks that communications theory is designed to accomplish. Habermas is truly Hegelian in that he holds that values possessing a truly universal, transhistorical validity were *recognized* during the period of the *bourgeois*-democratic revolutions. To be sure, these norms were enlisted by the bourgeoisie to rationalize the capitalist order. However, it is not so much, as the traditional Marxist interpretation has it, that the bourgeois presented its own particularistic values as universal, but that it adopted and distorted genuinely universal values for its own purposes. Intrinsically valid norms rode into historical recognition, so to speak, on the back of the emerging capitalist order. The *desideratum* is, therefore, to detach the instrinsically valid norms from their ideological functions, which means to disassociate "bourgeois right" from its economic interpretation. By reinterpreting the content of "bourgeois right" in terms of human communication, rather than on the contractual model of economic exchange as the modern natural right theorists had, Habermas hopes to separate its rational kernel from its ideological husk.[63] It should therefore not be surprising that the values which Habermas claims are counterfactually presupposed by all human communication are precisely the central values of the bourgeois-democratic revolutions.

As opposed to the critique that rejects modernity as a whole, Habermas's strategy has one distinct advantage: it does not have to enlist standards whose validity has not already (at least tacitly) been recognized. Once the values of bourgeois society have been "de-ideologized," social theory has critical norms at its disposal which are immanent in, and in principle generally recognized by, the existing order. It has a foothold in the world, on the basis of which it can proceed immanently. Political theorists from Hegel to Arendt have recognized how difficult, if not impossible, it is to make successful appeal—as the proponent of our second position would have to do—to standards that are totally external to the established mode of *Sittlichkeit*.[64] For in that case no implicit consensual basis exists for political argumentation and the establishment of such a basis itself appears to be a theoretical impossibility.

Habermas then can be located in the mainstream of the Enlightenment tradition. Accordingly, rational autonomy (*Mündigkeit*)—to be subject to laws of our own making rather than to the compulsions of inner or outer nature—constitutes the fundamental value that animates his project, as it did for such classical *Aufklärers* as Kant and Freud.

Also, in accordance with the classical conception of the Enlightenment, Habermas views the "disembedding from nature"—that is, the conquest of the external environment and the mastery of inner drives—as the process through which the goal of rational autonomy can be achieved. Shapiro has made clear the extent to which the Habermasian version of critical theory is conceived of as the culmination of the Enlightenment: "emancipatory thought, of which critical theory is one part, is not just a current in the history of ideas, or an ideological reflection of class struggle, but part of the process of sociocultural evolution itself. Critical theory is a component of what could be called the emancipatory subsystem of sociocultural evolution, whose differentiation into a separate subsystem marks a turning point in world history. This differentiation occurred in the context of the industrial and democratic revolution at the end of the 18th and the beginning of the 19th centuries. From the viewpoint of intellectual history, it can be dated at beginning with Kant, whose critical philosophy founded the distinctive emancipatory method that was developed in the Hegelian and Marxian dialectic. The specific function of the emancipatory subsystem is the control and direction of the evolutionary process itself. It articulates and makes reflexive the capacities for such control that are given in human nature but emerge autonomously only with the emancipatory subsystem. If the main obstacle to human freedom and rationality is the embeddedness in the past, then we can say that the emancipatory subsystem is a cultural learning mechanism for trying to solve the problem of escaping from and overcoming embeddedness."[65] Let it be stressed that the conception of emancipation articulated in this passage remains thoroughly within a disenchanted world view. The goal of emancipatory process is the completion of the disembedding of humanity from nature, and the assumption of scientific and rational management, by that disembedded humanity, over the course of evolution. Not only does this conception reject the resurrection of inner or outer nature, but the goal of this process, the disembedding from nature, is exactly what is viewed as the difficulty by proponents of the second position.

Habermas's vision lacks at the same time the pessimism as well as the utopianism of the earlier critical theorists. He does not assess the historical situation as being nearly so irremediable as they did, and is not compelled, therefore, to go to such radical lengths in thinking a solution. Thus, on the side of human nature, one finds none of the erotic or aesthetic utopianism in Habermas that was so characteristic of the early Frankfurt School. He takes rational autonomy as a perfectly adequate idea of selfhood, and feels no need to formulate an alternative emancipatory concept such as a new sensibility. Likewise, as we have repeatedly seen, since his quarrel is not with Galilean science as such, but with its hypostatization into the "proportion of a life form, of a 'historical totality,' of a life world,"[66] it is not incumbent on him to argue for an alternative science. All of this becomes very clear in Habermas's "discussion" with Marcuse, where Habermas tries to draw out the systematic interconnections within their respective positions. It is obvious that Habermas perceives a necessary connection in Marcuse between his sensuous conception of reason, his erotic and aesthetic utopianism, and his call for a new science, and he believes that, by moving to the standpoint of communicative rationality, these necessary connections can be broken and the more dubious aspects of Marcuse's theory avoided.[67]

As Habermas has often been criticized for his thoroughly disenchanted stance toward the natural world, Shapiro tries to defend him on this count by arguing that the disembedding of humanity from nature is somehow equivalent to the resurrection of nature: "The dialectic of history is resolved through the completion of the self-transcendence of nature that occurs when embeddedness in nature is overcome and human beings bring the historical process under control. *This self-transcendence is at the same time the 'resurrection of nature,' because it ends the conflict of nature with itself* that was manifest in the embeddedness and the abstract contradiction between universality and its limits."[68] It is difficult to see how the state of affairs described in this passage can be construed as the resurrection of nature. On the contrary, the situation described above is perfectly compatible with a totally disembedded and denaturalized humanity standing over against a thoroughly reified nature and manipulating it for its own purposes. This would represent the pinnacle of domination rather than its opposite. While the conflict between humanity and nature might no longer occur *within* nature, it would not, for that reason, be any less a conflict. Habermas's position may very well be correct, and we may have to give up the idea of the resurrection of nature as a romantic illusion, but in that case, let us at least be perfectly clear about what is at stake.

The genealogy of the second view I enumerated above can be traced to the more messianic and eschatological traditions rather than to the modern, bourgeois Enlightenment. Indeed, instead of conceiving of an emancipated society as somehow constituting the fulfillment of modernity, the advocates of this position tend to see modernity, "this nullity [which] imagines that it has attained a level of civilization never before achieved," as the original evil which is to be extirpated. The dissolution of more organic forms of solidarity and their replacement by a ubiquitous bureaucracy, the decay of great cities and the simultaneous urbanization of the countryside,[69] the atomization of the individual, the substitution of formal for substantive justice, the ever-increasing danger to the earth's life-support systems, the constant threat of major war, not to mention the general banality of everyday life—these, and not the nobler sentiments of the Enlightenment, are held to comprise the essential reality of modern society. It is argued, therefore, that a social revolution must involve a radical hiatus in history so that a new social order can be established which differs from the existing historical constellation. Central to the new constellation would of course be a renewed relation to the natural world which would close the schism between society and nature that was opened with the rise of modernity and the methodical conquest of nature. Since this position entails such a radical discontinuity in history, it is extremely difficult, if not impossible, to identify "objective tendencies" striving in the direction of the type of society which is envisioned.[70]

In this context, it is interesting that Wellmer explicates the difference between Habermas and the early Frankfurt School in terms of the question of historical continuity. For Horkheimer, Adorno, and Marcuse, liberation "as the negation of instrumental reason . . . would be the resurrection of external and internal nature as well as the beginning of a new history of man. While in Marx's theory there is a tendency to blur the historical *discontinuity* which would separate a liberated society from the universe of instrumental reason, the philosophy of the Frankfurt School is in danger of losing the historical *continuity* which alone would make socialism a *historical* project: liberation becomes an eschatological category."[71] Because he introduces the categorical distinction between instrumental and practical reason, Habermas can determine the proper amount of discontinuity, as it were, between capitalism and a free society. Against Marx, who tends to see socialism as continuous with capitalism on the technical level insofar as he sees a socialist society as the nearly automatic outcome of the development of the forces of production, Habermas insists on the need for a practical enlightenment and the formation of a new conception of the good life. And, against the early Frankfurt School, Habermas appreciates the sedimented norms of bourgeois society so that the gap between capitalism and socialism does not have to be so radical as to be impossible. *Habermas, in short, asserts the enlightenment heritage of practical reason against its legacy of technical reason in order to achieve the fulfillment of the former.*

The second version of radicalism we are examining, as it tends to view modernity as corrupt *in toto*, cannot locate points within the modern social order on which to base itself. It must, of necessity, seek values completely outside of the existing "reality principle" in terms of which to legitimate itself and to oppose the established order. Rather than understanding the values of the bourgeois order of reason existing in an irrational form, this position views those values simply as the rationalizations for a system of domination. We are all familiar with contemporary attempts to ground radicalism in myth, art, fantasy, eroticism, play, and even insanity, all of which, more or less self-consciously, derive from the sort of assumptions we are discussing. And some of our most creative cultural commentary and artistic productions have undeniably come from this sort of avant gardism. However, the attempt to uncover the *promesse de bonheur* in marginal phenomena has the unfortunate tendency of degenerating into a position which identifies marginality with progressiveness, in which case only the spiritual or material lumpen can become the agent of enlightenment.

In the 1960s one was more readily disposed to see the Enlightenment as the root cause of the contemporary social crisis and to recognize progressive tendencies in a much wider variety of cultural phenomena. The events of the intervening years, however, must have prompted every thinking person to reflect on his attitude toward the heritage of the West. Recent events in Iran, Cambodia, and Jonestown, to mention but a few, have dramatized the frailty of that tradition, the amount of resentment against it, and the very real danger of massive historical regression. Moreover, many of the cultural phenomena which seemed so progressive in the past have subsequently revealed their reactionary undersides. In this context, a program such as Habermas', which envisions the fulfillment of that tradition, despite its many difficulties appears as the appropriate item on the historical agenda. And yet, at the same time, one is plagued by the sense that the vision, however laudable, is inadequate to historical reality. One's "partiality for reason" notwithstanding, it is often difficult at the level of political reality to discern the cunning of reason at work in the historical events we witness daily. It may be that the scope and depth of the social and ecological crisis are so great that nothing short of an epochal transformation on the scale of world views will be commensurate with them. While talk of an epochal transformation may sound grandiose, it must be remembered that they have occurred before—for example, with the emergence of the Greek *polis*, the rise of Christianity, and the Reformation—and, however unsettling the prospect may be, we cannot rule out the possibility that we are in the midst of such a transformation today.

Notes

1. See Joel Whitebook, *Perversion and Utopia: A Study in Psychoanalysis and Critical Theory* (Cambridge, Mass.: The MIT Press, 1995).

2. Habermas replied to my arguments in Jürgen Habermas, "A Reply to My Critics," in *Habermas: Critical Debates*, ed. John B. Thompson and David Held (Cambridge, Mass.: The MIT Press, 1982), pp. 238ff.

3. Of the two, Bookchin had by far the greater impact on my development. Politically, he helped me to extricate myself from the authoritarian and violent deterioration of the New Left and to develop an appreciation of the importance of the ecological crisis.

4. One notable exception has been a recent paper given by Peter Dews in the Philosophy Department of the New School for Social Research on 3 February, 1994.

5. See Axel Honneth, "Communication and Reconciliation," *Telos* 39 (Spring 1979): 45–61.

6. Max Horkheimer and Theodor W. Adorno, *Dialectic of Enlightenment*, trans. John Cumming (New York: The Seabury Press, 1972), p. 57.

7. According to Horkheimer and Adorno, however, myth already contains the dialectic of enlightenment *in nuce* and the Enlightenment notion of rationality is not without mythical elements.

8. William Leiss attempts to defend Marcuse on the criticisms that have been raised against his notion of a new science. However, as D'Amico has pointed out, Leiss's defense tends to be more textual than substantive. That is, by the time Leiss is through reconstructing Marcuse's position on the necessity of a new science, it appears indistinguishable from Habermas's so that Leiss never takes up the substantive questions about a new cognitive mediation between society and nature in an adequate fashion. See "Appendix," in *The Domination of Nature* (New York: George Braziller, 1972); and Robert D'Amico's review of that work in *Telos* 15 (Spring 1973). Murray Bookchin, on the other hand, argues that we already have the makings of a new science in ecology. However, while his argument is extremely suggestive, Bookchin has yet to show that his position can avoid the theoretical difficulties that confront such an approach and which are discussed below. See Murray Bookchin, "Toward an Ecological Philosophy," in *Philosophica: Ecology and Philosophy*, ed. F. L. Van Damme, 13 (1974).

9. Albrecht Wellmer observes that, their criticism of Marx notwithstanding, Adorno and Horkheimer were the heirs of his monism: ". . . . the latent reductionism . . . of Marx's philosophy of history has survived in the philosophy of the Frankfurt School, although, as it were with inverted signs. This comes to the fore, I believe, in Adorno and Horkheimer's *Dialectic of Enlightenment*. For here 'instrumental reason' becomes the category by which both dimensions of the world-historical process of civilization are conceived, namely the transformation of external nature (technology, industry, domination of nature) as well as the transformation of internal nature, individuation, repression, forms of social domination)." Wellmer, "Communication and Emancipation: Reflections on the Linguistic Turn in Critical Theory," in *On Critical Theory*, ed. John O'Neill (New York: The Seabury Press, 1976), p. 245.

10. See Karl-Otto Apel, "The A Priori of Communication and the Foundation of the Humanities," in *Man and World* 5 (1972).

11. Horkheimer and Adorno, *Dialectic of Enlightenment*, p. 33.

12. See Serge Moscivici, "The Reenchantment of the World," in *Beyond the Crisis*, ed. Norman Birnbaum (New York: Oxford University Press, 1977).

13. Both thinkers want to (1) criticize the positivism of their day, in large part (2) to avoid the skepticism that inevitably accompanies it. Where Kant had the classical British empiricists as the object of his critique, Habermas concerns himself with the contemporary varieties of scientism (including Marxist), especially as they affect our conception of the social sciences. The transcendental critique of positivism, in both cases, proceeds through (3) a *reflection* on the conditions of the possibility of knowledge which is meant to serve the dual purpose of (4) exposing the dogmatically held assumptions of positivism and of validating the fundamental principles of our knowledge in the various epistemological regions. Ultimately, Habermas's transcendental reflec-

tion, like Kant's, seeks to determine the objects of possible experience in order to establish the scope and validity of our knowledge.

14. Jürgen Habermas, "Some Difficulties in the Attempt to Link Theory and Praxis," in *Theory and Practice*, trans. John Viertel (Boston: Beacon Press, 1973), p. 14.

15. It is interesting to note that, evolutionarily, the very fact which made human intelligence and language physiologically possible, larger brains and heads, also necessitated the extended period of dependency which makes the acquisition of culture possible: "Language and culture . . . select for bigger heads. Bigger heads mean greater difficulty in parturition. Even today, the head is the chief troublemaker in child birth. The difficulty can be combatted to some extent by expelling the fetus relatively early in its development. There was therefore a selection for such early expulsion. But this, in turn, makes for a longer period of helpless infancy—which is, at the same time, a period of maxium plasticity, during which the child can aquire the complex extra-genetic heritage of the community." Quoted in Theodosius Dobzhansky, "Evolution and Transcendence," in *The Problem of Evolution: A Study of the Philosophical Repercussions of Evolutionary Science*, ed. John Deely and Raymond Nogar (Englewood Cliffs, N.J.: Prentice-Hall, 1973), p. 98.

16. Jürgen Habermas, *Knowledge and Human Interests*, trans. Jeremy Shapiro (Boston: Beacon Press, 1971), p. 314. See also C. H. Waddington, *The Ethical Animal* (Chicago: University of Chicago Press, 1960).

17. See Jürgen Habermas, "On Systematically Distorted Communication," and "Towards a Theory of Communicative Competence," in *Inquiry* 13 (1970).

18. Wellmer, "Communication and Emancipation," p. 251.

19. Habermas, *Knowledge and Human Interests*, p. 26.

20. See Habermas, "Appendix," in *Knowledge and Human Interests*, op cit.; and Theodor W. Adorno, "Metacritique of Epistemology," *Telos* 38 (Winter 1978–1979).

21. See Habermas, "Some Difficulties in the Attempt to Link Theory and Praxis," note, p. 285.

22. Habermas, *Knowledge and Human Interests*, p. 312, emphasis added.

23. See Habermas, "Some Difficulties in the Attempt to Link Theory and Praxis," pp. 21–22.

24. See Habermas, "A Postscript to *Knowledge and Human Interests*," in *Philosophy of the Social Sciences* 3 (1975): 164–165.

25. See Habermas, *Knowledge and Human Interests*, p. 268; and Thomas McCarthy, *The Critical Theory of Jürgen Habermas* (Cambridge, Mass.: 1978), pp. 115ff.

26. McCarthy, *The Critical Theory of Jürgen Habermas*, p. 125.

27. See Habermas, *Knowledge and Human Interests*, p. 41.

28. Habermas, "A Postscript to *Knowledge and Human Interests*," p. 165.

29. While I have not considered Habermas's more recent theory of "reconstructive science," which has tended to replace transcendental anthropology as his foundational theory, the same sort of objections could be raised against it as well. See McCarthy, *The Critical Theory of Jürgen Habermas*, pp. 278–279.

30. Habermas, "A Postscript to *Knowledge and Human Interests*," p. 165. Transcendental philosophy's goal of providing validation of our fundamental epistemological principles, as Hegel so brilliantly saw, seems to involve the dilemma of requiring, as he put it, that one be able to swim before entering the water. The act of claiming validity for epistemological principles is already a claim to knowledge, albeit of a very peculiar sort. As such, its own standards must themselves already have been validated which is, in principle, impossible. At the heart of every transcendental philosophy in the strict sense there must therefore be a theoretical device to overcome this dilemma. Thus, for example, Kant's transcendental deduction is meant to be a unique form of argumentation

which can avoid the problem of self-reference. And, Husserl's phenomenological *epoche* is designed as a propaedeutic to place one in a privileged position from which basic epistemological principles can be clarified. There exists no such theoretical device in Habermas; he seems to want the advantages of transcendental philosophy without confronting the dilemmas.

31. Habermas, *Knowledge and Human Interests*, p. 9.

32. Ibid., p. 311.

33. McCarthy, *The Critical Theory of Jürgen Habermas*, pp. 109–110.

34. "The Conflicts of Our Time and the Problem of Political Ethics," in *From Contract to Community: Political Theory at the Crossroads*, ed. Fred Dallmayr (New York: Marcel Dekker, Inc., 1978), pp. 81–102.

35. See Hans Jonas, "Technology and Responsibility: Reflections on the New Tasks of Ethics," in *Philosophical Essays: From Ancient Creed to Technological Man* (Englewood Cliffs, N.J.: Prentice-Hall, 1974).

36. Jonas, "Technology and Responsibility," pp. 10–11.

37. Apel, "The Conflicts of Our Time and the Problem of Political Ethics," pp. 3–4.

38. By introducing a dualistic framework to overcome the shortcomings of his predecessors Habermas is following Lukács earlier in the century. Whereas Habermas introduces his dualistic framework to correct the monism of Horkheimer and Adorno, Lukács introduced his to correct the monism of Engels and the Second International. In both cases the goal is to "save the subject." Despite their intentions, Horkheimer and Adorno did not have the categories needed to formulate an adequate notion of subjectivity and Habermas introduces the distinction between instrumental and communicative rationality for that reason. Likewise, Lukács, who was concerned to save the subject in the face of the "reification of consciousness"—a reification which ran counter to the expectations of Marxian theory, criticized the monism of Engels's dialectics of nature. Lukács, who came out of neo-Kantian milieu, argued that dialectics did not pertain to the realm of nature, but only to the realm of history. Nature could be properly treated by natural science. Marxism, insofar as it treated subjectivity as an epiphenomenon of the quasi-natural *(Naturwüchsig)* process of production, Lukács argued, could not adequately deal with the problem of the reification of consciousness. See Georg Lukács, "Reification and the Consciousness of the Proletariat," in *History and Class Consciousness*, trans. Rodney Livingstone (Cambridge, Mass.: The MIT Press, 1971); Alfred Schmidt, *The Concept of Nature in Marx*, trans. Ben Fowkes (London: New Left Books, 1971), pp. 51ff.; and Andrew Arato, "Lukács's Theory of Reification," *Telos* 11 (1972): 41ff.

39. See Alfred Schmidt, "Appendix," in *The Concept of Nature in Marx*, op cit.

40. Habermas, *Knowledge and Human Interests*, pp. 32–33. See also "Technology and Science as 'Ideology,'" in *Toward a Rational Society: Student Protest, Science and Politics*, trans. Jeremy Shapiro (Boston: Beacon Press, 1970), pp. 85ff. While I shall continue to use the term "resurrection of nature," only because it has already entered the literature, it *prejudices the discussion*. It prejudges the issues whether we can *rationally* get beyond Galilean science by insinuating that all such attempts are necessarily quasi-religious.

41. Habermas, "Technology and Science as 'Ideology,'" p. 87.

42. Murray Bookchin makes an even stronger claim: he argues that not just biology, but the science of ecology itself constitutes a major anomaly for Galilean science. The generalizing tendencies of deductive-nomological science will always violate the specificity of "eco-systems," each of which is a unique entity. It is therefore necessary, to use Habermas's language, to enter into a quasi-communicative relationship

with each of them. See Murray Bookchin, "Ecology and Revolutionary Thought," in *Post-Scarcity Anarchism* (San Francisco: Ramparts Press, 1971).

43. E. A. Burtt, *The Metaphysical Foundations of Modern Science* (Garden City, N.Y.: Humanities Press, 1954), p. 33.

44. See Immanuel Kant, *The Critique of Judgement*, trans. James Meredith (Oxford: Oxford University Press, 1973), pp. 18ff.

45. See Hegel, *Logic*, trans. William Wallace (Oxford: Oxford University Press, 1975), p. 88; *History of Philosophy*, vol. 3, ed. and trans. E. S. Haldane and Frances Simson (New York, 1968), pp. 467–468; and *Aesthetics*, vol. 1, trans. T. M. Knox (Oxford: Oxford University Press, 1975), pp. 57–78.

46. See Ernst Cassirer, *The Problems of Knowledge: Philosophy, Science and History since Hegel*, trans. William Woglom and Charles Hendel (New Haven, Conn.: Yale University Press, 1974), pp. 188ff.

47. Ronald Munson, "General Introduction," in *Man and Nature: Philosophical Issues in Biology*, ed. Ronald Munson (New York: Delta Books, 1975), p. 8.

48. As one eminent biologist has noted: "What finally produced a breakthrough in our thinking about teleology was the introduction of new concepts from the fields of cybernetics and new terminologies from the language of information theory." See Ernst Mayr, "Teleological and Teleonomic: A New Analysis," in *Evolution and the Diversity of Life* (Cambridge, Mass.: Harvard University Press, 1976).

49. The *locus classicus* of this position in Arturo Rosenbleuth, Norbert Wiener, and Julian Bigelow, "Behavior, Purpose and Teleology," *Philosophy of Science* 10 (1943).

50. This attitude is evident in Ernst Mayr's suggestion that we should, in light of current knowledge, return to Aristotle's biological texts and replace the term *eidos*, wherever it appears, by the term "DNA formula." "Teleological and Teleonomic," p. 400.

51. Ludwig von Bertalanffy, *General System Theory* (New York: George Braziller, 1968), pp. 144–156.

52. See Ervin Laszlo, *Introduction to Systems Philosophy: Toward a New Paradigm of Contemporary Thought* (New York: Gordon and Breach Science Pubs., 1972), p. 49. Cf. Marjorie Grene, "Aristotle and Modern Biology," in *Topics in the Philosophy of Biology*, ed. Marjorie Grene and Everett Mendelson (Boston, 1976), p. 8: "What is finalistic about information theory? Admittedly, both teleological explanation and cybernetic information both resist a one-level Democritean approach. But that does not make them equivalent."

53. See Hans Jonas, "Is God a Mathematician? (The Meaning of Metabolism)," in *The Phenomenon of Life: Toward a Philosophical Biology* (New York: Delta, 1968).

54. Habermas, *Knowledge and Human Interests*, p. 33.

55. See Richard Bernstein, *Praxis and Action* (Philadelphia: University of Pennsylvania Press, 1971), Chapter 3.

56. See Jürgen Habermas, *Zur Logik der Sozialwissenschaften* (Frankfurt: Surkamp Verlag, 1970), pp. 162ff.; and McCarthy, *The Critical Theory of Jürgen Habermas*, pp. 151ff.

57. Boris Frankel, "Habermas Talking: An Interview," *Theory and Society* I (1974): 44–45.

58. Habermas, "Technology and Science as 'Ideology.' " p. 95

59. Ibid., p. 96.

60. Ibid., p. 87.

61. See Albrecht Wellmer, *Critical Theory of Society*, trans. John Cumming (New York: The Seabury Press, 1971), pp. 80ff. and 99ff.

62. Ibid., p. 99.

63. See Jürgen Habermas, *Legitimation Crisis*, trans. Thomas McCarthy (Boston: Beacon Press, 1975), p. 105.

64. See Hegel, *Philosophy of Right*, remark to Paragraph 93, remark to Paragraph 139, addition to Paragraph 94, and addition to Paragraph 158; and Hannah Arendt, *On Revolution* (New York: Penguin Books, 1965), pp. 158ff. and 205ff.

65. Jeremy Shapiro, "The Slime of History," in *On Critical Theory*, ed. John O'Neill, p. 157.

66. Habermas, "Technology and Science as 'Ideology,' " p. 90.

67. See Herbert Marcuse, Jürgen Habermas, Heinz Lubasz, and Telman Spengler, "Theory and Politics: A Discussion," *Telos* 38 (Winter 1978–1979).

68. Shapiro, "The Slime of History," pp. 149–150, emphasis added.

69. See Murray Bookchin, *The Limits of the City* (New York: Harper and Row, 1974), Chapter 3; and "The Myth of City Planning," *Liberation* 18 (Sept.–Oct., 1973).

70. See Walter Benjamin, "Theologico-Politico Fragment," in *Reflections*, trans. Edmund Jepheatt and ed. Peter Demetz (New York: Harcourt Brace Jovanovich, 1978).

71. Wellmer, "Communication and Emancipation," p. 245.

Chapter Thirteen

Social Ecology and Social Labor: A Consideration and Critique of Murray Bookchin

Alan Rudy and Andrew Light

Introduction

Murray Bookchin has been an important figure in social movements and ecological thought for the past forty years, most notably as the leading proponent of social ecology for three decades. Additionally, he has been much more than an activist; Bookchin is a theorist who grounds social ecology and anarchocommunism in scholarly traditions ranging from evolutionary biology to political philosophy and ecology. During the past decade, his works have been increasingly widely read by people in the ecology movement and academia.

This paper is a critique of social ecology and Bookchin's form of political anarchism out of an emerging socialist ecological tradition, something Bookchin once claimed was a contradiction.[1] We begin with an exposition of some of Bookchin's central ideas: his theory of the development of domination in the early evolution of human society, the importance of the transition to capitalism for human society, and his exposition of a new future for modern society rooted in his Kropotkin-inspired evolutionary anarchism. The next section criticizes this work. Our central critique is that Bookchin powerfully underplays the importance of labor as a mediating force within and between the social relations of humans, and within and between humans and the nonhuman natural world. This critique is advanced in relation to some of Bookchin's positions relative to capitalism, most notably that Bookchin fails to see the importance of the qualitative change evidenced by the development of social labor under and through capitalism's uneven and combined development. The primary consequences of Bookchin's neglect of labor as a category of analysis, and of social labor as a defining characteristic of capitalism and its contradictions, are that Bookchin's natural histories are incomplete and produce a problematic analysis of historical change. These problems derive from the theoretical rigidity of his political project, which results in a skewed interpretation of material, global, and political problems. Prior to beginning, some brief biographical notes on Bookchin are in order.

Well before the development of a public environmental consciousness, Bookchin wrote about the social, psychological, and health consequences of urbanization, the use of industrial chemicals in food production, and a variety of other antiecological consequences of modern industrial society. Some key texts were *Our Synthetic Environment* (1962)[2] and *Crisis in Our Cities* (1965),[3] both published under the pseudonym Lewis Herber.[4] These books represent the initial development of Bookchin's ecological and anarchist perspectives during his involvement in a variety of movements and ecological debates in the 1950s and early 1960s. Since the mid-1960s, Bookchin has written widely on ecological and social issues and the ways each interpenetrates the other. He has published theoretical works and debates with activists, including influential collections of articles in 1971, 1980, and 1986.[5]

In 1982 Bookchin published perhaps his best-known book, *The Ecology of Freedom*,[6] which had taken more than a decade to write and research. This book has been widely read in the ecology movement by both theorists and practitioners. It is here that Bookchin most extensively develops his view that the social domination of human by human leads to ideas about, and practices designed to produce, the domination of nature, and, then, to the destruction of nature. Social domination and attempts to dominate nature then degrade communal and ecological evolution and reproduction.

Bookchin's social ecology takes the position that evolution is primarily grounded in spontaneous mutualism, and that fecund, diverse, and harmonious ecologies existed before the introduction and spread of social relations of domination. In prehistory, the emergence of relations of domination within human societies proved destructive to naturally evolved, spontaneous, and mutualistic human activity, and led to relations with nature predicated on hierarchical ideas derived from the practices of social domination. The expansion of institutionalized domination culminated in the modern nation-state, capitalism, and the present ecological crisis.[7]

Actions predicated on belief in the possibility of dominating nature result in the destruction of ecological and social spontaneity, which is the predicate of flexible evolution. This novel perspective has made Bookchin one of the most widely read ecological thinkers in the past thirty years, linking modern social relations of domination with global ecological destruction. His approach has been developed further in *The Rise of Urbanization and the Decline of Citizenship*,[8] summarized in *Remaking Society*,[9] and extended theoretically in *The Philosophy of Social Ecology*.[10]

Bookchin is regarded by many activists as important not only for his emphasis on the interaction between deep-seated social relations and their effects on the external, "natural" world, but also for his utopian visions for the future. His concerns encompass far more than those of either liberal environmentalists or traditional leftists. His libertarian vision of confederalized, independent, human-scale, and self-reliant communities has been attractive to many. His critique of domination, which includes theories of technology, politics, and social conflict, strikes a deep chord among progressive and radical ecologists in many countries, including the West German Greens and the Left-Green Network in the United States.

Additionally, Bookchin has offered significant critiques of the Left, as well as nonleft strands of ecological and environmental thought and activism. His broadsides against vulgar Marxism have helped force Marxists to address ecological theory. His exchanges with deep ecologists have created lively (and often acrimonious) debate within the environmental movement, especially regarding his insistence that deep ecologist address the social roots of ecological destruction. For these reasons and more, including his support of feminism, gay rights, and struggles against racism, Bookchin's contributions to left ecology must not be understated. However, while his work has been and remains important to the development of left ecology, his theory (and, too often, his polemical style) have serious shortcomings, which this paper is intended to address.

The Evolution of the Idea of Domination

Bookchin's social ecology posits a spontaneous and teleological evolution of matter towards increasing complexity and consciousness. "The universe bears witness to an ever-striving, *developing*—not merely 'moving'—substance, whose most dynamic and creative attribute is its ceaseless capacity for self-organization into increasingly complex forms. Natural fecundity originates primarily from growth, not from spatial 'changes' in location."[11]

This fecund self-organized growth of planetary life, the atmosphere, and the land and sea is based on symbiosis,[12] or what might be called evolutionary cooperation. Following and adapting the "mutual aid" thesis of the nineteenth-century philosophical anarchist Peter Kropotkin, Bookchin claims, "Mutualism, not predation, seems to have been the guiding principle for the evolution of [the] highly complex aerobic life forms that are common today."[13] Although he does not say that predation, competition, and adaptation to environmental change do not occur he very strongly privileges mutualism and symbiosis as central evolutionary forces.

While this teleological approach is not determinate,[14] Bookchin does see nature's evolution generating "its own natural philosophy and ethics."[15] Healthy ecological and social differentiation is possible only under conditions of natural and social "spontaneous development" which "unfold and actualize [universal nature's] wealth of possibilities."[16] Social ecologists do not look "upon (nature) as a necessitarian, withholding, or 'stingy' redoubt of blind 'cruelty' and harsh determinism, but rather as the wellspring for social and natural differentiation."[17]

For social ecology, the link between the evolution of external and social nature is profound. The "very natural processes that operate in animal and plant evolution along the symbiotic lines of participation and differentiation reappear as social processes in human evolution, albeit with their own distinctive traits, qualities and gradations or phases of development."[18] As humans and human societies emerge, Bookchin believes that "it is the logic of differentiation that makes it possible to relate the mediations of nature and society into a continuum."[19] For him, what "makes unity and diversity in nature more than a suggestive ecological metaphor for unity in diversity in society, is the underlying fact of wholeness,"[20] which represents the universal self-organizing and self-maintaining attributes of all matter.

Though a continuation of natural evolution, human societies are quite different from the animal communities from which they evolved. For Bookchin, animal "communities are not societies . . . they do not form those uniquely human contrivances we call institutions . . . [they have] genetic rigidity . . . not contrived rigidity."[21] The initial evolution of human institutions is seen by Bookchin to have generated "organic," preliterate societies in which internal social relations and relations with the external world were organized around mutualistic practices supporting social and ecological differentiation. Historically, the conditions within which these societies existed were disturbed by nascent, increasingly institutionalized social relations of domination and hierarchy. These relations of hierarchy and domination are inherently *social* and

> must be viewed as *institutionalized* relationships, relationships that living things literally institute or create but which are neither ruthlessly fixed by instinct on the one hand nor idiosyncratic on the other. By this, I mean that they must comprise a clearly *social* structure of coercive and privileged ranks that exist apart from the idiosyncratic individuals who seem to be dominant within a given community, a hierarchy that is guided by a social logic that goes beyond individual interactions or inborn patterns of behavior.[22]

As components of institutions, hierarchy and domination can be found only in human societies and, for Bookchin, cannot be said to exist in animal communities. These institutions are defined by Bookchin to represent complex "cultural, traditional and psychological systems of obedience and command . . . in which elites enjoy varying degrees of control over their subordinates without necessarily exploiting them."[23]

As a coercive social relation, domination is seen to work against spontaneity and, unlike social relations in "organic" societies, to counter the "natural processes" of evolution. Given Bookchin's distinction between animal and plant communities and human societies, domination becomes *the* primary destructive relationship within society in that it destroys social "participation and differentiation."[24]

The ecologically destructive character of hierarchy and domination emerge from ideologies of the domination of nature, which themselves spring from the real domination of human by human. Social domination begins, then, with the destruction of naturally spontaneous social relations within "organic" societies. Domination diminishes the variety of the natural world and imposes on it a regression to its natural state that ultimately may be life-threatening.[25] In Bookchin's scheme, the history of social and natural evolution is that of two competing logics: the logic of spontaneous, mutualistic, ecological differentiation, and the logic of domination.

Bookchin's historical work explores how these two logics work themselves out as spontaneous organic societies are transformed into cities, city-states, nations, nation-states, and capitalist political economies increasingly organized through domination and hierarchy. History, under Bookchin's precepts, becomes largely the story of the battle between communities committed to freedom and elites committed to domination.[26]

Modern Capitalism

Bookchin's historical thesis tracing this division through time is extraordinary. Important for us here is Bookchin's account of the advent and effect of capitalism in history. By the seventeenth century, Bookchin finds Europe at a juncture where "development" (not to be conceived of as "progress") becomes dependent on the expansion of technology more than on the growth of a centralized state.[27] For the first time in Bookchin's analysis, developments in "the instruments of production" become more important to the expansion of domination than developments in "the instruments of administration." This emphasis on machinery is part of Bookchin's belief that capitalism "did not 'evolve' out of the feudal era, [it] literally exploded into being in Europe, particularly in England, during the eighteenth and especially nineteenth centuries, although [capitalism] existed in the ancient world, the Middle Ages, and with growing significance in the mixed economy of the West from the fourteenth century up to the seventeenth."[28] The explosive expansion of domination from the political to the economic realms of European society, in large part by way of mechanical technological innovation, is considered by Bookchin to be metastatic.

For Bookchin, the capitalist conversion of human beings and nature into commodities[29] is destroying society.[30] Society is experiencing the "social dialectic and the contradictions of capitalism" as they expand from economic to hierarchical realms of society.[31] These expanding forms of domination shape social life so that the "suffocating impersonality" of modern market-oriented life produces nothing that can serve as a normative guide for social interaction.[32] The combination of the domination of the nation-state—which "makes us less than human . . . towers over us, cajoles us, disempowers us, bilks us of our substance, humiliates us—and often kills us in its imperial adventures"[33]—and capitalism—which "is unique in contrast with other societies in that it places no limits on growth and egotism"[34]—has destroyed all but the very last vestiges of "organic" society. For Bookchin, the nation-state and capitalism are simplifying both human social life and the natural world to the detriment of both.[35] This simplification of society and ecology is inseparable from capitalist technologies and "resources of abundance" which "reinforce the monopolistic, centralistic and bureaucratic tendencies in the political apparatus. In short, they furnish the state with historically unprecedented means for manipulating and mobilizing the entire environment of life—and for perpetuating hierarchy, exploitation and unfreedom."[36]

Despite all this, the "foremost contradiction of capitalism today is the tension between what-is and what-could-be—between the actuality of domination and the potentiality of freedom."[37] Modern technology, "by its continual development . . . to expand local possibilities"[38] has the potential to reinforce the "realm of freedom." Initially, Bookchin's hopes for modern technology were initially extraordinarily high.[39] The revolutionary resolution of the apocalyptic consequences of capitalism was seen to arise from modern technology, producing a "a turning point . . . that negates all the values held by mankind throughout all previous recorded history."[40] This techno-social juncture was seen to provide the ground for the resolution of the historical conflict between the "realm of freedom" and the "realm of domination," to allow for a new, ethical reconstruction of society.[41]

While he has maintained a messianic vision of the future,[42] Bookchin's more recent ecotopic visions have become increasingly low-technology affairs.[43] In the early 1980s, his view of technology had evolved to the point where his concerns were focused on "how we can contain (that is absorb) technics within an emancipatory society."[44] In 1986, in his introduction to the second edition of *Post-Scarcity Anarchism*, Bookchin wrote that, if he were to rewrite the book, he would "temper the importance [he gave] to the technological 'preconditions' for freedom."[45] Similarly, his perspective on scarcity, viewed "as a drama of history that our era has resolved technologically," has changed to the point "that such an interpretation is now unsatisfactory."[46] Most recently, Bookchin has said that what must be overcome is not the contradiction between the modern potential for post-scarcity and its lack of realization but rather the "gravest most single illness of our time . . . disempowerment."[47]

As Bookchin's views on technology have changed, so have his views on revolutions in developed countries and on the political tactics demanded by the modern situation. Bookchin once held that the "problems of necessity and survival . . . cease to require any theoretical mediation, 'transitional' stages, or centralized organizations to bridge the gap between the existing and the possible,"[48] and that in the past, "nearly all the major revolutionary upheavals began spontaneously . . . [and] from *below*."[49] His position has become very different. Whereas he would likely still insist that a social ecological revolution is an "absolute necessity," the achievement of such a goal "can no longer be seen as a sudden 'revolution' that within a brief span of time will replace the present society with a radically new one. Actually, such revolutions never really happened in history as the litany of their failure so dramatically reveals."[50]

However, while the centrality of technology and specific political techniques for revolutionizing society have evolved in Bookchin's thought, his position on the agents of social transformation has not changed. As in the 1960s, when he wrote that the making of a revolutionary population would come from class decomposition and the "emergence of an entirely new class, *whose very essence is that it is a non-class*,"[51] he has recently written of the "reemergence of 'the People,' in contrast to the decline of 'the Proletariat.' "[52] For Bookchin, such a constituency has produced "entirely new issues, modes of struggle and forms of organization and calls for an entirely new approach to theory and praxis"[53] than those of any other revolutionary theory, but as in the past, such a movement organized around liberatory struggles continues to require "a radical intelligentsia (in no way to be confused with 'intellectuals' and academics) to catalyze revolutionary change."[54] The "non-class" or "transclass constituency"[55] is constructed in opposition to what Bookchin calls Marxian (specifically Lenin's and Trotsky's) "scientific socialism." In his formulation, political and cultural ties bind, rather than economic unions.

Another aspect of Bookchin's writing that has not changed is that the intelligentsia and transclass constituencies necessary for truly liberatory social movements are situated almost exclusively in the North. For Bookchin, what "the Third World needs is a revolution in America, not isolated sects that are incapable of affecting the course of events."[56] In the 1960s he wrote:

> The attempt to describe Marx's immiseration theory in international terms instead of national (as Marx did) is sheer subterfuge. In the first place, this theoretical legerdemain simply tries to sidestep the question of why immiseration has not occurred within the industrial strongholds of capitalism, the only areas which form a technologically adequate point of departure for a

> classless society. If we are to pin our hopes on the colonial world as "the proletariat," this position conceals a very real danger: genocide. America and her recent ally Russia have all the technical means to bomb the underdeveloped world into submission.[57]

More recently, Bookchin's position on Third World revolutions has focused on the "material underdevelopment"[58] and limited history of "republican institutions"[59] in the South.

> "Third World revolutions" are nationalist rebellions, not "patriotic" revolutions that speak for humanity. One cannot fault them for their parochial qualities. They have lived under the lash of alien exploiters for centuries. . . . Whatever good they bring their own people, their solidarity is local or regional, not international or guided by a world vision of freedom. . . . Their economic emphasis . . . speaks to real material needs—but it permits them to elude a respect for person, freedom to dissent, and a genuine sense of fraternity and happiness based on human solidarity rather than on class interest.[60]

Additionally, he writes that

> the "law" of uneven and combined development makes it possible for people who used bows and arrows a decade ago to leap "forward" to automatic rifles and shoulder rockets without undergoing the intervening evolution of arms which led from one group of weapons to another . . . [and to] leap directly from tribal and village cultures to viciously totalitarian and technocratic cultures. The political freedom and republican institutions which might restrain these new governments, as they restrain those of their western would-be masters, are never allowed time to develop.[61]

In Bookchin's analysis, perhaps the only thing that has kept the United States from militarily annihilating Third World revolutionary movements is "its [the United States's] own institutional structure, the burdens—ideological and political—of its own great Revolution."[62]

In opposition to what he sees as the overly material orientation of Third World revolutions, Bookchin emphasizes the "utopian side of the American Dream"[63] when speaking about the American Revolution. Bookchin attributes to immigrant socialists and syndicalist anarchists the error of moving away from this utopianism within the Left in the United States. Bookchin has repeatedly chastised these immigrants for being "largely unionists rather than revolutionary utopians."[64] For him, their error lay in the focus on material conditions without making appeals to utopian visions and a transclass constituency. It is in this context that Bookchin most emphatically differentiates himself from Marxists. Not only is the Third World not a locus for truly liberatory revolutions, but for Bookchin, the "very class nature of the proletariat, its existence as a creature of a national division of labor and its highly particularistic interests that rarely rise to the level of general interest, belie Marx's claims for its universality and its historic role as a revolutionary agent."[65]

Bookchin believes that the majority of the proletariat are complicit in their own subjugation and that their political activities materially support the system of capitalist domination. The proletariat simply takes part in the larger social system's rejection of the transclass revolutionaries of the new social movements (such as feminists, gays, and environmentalists).[66] Bookchin writes that Marx gave to proletarians "no personal will but only an historical one."[67] In his presentation, the key statement by Marx on proletarian agency is found in *The Holy Family*, where Marx suggests that the key questions do not concern individual workers but the proletariat as a class: what it is and what it is compelled historically to do.[68]

For Bookchin, an emphasis on the individual and group agency of a transclass constituency vastly alters politics. Revolution can be anchored "in sexual relations, the family, community, education and the fostering of a truly revolutionary sensibility and ethics,"[69] but in his view Marxist approaches preclude this. There "can be no Marxian theory of the family, of feminism, or of ecology because Marx negates the issues they raise, or worse, transmutes them into economic ones.[70] There is a "reactionary aspect" to the Marxist and socialist projects in that they retain "the concepts of hierarchy, authority and the state as part of humanities' post-revolutionary future."[71] As such, Bookchin believes that Marxism "is not revolutionary enough" and the "problem is not to 'abandon' Marxism or to 'annul' it, but to transcend it dialectically."[72]

Foremost among the steps necessary for the reconstitution of organic society and nature is, for Bookchin, a rebuilding of local human association and the reconstruction of human beings within a libertarian political ecocommunity. Reempowering communities and people demands the abolition of the state, and the nation-state in particular. Politically, social ecology emphasizes decentralization and the return to the local level of the "resources and . . . potential for development" within society,[73] though Bookchin believes that these communities should be linked in confederations.

For him, there "can be no politics without community"[74] and democratic politics, as distinguished from logistical administration, is the social relation by which ecological decisionmaking is achieved. Bookchin envisions human-scale communities encompassing a political and productive division of labor which is participatory and skilled. The community makes its own decisions for which he advocates a strong form of direct democracy.

Programatically, Bookchin arrives at four basic principles for the regeneration of society and ecology. These coordinating principles focus on (1) "the revival of the *citizens assembly*,"[75] (2) the confederation of assemblies,[76] (3) the construction of communal and confederal politics as a "school for genuine citizenship,"[77] and (4) the economic empowerment of communities through the "municipalization of property"[78] and the formulation of communal productive policy by public democracy.[79] For Bookchin, it is only through social institutions such as these that the social and ecological simplification resulting from the centralizing and dominating tendencies of the state, modern market economies, and the technologies they spawn, can be overcome and the conditions for free social and ecological differentiation reestablished.

Social Labor and the Contradictions of Bookchin's Capitalism

Our central criticism of Bookchin's work on captalism and its contradictions is that he does not investigate what Marxists call social labor and, consequently, that his anthropology, history, and politics fall short in their criticism and analysis. Within all modes of production, labor is a social process which encompasses the ways in which human beings actively organize the relations by which they (re)produce themselves day to day within complex, dialectical, historical, and ambiguous social structures. Further, within the socially organized labor of each mode of (re)production is included the (re)production of ethics, culture, gender, politics, economy, art, and geographical and ecological space.

A full treatment of this critique would involve a careful reading of Bookchin's position on the nature of global evolution, his explanation of the origin of organic preliterate societies and hence the origins of domination, and his historical account of domination through human history. Given space limitations, we will restrict our analysis to a limited examination and critique of Bookchin's views on the final stage of this process so far—capitalism. We will use this analysis to critique some aspects of the utopianism and insularity of his political program.

Following a controversial account of the early mutualistic tendencies of human societies, and a history of the destruction of the capacity for mutual aid in cities, Bookchin finds that the global expansion of capitalist economic domination arises out of feudalism and results in economicized social relations. This change derives from a successful move by social elites to dominate what had been more or less organic, free communities. Their domination is accomplished through technical administration and the development and extension of machinery designed to an increasingly nonhuman scale. These processes encourage the tendencies towards centralization and domination of the nation-state and capitalism. Thus, even vestigal legacies of mutual aid and cooperation were all but been destroyed under the nation-state and capitalism.

Let us assume another view, one that sees "freedom" and "domination" as being linked with one another. This is not to say that domination is a necessary precursor to freedom, as Bookchin accuses Marx and Marxists of thinking, but rather that institutionalized social relations have been composed of dialectically intertwined degrees of freedom and domination throughout history. Neither arises from society nor nature. Both evolve as part of the dialectical, contextual, contingent, and largely unintended activities and consequences of human actors within environmentally and socially constrained circumstances.[80]

Specifically, the contradictory evolution of social labor during the rise of capitalism has encouraged both homogenization and differentiation, neither of which is unambiguously suffused with domination or freedom. Thus, our hopes for ecocommunism (a goal we share with Bookchin despite the differences in our theoretical bases) are informed by the material and social freedoms of modern society, unknown to any prior epoch, as well as by the often suffocating exploitation and domination under which we live.

The heart of our critique of Bookchin lies in our understanding of historical change, and therefore our understanding of the processes of future transformations. Social ecology presents history as the ongoing struggle between the realms of freedom and domination. Historical change appears as change in the amount or quantity of institutional relations of domination relative to mutualistic relations of freedom. Thus, the processes by which Bookchin suggests feudal relations in Europe were transformed into capitalist relations are presented as the expansion of institutional political domination in the form of absolutist kings, followed by domination's further expansion into production through technical administration and machinery. In short, the transition from feudalism to capitalism is presented as the quantitative expansion of existing social relations rather than their qualitative transformation.

This may seem unfair, as Bookchin clearly understands the extreme difference between life in feudal cities (which he presents as actively egalitarian and participatory) and life under the capitalist nation-state. However, he also argues that capitalist exploitation is a form of domination (one which existed in the prefeudal Mediterranean and the "mixed economy" of feudal Europe), rather than viewing capitalist exploitation as arising as the central moment of a qualitatively different mode of production.[81] For Bookchin, domination expands from the civil to the political to the economic sphere during the transition from feudalism to capitalism. For Marxists, and perhaps especially for ecological Marxists, a new dimension must be added to this picture. In our account, it is not enough to say that a transition has occurred whose roots are qualitatively the same. Instead, the qualitative transformation of the social forces and relations of production and reproduction (which have characterized the development of mercantilism, manufacture, modern industry, monopoly capitalism, and the present hegemony of international conglomerates) must be recognized in order to understand *why*, as well as *how*, economic, political, and ecological crises occur under capitalism. Consequently, ecological Marxists are developing refined explanatory strategies to make possible a deeper analysis of the varieties of historical change and the development of countervailing struggles.

By understanding capitalism as having generated a qualitatively new social relation, social labor, both the power and contradictions of capitalist political economy can be structurally assessed, socially resisted, and politically overthrown.[82] Recently, these insights have been elaborated to suggest the manner by which the political (and) economic structures and contradictions of capitalist modes and forms of production codetermine the processes by which the destruction of human health, ecological sustainability, and community reproduction occurs under capitalism.[83]

By understanding social labor as a qualitatively new social relation in production, one that conditions reproduction—the means by which people make their own lives under capitalist political economic and environmental conditions not of their own choosing—ecological Marxists possess an analytic tool which Bookchin has thrown away. This tool permits an understanding of how political, economic, and environmental resistance in the capitalist West must be different than other sites of resistance: whether Athenian, feudal, or Third World. Rather than subsuming all resistance to struggles against the same social relation—institutionalized domination—our approach suggests that struggles must be generalized and localized in ways which assess opportunities, contradictions, and consequences in relation to the common and particular moments and places in which resistance is organized. The possibilities for and forms of resistance must be tailored to the characteristics of immediate economic, political, and ecological crises and the ways in which those crises are intertwined. Thus, the combined and uneven development of capitalist political economies, constrained and enabled by local and global conditions of production, is understood in a manner more nuanced and flexible than Bookchin's.

Further, Bookchin's concept of capitalism as an uncontrollable cancer is problematic. A good part of what is wrong with his imagery has to do with his failure to investigate the particularity of capitalist production relations and their generation of a qualitatively different (re)productive relation, social labor. Capitalism must grow in order to survive, much less prosper, but it cannot grow indefinitely. There are many obvious constraints, both structural and contingent, to capital accumulation which can be analyzed so as to understand how to take advantage of the freedoms that the uncertainty of economic, political, and environmental crises make available. Among other ways, these constraints are expressed in the opposition and organization of labor as a class, in associations, and as individuals. In addition, constraints exist in the cyclical valorization crises within particular markets; in the global economy; and most important for this discussion, in the destruction of the conditions for capital's own (re)production: human beings and their reproductive health; global, regional, and local ecosystems; and the organization of communities and social spaces in and through which people interact with their ecosystems.

Bookchin's offhanded assertion that capitalism "is not a social phenomenon but rather an economic one"[84] is clearly false even if meant only as a polemic. The inherently social nature of all human (re)production can be denied only by reifying both society and economy. In fact, Bookchin treats capitalist economics as "the buyer-seller relationship," failing to mention the centrality of production. Instead of analyzing social labor, he emphasizes exchange and consumption.

This follows from Bookchin's sense that "the grand secret from which [the market] draws its power" is "the power of anonymity."[85] In fact, the key to capitalism after mercantilism is not anonymity in the marketplace, but the wage labor that is at the root of the production- and class-based extraction of a social surplus treated as abstract exchange-value. While the capitalist mode of production is inseparable from market exchange and market manipulation, market conditions and relations represent the valorization of capital based on commodity production. The impersonal relations associated with modern market exchanges do not lie at the root of economic crisis, political delegitimation, or ecological destruction, despite their real social-psychological costs.

Without the historical development of "free" labor, there would remain a much more parochial interchange of use-values rather than production of commodities as exchange-values bought and sold through global markets. The grounds for this transition in production were laid by mercantile exchange and the coevolved interests in accumulation of bourgeois and feudal social elites at the expense of an increasingly displaced and dispossessed labor force. The grand ideological secret of capitalism has to do with the reification of exploitation, not the anonymity of the market.[86]

Depending where one looks or at what level level of analysis, capitalist social relations are spontaneously and dynamically evolving more complex forms of production and distribution for accumulation *and* are developing a thoroughly homogenized social milieu in which everything is subjugated to the market. Capitalism is not unambiguously simplifying human social relations, but at the same time and in contradictory ways, evolving a social complexity infused with social potential as well as destructive tendencies.[87]

When Bookchin writes about the modern world, he has a tendency to contrast the domination of the market with libertarian practice within highly autonomous communities. He generally does not discuss capitalist production in terms other than those related to the overwhelming and inherent domination present in the state and the factory as well as their nonhuman scale.[88] He emphasizes the impersonality of the market and the factory and what he views as the total lack of ethical interaction in either location,[89] but he does not look at the ambiguous complexity of site-specific or more general divisions of labor. The myriad cultures and individual acts of resistance within the capitalist workplace, attempting to overthrow and undermine corporate cultures and divisions of labor, are rejected out of hand or, more often, omitted altogether from Bookchin's analysis.

In this context, Bookchin's approach to "technics" is salient. His position (since *The Ecology of Freedom*) that a "liberatory technology presupposes liberatory institutions"[90] again privileges the capacity of good social relations to encompass "adaptive" rather than "innovative" movements in craftlike "instruments of production."[91] Thus, he misses the basic point of Marxist analysis. The division of laborers (the cooperation necessary for production within a particular division of labor involved with specific technologies) is *both* a force and relation of production. Therefore, production in class societies creates an ambiguous intersection of social relations of freedom and domination, exploitation and cooperation, friendship and ageism, love and sexism. As O'Connor has argued:

> Modes of cooperation are not determined by technological developments or . . . power relations alone. The view that work relations are governed by technology gives you productive force determinism. The view that work relations are determined by . . . ideology and power gives you productive relations determinism. The first puts us in the world of Engels and Habermas; the second in the world of Lukács. It seems . . . that these two themes have theoretically played themselves out. A better view is that work relations combine, in contradictory ways, "technical relations," "relations of power," including ideology, and *also* self-organized relations by the direct producers governed by working class concepts of morality or "what is right."[92]

Thus, in both feudal and capitalist societies, the ethical demands of hierarchical organizations, from the most democratic guilds to the most stultifying corporate hierarchies, are mediated and countered by historically conditioned worker moralities and struggles, but Bookchin fails to see these sorts of relationships in his work.

By looking at relations within capitalist production as suffused with working-class morality, Marxists gain access to the working class as a political force and can elaborate a theory of new social movements absent in Bookchin's work.[93] Capitalism becomes a set of social relations in which there is the potential to act in forward-looking and positively self-negating ways. Just as the state has ambiguously served the interests of capital, its citizens, and the environment (though leaning strongly toward instrumental relations with capital especially since the Second World War), capital's contradictory moves towards universalization and differentiation work both to homogenize the global populace and to fill it with revolutionary potential.[94]

Since Bookchin disagrees with this picture, it is important to ask which forces and relations would be needed to move his transclass constituency to gradualist revolutionary activity. During the 1960s, Bookchin's position on the potential within modern technology for the production of postscarcity societies, in combination with an apparently highly dissatisfied population, led him to argue that "all the values held by mankind throughout all previous recorded history" had been negated and he anticipated a spontaneous ecocommunist revolution.[95] Since the publication of *The Ecology of Freedom*, he has grown more skeptical of technology's potential to motivate social change. Having eschewed materialist analysis of the division of social labor under capitalism, Bookchin's utopian philosophy now encounters its greatest problems.[96] As he no longer believes modern technology to have the potential to eliminate scarcity, and since the populations of the advanced democratic and technological countries of the world are not up in arms, it is difficult to ascertain the forces or relations he thinks would move people to support social ecology.

Given his skepticism about the existence of the working class in any meaningful form, the cross-class constituency on which he hopes to build his politics would presumably be flush with the much-discussed new social movements.[97] There is clearly a great deal of political potential in such movements, and the criticisms of determinist Marxism which have arisen out of these movements (including Bookchin's) often have been excellent. However, there remains massive potential within the working classes of both the North and the South and Bookchin's rejection of the proletariat's revolutionary potential (including his misinterpretation of the agency Marx recognized in the proletariat) is only the product of his insistence on domination as the absolute central term in social and ecological revolutionary theory and practice.[98]

Bookchin's focus on the category of domination treats the spatial, social, and economic expansion of the nation-state and capitalism as contradictory to earlier forms of social relations and evolved ecologies, but he does not explore internally contradictory tendencies within the state or capitalism, or in the relation between the two. This is most clearly evidenced in his repeated denigration of the revolutionary potential of the South, or the "Third World." In fact Bookchin takes a highly problematic approach to the uneven development of capitalism.

It is very difficult to know how to take Bookchin's comments about "bows and arrows" and "shoulder rockets" (quoted above). If these references are meant to evoke a sense of the scale of the problem of transition (which is by no means obvious from the text) a more sympathetically applied or historically situated manner of explication would have been less objectionable. If Bookchin believes that these revolutions are driven primarily by members of "tribal and village cultures" inherently unable to build "republican institutions" because revolutions "accomplished in a decade" do not allow these institutions "time to develop," then he clearly is not a scholar of such revolutions.[99]

That Bookchin appears to desire that Third World countries "undergo the intervening evolution of arms," technology, and political institutions which he apparently finds necessary for true revolution, indicates the depth of his misunderstanding of international relations and the combined and uneven development of global capitalism.[100] The problems of attempting to transform the internal and external institutional structure of any country politically, economically, and socially, but particularly those subjugated to the international capital of the superpowers, are all but insuperable. Bookchin lays the blame for the failure of these struggles to produce fully successful, democratically egalitarian social arenas at the feet of the revolutionaries, which is ahistorical, unfair, and unsupported textually. Equally hard to understand is that Bookchin speaks of these struggles as though they occurred in a decade, led by people who recently gave up tribal life, ignoring the fact that in many cases these struggles are theorized and led by educated members of urban areas who proclaim their "international solidarity" and a "world vision of freedom." If Bookchin wants to discuss these revolutions, he desperately needs to give more historical presentation, analysis, and contextualization.

These struggles are for self-determination in the face of a long history of subjugation, deeply tied to capital's ability to displace economic and ecological crisis by the exercise of its geographic mobility. Among other ways, capitalist economic crisis is expressed as crises in society and the environment. For capital, the only way accumulation crises can be successfully displaced or resolved is through increased exploitation or cost-cutting, but both lead to crises in social relations. For example, with cost-cutting, worker health declines, unemployment rises, resources are depleted, and pollution increases. Both the ecological and social crises in the Third World, as in the First World, are inextricable from capitalist displacement of crisis through uneven and combined development, increased exploitation, and capital's efforts to cut costs. If Bookchin agrees with this critique of imperialism, then it must become part of his analysis of the international balance of power and his assessment of politics in the South, North America, and Europe. Without such analysis, his understanding of social labor and political power produces a political program that suggests localism in the face of an increasingly powerful, internationally coordinated capitalist world system.

As we have seen, for Bookchin the national divisions of labor within the capitalist world economy are merely extensions of domination which grow out of the nation-state and capitalism, effectively destroying all organic relations across the globe. He believes that there is nothing spontaneous, symbiotic, differentiating, or free in the international division of labor and, for him, the future must be constructed without the institutions that have developed as a part of this economic system. The state, capitalism, and the international division of labor apparently have nothing social within them that is worth saving in Bookchin's utopian world of the future.

If one wants to claim that technology has a social dimension, then perhaps it is true that Bookchin's work once embraced the idea of retaining certain forms of technology that had emerged under capitalism.[101] Even then, however, he analyzed technology more for its practical impact than for its rich *social* dimension. There is little in his work that really attempts to theorize the problems of technological manipulation and control under capitalism as properties of a unique form of determinism. Even in his work on municipal communities, criticizing the reduction of community to forms of electronic participation, there is little to explain the politics and rationality behind the proposals for such technological choices.[102]

Bookchin's more recent, thinner views on technology fit better with the traditional view that the push for liberation emerges finally from the people, but again he gives short shrift to social labor. For example, he is a great fan of solar-energy sources and some other advanced, labor saving technologies. Yet, in the transition from deep integration in the international division of labor to ecocommunist confederalized communities it is highly unlikely that locally organized egalitarian communities would have access to high technology during their struggle with the state and international capital. If U.S. capital and international political power can destroy Nicaragua's economy in just a few years, what would be the likely response to confederated small communities in Vermont? It is inconceivable that that Hewlett-Packard or some other multinational producer of long-lasting solar cells made from materials mined, processed, manufactured, and transported across the globe through the international division of labor would trade or barter solar assemblies for the goods produced by a small city or agrarian community or that such communities could amass the economic resources to purchase such goods.

With respect to Bookchin's revolutionary program and its more recent gradualism, these criticisms have serious ramifications. If the state and capitalism are as powerful and hegemonic as he often says they are, then a gradualist movement in which communities guard themselves with militias against "the ever-encroaching power of the state" seems hopelessly utopian.[103] Again, if the state and capitalism are as all-pervasive in their domination as Bookchin often writes, how is it that communities are going to be "allowed" to "play . . . democratic institutions against the state" to the point where "state power itself will have been hollowed out institutionally by local or civic structures" and "an aggressive attempt can be made to replace the state with municipally based confederal structures?"[104]

Bookchin's attempts to deal with the human and natural stakes in an ecocommunist revolution are hopefully utopian, providing a vision toward which people can work (despite his intolerance for other visions).[105] However, his unrealistic analysis of the national and international balance of power, and of the political strategies by which that order might be overthrown, is utopian to the point of being incredible.

Bookchin is perfectly correct that "we must try to democratize the republic," but in the process we must also democratize production and (initially) the state, or whatever transgeographic political institutions that are to regulate and mediate nonlocal activities from trade to cultural exchange. He is equally correct that we must have a vision of where we are going with our democratization, but that vision should not blind us to social relations in the world today. Neither the state nor the international division of labor is an unambiguous social construct. Both serve the interests of the public as well as those of political and economic elites which is one reason that it appears to Bookchin that the "litany" of revolutionary failures indicates that effective spontaneous revolutions do not happen. In short, his vision of modern society as so suffused with social and ecological domination as to demand unmediated moves towards the reconstruction of ecocommunities, has led him to demand of revolutions impossible measures of success. He makes similarly unreasonable demands of modern radicals and their programs.

Conclusions

Although Bookchin has a theory of ecological dialectics, a social ethics, and an anarchocommunist politics, he does not have a rich materialist theory to link his dialectics and ethics with practical politics. Thus, he promotes not only a utopian goal for society, but also an unreasonable political program. A materialist approach to society and nature includes theoretical analysis of labor and social labor; of the qualitative alteration of relations of production, distribution, and consumption; and of specific historical structures, tendencies, and crises which generate change. There is in Bookchin no such theory only a rejection of the most sophisticated extant theory, late-modern Marxism. Our argument is that in emphasizing domination and freedom, rather than the crisis-ridden and ambiguous nature of social institutions, Bookchin gives insufficient attention to labor in his historical work and analysis of the present. By marginalizing the pivotal concept of social labor, Bookchin is able to tell his story of history in support of utopian politics.

Finding both the nation-state and the international division of labor to be the products of hierarchy and domination, Bookchin sees increasingly little worth saving or working to transform in either form of social organization. For him, neither is spontaneous or part of the "ever-striving *developing* . . . substance whose most dynamic and creative attribute is its ceaseless capacity for self-organization into increasingly complex forms."[106] While his position is far preferable to uncritical acceptance of capitalist social relations, it is nonetheless too rigid to ground an oppositional politics.

In order to understand the role of domination in the modern capitalist division of labor in its myriad of historically and geographically specific forms, it is absolutely central that one have a sophisticated theory of capitalist production, distribution, consumption, accumulation, and crisis. Without such a theory, understanding the specific means by which capital destroys the conditions of its own (re)production becomes impossible, and it is extremely difficult to assess the likely or unintended effects of political action aimed at transforming ecological, social, and political conditions.

Bookchin's insistence that political programs maintain a vision of future goals is important, but it is certainly no more important than Marxists' insistence on materialist, political, and economic analysis. A reductive materialist theory can of course be stultifying, but without due regard to material conditions there is a tendency to develop unrealistic politics. In the first case, one gets the depressing sort of academic Marxism extant today; in the second case, one gets Bookchin's advocacy of a move from international capitalism directly to ecocommunism. Neither is particularly helpful in the search for utopias that might be imagined out of the present and that might ground political strategy for the future.

What Marxist political economy provides (that is not found in Bookchin's work) is a thoroughgoing investigation of the contradictions within capital accumulation on the many levels of its operation. Among these are "the detail division of labor within the workplace," "the division of the social capital between different individual capitals," "the division of labor (and capital) in particular different sectors" of capital, and "the general division of labor (and capital) into different departments"[107]; the division of labor between capital and the state and its ramifications for the state[108]; and the development of international capital.[109] Perhaps most important, are the ways in which capitalism both constrains and enables the ecological, personal, and communal conditions of production which, in turn, constrain and enable capital accumulation.[110] Rather than relying on a relatively disembodied analysis of domination in a universal marketplace to understand the system against which one struggles collectively and individually, one can assess the crises internal to capitalism, during which there is the greatest political and social latitude to act.

The effect of Bookchin's decision to analyze domination and hierarchy, without reckoning with exploitation and accumulation, results in a political project that effectively suggests global ecological transformation of society and social relations with the nonhuman world through a localist reconstruction of community. In not fully analyzing the specific conditions of state and capitalist power and hegemony, such an approach is unlikely to anticipate the responses of either the state or capital to the actions of such communities, much less the unintended consequences of the actions of such communities as their activities reverberate through the global division of labor. We feel that analysis informed by Marxist investigation of economic, political, and environmental crises is far more capable in this respect.

Bookchin's writings have had a tremendous impact on the left-green community in the West, but left-green politics falls short without red-green analysis. It is only through an internationalist approach—(which investigates the specific ways in which social labor both develops and destroys social and ecological conditions under global capitalism)—along with a good analysis of localities, that political solidarity can be generated in a liberatory way. Internationalism and a focus on social labor are central to Marxist analysis, from Marx himself through contemporary Marxist-feminist deconstruction and socialist ecology. It is by these routes and through a wide variety of analyses, among them capital accumulation strategies, forms of exploitation, sexism, racism, ecological destruction, and enforced heterosexuality, that a socialist ecology may offer a deeper utopian moment than social ecology.

Notes

1. "A 'socialist' ecology, a 'socialist' feminism, and a 'socialist' community movement . . . are not only contradictions in terms; they infest the newly formed, living movements of the future with the maggots of cadavers from the past and need to be opposed unrelentingly." Murray Bookchin, *Toward an Ecological Society* (Montreal: Black Rose Books, 1980) p. 16.

2. Lewis Herber, *Our Synthetic Environment* (New York: Harper and Row, 1962), reissued under the author's name in 1974 (New York: Knopf).

3. Herber, *Crisis in Our Cities* (Engelwood Cliffs, N.J.: Prentice-Hall, 1965).

4. Many of Bookchin's early articles were published under pseudonyms, including a series under the names of M. S. Shiloh, Harry Ludd, and Robert Keller, which appeared primarily in the journal, *Contemporary Issues: A Magazine for a Democracy of Content*, between 1950 and 1958. A few articles which would later appear in the journal *Anarchos* were published in other places under the name Lewis Herber in the early sixties, including the influential and much republished "Ecology and Revolutionary Thought," in *Comment* (1964).

5. Bookchin, *Post-Scarcity Anarchism*, (1971; 2nd ed., Montreal: Black Rose Books, 1986); *Toward and Ecological Society*; and *The Modern Crisis* (Philadelphia: New Society Publishers, 1986).

6. Bookchin, *The Ecology of Freedom* (Palo Alto, Calif.: Cheshire Books, 1982).

7. It should be made clear that Bookchin is not a romantic antimodernist. He believes that parochial localism remains unacceptable and that social ecology must be infused with rational and universalistic forms of humanitas in participatory local democracies confederalized with other municipalities and regions. Nevertheless, he argues, "After some ten millenia of a very ambiguous social evolution, we must reenter natural evolution again—not merely to survive the prospects of ecological catastrophe and nuclear immolation but also to recover our own fecundity in the world of life." *The Ecology of Freedom*, p. 315.

8. Bookchin, *The Rise of Urbanization and the Decline of Citizenship* (San Francisco: Sierra Club Books, 1987), later republished as *Urbanization without Cities* (Montreal: Black Rose Books, 1992).

9. Bookchin, *Remaking Society* (Montreal: Black Rose Books, 1989).

10. Bookchin, *The Philosophy of Social Ecology* (Montreal: Black Rose Books, 1990).

11. Bookchin, *The Ecology of Freedom*, p. 357.

12. Ibid., p. 358. Bookchin does not specifically define symbiosis or mutualism.

13. Bookchin, *The Modern Crisis*, *p*. 359.

14. "Our notion of teleology need not be governed by any 'iron necessity' or unswerving self-development that 'inevitably' summons forth the end of a phenomenon from its nascent beginnings. Although a specific phenomenon may not be randomly self-constituted, fortuity could prevent its self-actualization" (Bookchin, *The Ecology of Freedom*, p. 355).

15. Ibid., p. 355.

16. Bookchin, *Toward an Ecological Society*, p. 59.

17. Bookchin, *The Modern Crisis*, p. 11.

18. Ibid., pp. 42–43.

19. Ibid., p. 60.

20. Ibid.

21. Ibid., pp. 16–17.

22. Bookchin, *The Ecology of Freedom*, p. 29.

23. Ibid., p. 4. For Bookchin, historically and today, the fact that hierarchy exists "is an even more fundamental problem than social classes, that domination exists today is an even more fundamental problem than economic exploitation" (Bookchin, *The Modern Crisis*, p. 67).

24. Bookchin, *The Modern Crisis*, p. 42.

25. Bookchin, *Post-Scarcity Anarchism*, p. 98.

26. This can be seen in his discussion of "the trickery and cunning of elites and power-brokers" who induced "our distant ancestors . . . to chain themselves in servitude" (Bookchin, *The Modern Crisis*, p. 120); the purposes behind Athenian democracy (Bookchin, *The Rise of Urbanization*, p. 70); the elaboration of patrimony among and between elites and communities during the transition to the nation-state (ibid., p. 141); and in the technological control and military power of modern elites (Bookchin, *The Modern Crisis*, pp. 111, 126).

27. Bookchin, *The Rise of Urbanization*, pp. 200–201.

28. Ibid., p. 201.

29. Bookchin, *Post-Scarcity Anarchism*, p. 85.

30. Bookchin, *The Modern Crisis*, p. 30.

31. Bookchin, *Post-Scarcity Anarchism*, p. 60.

32. Bookchin, *The Modern Crisis*, p. 85.

33. Ibid., p. 44.

34. Ibid., p. 49.

35. "Like the biotic ecosystems we have simplified with our lumbering and slaughtering technologies, we will have simplified the psychic ecosystems that give each of us our personal uniqueness. . . . The process of simplification, even more significantly than pollution, threatens to destroy the *restorative* powers of nature and humanity—their common ability to efface the forces of destruction and reclaim the planet for life and fecundity." Ibid., p. 108.

36. Ibid., p. 57.

37. Ibid., p. 14. This was particularly important as, for Bookchin, the "tension between actuality and potentiality, between present and future, acquires apocalyptic proportions in the ecological crisis of our time" (ibid., p. 18). These "apocalyptic proportions" then made ecotopic revolution "not merely desirable or possible, but an absolute necessity, . . . the very development of the material preconditions for freedom makes the achievement of freedom a social necessity" (ibid., p. 29).

38. Ibid., p. 142.

39. For example, he hoped during the 1960s, that "communities would stand at the end of a cybernated assembly line with baskets to cart the goods home" as material scarcity disappeared (ibid., p. 155).

40. Ibid., p. 115.

41. This modern juncture was pivotal in that "lacking the material resources, the technology and level of economic development to overcome class antagonisms . . . Athens and Paris [representatives of prior historical revolutions oriented to the 'realm of freedom'] could achieve an approximation of the forms of freedom only temporarily—and only deal with the more serious threat of complete social decay" (ibid., p. 185).

42. See Bookchin, *The Ecology of Freedom*, pp. 70–71; *The Modern Crisis*, p. 55; and *The Rise of Urbanization*, p. 288.

43. Bookchin presents utopian ecocommunities in which "work, more craft-like than industrial, would be rotated as positions of public responsibility; that members of the community would be disposed to deal with one another in face-to-face relationships rather than by electronic means," and in which industry would likely be limited to entities "such as (a) small-scale foundry, machine shop, electronic installation, or utility" (Bookchin, *The Ecology of Freedom*, p. 345). Technology gets little mention though it is hoped that it will be seen "as a way of fostering natural fecundity rather than exploiting or vitiating it" (Bookchin, *The Rise of Urbanization*, p. 267). In his most technologically conservative statement, he emphasizes "that the 'means of production' have now become too powerful—too manipulable by small, idiosyncratic if not crazed elites, too prolific and cancerous in their metastatic growth—to be designed, much less used, as a means of destruction. . . . We are passing the point where technical and scientific advances, apart from mere marginalia, hold any promise for human survival and well-being" (Bookchin, *The Modern Crisis*, p. 111).

44. Bookchin, *The Ecology of Freedom*, p. 240; parentheses in original.

45. Bookchin, *Post-Scarcity Anarchism*, p. 44.

46. Ibid., p. 45.

47. Bookchin, *The Modern Crisis*, p. 123.

48. Bookchin, *Post-Scarcity Anarchism*, p. 62.

49. Ibid., pp. 188–189, emphasis in original.

50. Bookchin, *The Rise of Urbanization*, p. 286.

51. Bookchin, *Post-Scarcity Anarchism*, p. 185, emphasis in original.

52. Bookchin, *The Modern Crisis*, p. 153.

53. Bookchin, *Post-Scarcity Anarchism*, p. 209.

54. Bookchin, *The Modern Crisis*, p. 150.

55. Bookchin, *The Ecology of Freedom*, p. 215.

56. Bookchin, *Post-Scarcity Anarchism*, p. 28.

57. Ibid., pp. 206–207.

58. Ibid., pp. 128–129.

59. Ibid., p. 134

60. Ibid., p. 132.

61. Ibid., pp. 133–134.

62. Ibid., pp. 134–135.

63. Ibid., p. 130.

64. Ibid., p. 139. Bookchin has also blamed immigrants for derailing American utopianism, asking, "can we perhaps look back and ask where they erred and formed a branch of their own away from the historical flow of the utopistic American Dream as distinguished from the material one they cherished?" Bookchin's account of immigrant "errors" leaves out supporting data showing that the American Dream was primarily utopian and that the immigrants eschewed utopianism in their materialism. Further, the complexity of the many forms of oppositional politics across the many regions of the United States initiated by people of many races in contradictory and evolving class positions is omitted from his account. Finally, Bookchin leaves out an investigation of

the material conditions which enabled and constrained materialist and unionist movements of the sort he does not appreciate.

65. Bookchin, *Toward an Ecological Society*, p. 218

66. Ibid., p. 196.

67. Ibid., p. 124.

68. Karl Marx and Frederick Engels, *The Holy Family*, in vol. 4 of *Collected Works*, (New York: International Publishers, 1975), p. 37. For Bookchin's most emphatic criticism, see *Toward an Ecological Society*, p. 124. We find it problematic that Bookchin's work on Marx focuses so exclusively on the pre-*Capital* works, but cannot pursue the implications of this omission adequately here.

69. Ibid., p. 209.

70. Ibid., p. 209.

71. Bookchin, *Post-Scarcity Anarchism*, pp. 20ff.

72. Ibid., p. 199.

73. Bookchin, *The Rise of Urbanization*, p. 228.

74. Ibid., p. 245. By community, Bookchin means "a municipal association of people reinforced by its own economic power, its own institutionalization of the grass roots, and the confederal support of nearby communities organized into a territorial network on the local and regional scale."

75. "The most important of these coordinates is the revival of the *citizens assembly*, be it in the form of town meetings in humanly scaled communities or neighborhood assemblies in large, even metropolitan urban entities" (ibid., p. 257).

76. "The second of these major coordinates is the need for the assemblies to 'speak' to each other, literally, to confederate. Leagues of towns and cities, as I have argued earlier, have always surfaced, however temporarily, as centripetal forms of municipal associations" (ibid., pp. 257–258).

77. Ibid., p. 258.

78. Ibid., p. 262.

79. Ibid., pp. 261–263. "Economic policy can be formulated by the *entire* community—notably its citizens in face-to-face relationships working to achieve a general interest that surmounts separate, vocationally defined specific interests" (p. 263).

80. "Men [*sic*] make their own history, but they do not make it just as they please; they do not make it under circumstances chosen by themselves but under circumstances directly encountered, given and transmitted from the past." (Marx, *The Eighteenth Brumaire of Louis Bonaparte*, in vol. 11 of *Collected Works*, 1979, p. 103).

81. Ibid., pp. 198, 201. Capitalist exploitation appears tied to wage labor in Bookchin's historical presentation. Because there were instances of wage labor in precapitalist modes of production, Bookchin sees only their quantitative expansion with the emergence of capitalism. The extraordinary power of Marxist analysis to explicate the qualitatively new forces and relations associated with the capitalist mode of production is lost to Bookchin. He also loses the ability to see its internal contradictions and various forms, and loses access to the breadth and extent of immanent possibilities for change within the economically, socially, and ecologically contradictory structure of the system and its transformation. The analysis of structural contradictions, associated as much with overproduction as with fiscal crisis and ecological destruction, is rejected by Bookchin in favor of a chaotic search for and attempts to organize revolutionary constituencies.

82. Marx is extremely specific about the processes through which the social, spatial, and technical restructuring of work in the transition from craft production to manufacture to modern industry generates new productive relations and forces as the modes of cooperation and divisions of labor and laborers are elaborated. It is in the

discussion of the social, productive, and newly cooperative characteristics of manufacture, in the context of elaborating the differences within and between absolute and relative surplus value, that Marx introduces the term "social labor." The mode of exploitation (and forms of domination) specific to the conditions generated prior to and necessary for the elaboration of social labor and extraction of relative surplus value is qualitatively different from previous forms of class-based society and exploitation. These conditions of and for social labor utterly transform the structural possibilities and contradictions of social and economic life, political forms and relations, and ecological relationships. See Marx, *Capital*, vol. 1, trans. Ben Fowkes (New York: Vintage Books, 1977) especially the appendix; and James O'Connor, "Historical Materialism Reconsidered," paper presented at the Gramsci Institute's Conference Commemorating the 100th Year of Marx's Death, Rome, 16 November 1983.

83. See the ongoing exchange in *Capitalism, Nature, Socialism*, by James O'Connor and others, on the thesis of the second contradiction of capitalism, especially issues 11–16.

84. Bookchin, *The Modern Crisis*, p. 30. Clearly he believes that capitalism has many social *consequences* ("capitalism has shown that it can grow indefinitely and spread into every social domain that harbored ties of mutuality and collective concern"). But when capitalism is said to destroy the social, the implication is that it is not itself a social system, or at least that it is one predicated only on domination.

85. Ibid., p. 84.

86. Georg Lukács, *History and Class Consciousness*, trans. Rodney Livingstone (Cambridge, Mass.: The MIT Press, 1971); Andrew Feenberg, *Lukács, Marx, and the Sources of Critical Theory* (Oxford: Oxford University Press, 1981).

87. Of course, this is the main thesis of Marxism. For its standard development in environmental political theory, see James O'Connor, "Capitalism, Nature, Socialism: A Theoretical Introduction," *Capitalism, Nature, Socialism* 1 (Fall 1988): 11–38.

88. See Bookchin's discussion of Engels's *On Authority in Towards an Ecological Society*, pp. 126–129.

89. Bookchin, *The Modern Crisis*, p. 85.

90. Bookchin, *The Ecology of Freedom*, p. 242.

91. Ibid., p. 244. Bookchin does not explain the difference between these terms or why adaptive technologies should not be innovative.

92. O'Connor, "Historical Materialism Reconsidered," p. 10.

93. O'Connor "Capitalism, Nature, Socialism: A Theoretical Introduction"; and O'Connor and Daniel Faber, "The Struggle for Nature: Environmental Crisis and the Crisis of Environmentalism in the United States," *Capitalism, Nature, Socialism* 2 (Summer 1989): 12–39.

94. Neil Smith, *Uneven Development: Nature, Capital, and the Production of Space* (London: Blackwell, 1984).

95. Bookchin, *Post-Scarcity Anarchism*, p. 93.

96. It is important to note here that we are using "materialist" and "materialism" in the traditional Marxian sense. One of us has described Bookchin's approach to political ecology as "environmental materialism," meaning materialism in a very thin sense. Nothing in this chapter should be taken to contradict that analysis. See Andrew Light, "Rereading Bookchin and Marcuse as Environmental Materialists," *Capitalism, Nature, Socialism* 13 (March 1993): 69–98.

97. We are using the term "new social movements" in a very broad sense that would include environmentalists, whom we presume must be part of Bookchin's scheme. This is difficult to assume, however, as there is no clear identification in his work comparable to that found in work of Iris Young, Stanley Aronowitz, or Chantal Mouffe

and Ernesto Laclau. It is not clear that the force behind Bookchin's new municipalities would be the same political subjects as in their work. We maintain that a theory of production and social labor is necessary for a new theory of political identity, and that a simple Hegelian inversion of class politics to nonclass politics does not suffice. Bookchin is certainly on the way to a richer theory in his work on the relationship between municipal life and citizenship, but we would not dissociate workplace democracy so strictly from participation in the body politic. In *Participation and Democratic Theory* (Cambridge, England: Cambridge University Press, 1972), Carole Pateman makes a very strong case that better political participation is possible only in the context of better self-management by shop floor workers.

98. Bookchin, *Toward an Ecological Society*, p. 224.

99. Almost unbelievably, Bookchin writes that it is in fact the republican institutions of the United States, which "make the successes of 'Third World revolutions' possible" (*The Modern Crisis*, p. 134). It is hard to fathom how he can stress the restraint of the United States after the eco- and genocide of Vietnam, the bombing of Cambodia, and the massive loss of human and ecological life in Central America. For a more illuminating view of Third World revolutions, see Daniel Faber, *Environment Under Fire: Imperialism and Ecological Crisis in Central America* (New York: Monthly Review Press, 1993); and Eduardo Galeano, *We Say No: Chronicles 1963–1991* (New York: W. W. Norton and Company, 1992).

100. Bookchin, *The Modern Crisis*, p. 133.

101. Bookchin, "Toward a Liberatory Technology," in *Post-Scarcity Anarchism*.

102. Bookchin, *The Rise of Urbanization*, p. 250. For a theory of technology which takes social labor fully into acount, see Andrew Feenberg, *Critical Theory of Technology* (Oxford: Oxford University Press, 1991).

103. Bookchin, *The Rise of Urbanization*, p. 285. This is a little unfair to Bookchin, who writes two pages later that the arming of local militias would be only a "feeble step" more or less symbolically challenging the right of the nuclear state's monopoly on violence (ibid., p. 287). Yet, this caveat makes one wonder about Bookchin's belief in and commitment to this "feeble step."

104. Ibid., p. 287.

105. Recall Bookchin's claim that all of us who do not treat domination as the central term in revolutionary discourse are "unreliable" in our commitment to fully revolutionizing society. *Toward an Ecological Society*, p. 244.

106. Bookchin, *The Ecology of Freedom*, p. 357.

107. Smith, *Uneven Development*, p. 108.

108. James O'Connor, *The Fiscal Crisis of the State* (New York: St. Martin's Press, 1973).

109. Michael Redclift, *Sustainable Development: Exploring the Contradictions* (New York: Methuen, 1987).

110. Alan Rudy, "On the Dialectics of Capital and Nature," *Capitalism, Nature, Socialism* 18 (June 1994): 95–106.

Index